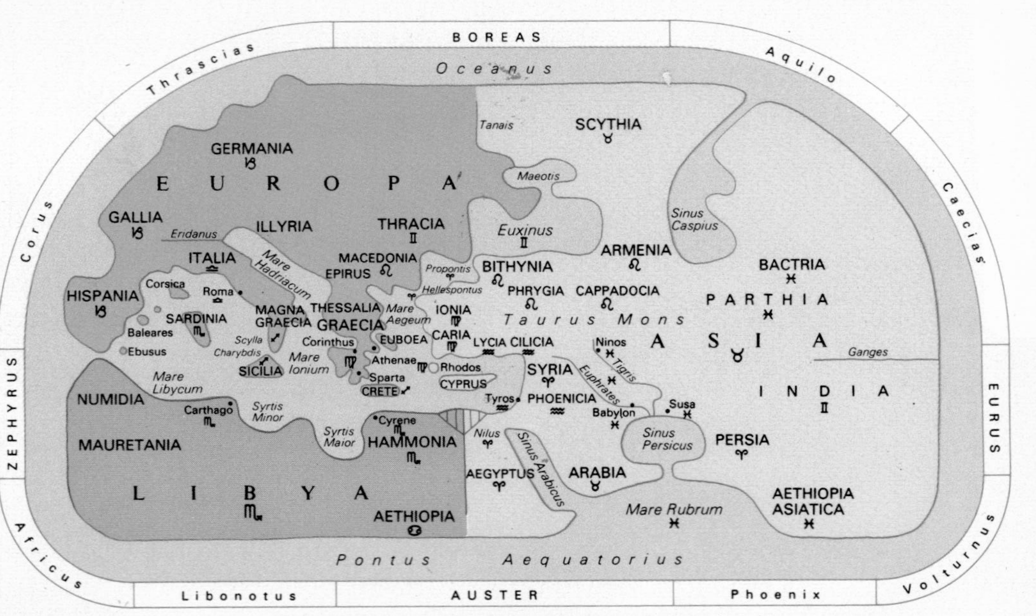

THE WORLD OF MANILIUS, 4.585-817

MANILIUS

ASTRONOMICA

WITH AN ENGLISH TRANSLATION BY

G. P. GOOLD
UNIVERSITY COLLEGE LONDON

CAMBRIDGE, MASSACHUSETTS
HARVARD UNIVERSITY PRESS
LONDON
WILLIAM HEINEMANN LTD
MCMLXXVII

American
ISBN 0-674-99516-3

British
ISBN 0 434 99469 3

Printed in Great Britain

CONTENTS

THE WORLD OF MANILIUS (*map*) . . *Frontispiece*

PAGE

PREFACE vii

INTRODUCTION:

About the Poet xi

A Guide to the Poem xvi

The Manuscripts cvi

Editorial Principles cx

Select Bibliography cxiii

THE ASTRONOMICA OF MARCUS MANILIUS:

Book One 3

Book Two 81

Book Three 161

Book Four 221

Book Five 299

APPENDIX (*readings not in Housman*) . . . 365

INDEX 367

THE SKIES OF MANILIUS (*star-charts*) . *following* 388

PREFACE

In preparing an edition of the *Astronomica* for the Loeb Classical Library I have felt compelled in the interests of simplicity and clarity to add rather more by way of commentary than is ordinarily desirable in the series. But Manilius is a difficult author. His text is unusually corrupt, his subject-matter highly specialized, and his Latin style so bizarre and bewildering that, as Garrod dolefully put it, "the helpfulness of a translation is not unlikely to be in inverse proportion to its readableness." Moreover, he frequently embarks on an audacious plan of rendering diagrams, tables, and maps in hexameter form ; and in these places even the best of translations would need visual aids to be readily comprehensible.

Perhaps then it is not surprising that only once before—as long ago as 1697—has the whole work been turned into English ; Creech's rhymed couplets are excellent in themselves, but all too often they leave the Latin far behind and are consequently of small value as an elucidation of it. My indebtedness to Garrod's translation of Book 2 and to other published selections will be obvious ; I have also made use of a model version of 1. 876-897 composed for his students by A. S. F. Gow. But in the matter of assistance I owe most to Percival Vaughan Davies, the translator of Macrobius, who after translating about two-thirds of the text of Manilius heard of my undertaking and at once generously presented me with all he had done and invited me to use it as I thought fit: I have gratefully revised my own version after diligent comparison with his, adopting from him many an apt word and felicitous turn of phrase; and often enough, where not moved to incorporate his words, I have recast my own or modified an earlier interpretation in the light of his.

Whilst the text and the interpretation given in this edition represent my own judgement of what the poet wrote and meant (and supersede any opinions to the contrary I may previously have expressed), my obligation to former editors and scholars in this field is practically total. But to none is it greater than to Housman. There cannot be a page of this book untouched by his influence. I differ from him more than I should have once thought reasonable, but if time has opened my eyes to his fallibility, it has also brought me a fuller awareness of his scholarship, for which my respect has grown and not diminished. Where I have asserted independence I have not done so lightly. On the other hand, hoping that many who study his great edition will find in this a convenient cicerone, as I hope also that many who use this will be stimulated to go on to his, I often shelter under his protection amid some blind uncertainty (as at 2. 935 ff.) sooner than embark in my frailer craft on some desperate argosy, and I regularly employ his example of a general principle (as at 3. 160 ff.) in preference to making up an alternative simply in order to differ from him. Housman's authority as a Latinist, like his distinction as a poet, the world has long since recognized ; but not everyone knows what a consummate astrological scholar he was. Indeed, his exposition of Manilius's astrology, largely free from the ill temper which obtrudes so much into his textual criticism, constitutes the most admirable feature of his work and merits special notice in these times, when exciting new vistas are being opened up by the editing, notably by Professor David Pingree, both of extant Greek texts and of oriental versions of others now lost.

I should like to express my warmest thanks to David and Pamela Packard for furnishing me with a computer-concordance to Housman's *editio minor* ; to Professor E. H. Warmington for carefully scrutinizing the translation ; and to Kevin Goold, my brother, for designing and executing the frontispiece, figures, and star-charts.

University College London
February 1976

G. P. GOOLD

INTRODUCTION

ABOUT THE POET

[Housman, ed. 1[2] lxix ff., 90 ff.; Goold[3]; Fletcher; Sikes][a]

HAD the archetype of the *Astronomica* not survived long enough to provide us with copies of the poem, we should have had no reason to suspect its existence or that of its author. Pliny, it is true, refers (*N.H.* 35. 199) to a Manilius of Antioch who was brought to Rome about 90 B.C. as a slave and became an astrological writer, but the date is incompatible with that of the poem. Several scholars have therefore toyed with the possibility that our Manilius was a son of the Antiochene, a possibility imaginatively embellished by Léon Herrmann; but the most that can be said is that the odds are against it.

The name to be elicited from the manuscripts is certainly Marcus Manilius; and hardly less certainly the *Boetius* attached to one occurrence of it in the chief manuscript is to be ascribed to confusion with the great Boethius. We need likewise entertain no doubt about the title of the poem, *Astronomica*: the poet obviously desired his didactic poem to stand

[a] For items in square brackets see the Select Bibliography, pp. cxiii ff.

comparison with Virgil's and named it "Astronomics" after "Georgics."

The following considerations determine the date of composition.

Book 1 : The disaster of the Saltus Teutoburgiensis (A.D. 9) has recently taken place (898 ff.); Augustus, to whom the poem is dedicated, is alive (7 f.; 385; 922 ff.).

Book 2 : Lines 507 ff., extolling the greatness of Capricorn as Augustus's natal sign (when it was occupied by the Moon), guarantee that Augustus is still on the throne.

Book 3, like Book 5, provides no clue to its date.

Book 4 : Capricorn has fallen from grace, and it is now Libra who is described (548 ff.) as presiding over the birth of the reigning emperor (to whose erstwhile domicile in Rhodes lines 763 ff. refer): evidently Tiberius (whose natal sign was indeed Libra) has succeeded Augustus, who died in A.D. 14.

The inferences to be drawn are plain: Books 1 and 2 were written while Augustus was still alive, 4 and 5 when he was dead.

Our ignorance of the personal circumstances of the poet is not much instructed by the contents of the poem. Since he hopes (1. 115) to live to old age so that he may complete his work, he cannot be an old man already, but the formulation of his wish would be distinctly odd if he were young. More debatable is his origin: for Scaliger he is a Roman, for Bentley (who thought some features of his language peculiar) an Asiatic. True, he identifies himself as Roman in speaking of the Latin language and in referring to Hannibal, but this need not exclude the possibility that he was of foreign descent. On the other hand, the singularity of his Latin consists not in incorrect or gauche or even unidiomatic language, but rather in his excessive striving after sententious utterance concentrated in lapidary brevity. He displays a thorough familiarity with Lucretius, whose philosophy he repeatedly seeks to rebut, and also with Virgil, from whom he repeatedly takes a word or phrase or idea. The early books of Livy he seems to have read, as also Cicero's *Somnium Scipionis*, and the great orator's *Verrines* have also inspired a curious borrowing at 5. 620. But against this we receive the impression that Manilius is much more steeped in Greek literature, not only astrological works or the great masters like Homer and the dramatists, but even minor authors as well. In particular his text seems to disclose several striking parallels (apparently borrowings) with phrases in the text of the *De Sublimitate*, which I therefore incline to date before A.D. 14 (2. 8 ff. < *De S.* 13. 3 ; 2. 58 < 13. 4 ; 3. 22 f. < 4. 2 ; 4. 158 f. < 35. 3 ; 5. 222 < 19. 1). Whether or no the author of the *De Sublimitate* is (as I have elsewhere urged) Gnaeus Pompeius, the friend of Dionysius of Halicarnassus, we seem to have a possible connection between Manilius and a Greek professorial circle at Rome.

It teases the imagination to ponder the precarious route followed by the text of our poet from his times to those of our oldest manuscript. Faint echoes for about a century, especially in Stoic circles, suggest that it enjoyed some sort of circulation ; it may have influenced Germanicus (184 < Man. 5. 23), Lucan (8. 365 < Man. 1. 655), Seneca (*Phoen.* 14 f. < Man. 5. 184), the author of the *Aetna* (591 < Man. 2. 3), Valerius Flaccus (3. 710 ff. < Man. 4. 570), and Juvenal (10. 175 < Man. 1. 776). One memorable line (4. 16) even appears in a verse inscription. Just once, at the end of the fourth century, the text comes into full view, for Firmicus Maternus quite certainly had Book 5 of Manilius open before him when he composed Book 8 of the *Mathesis.* Thereafter the *Astronomica* disappears from sight and next emerges in a tenth-century Bobbio catalogue, which lists as one of its items the Manilian archetype, or, if not that, certainly some ancestor of our extant manuscripts.

We need not doubt that, except for the lacuna in Book 5, we possess the whole poem ; nor need we entertain the thought that Manilius broke off his design. Of course Housman is right to insist that a nativity cannot be cast from the poem as it stands. But a didactic poem is seldom an exhaustive treatise, and various features of Manilius's should warn us against expecting a completely logical and systematic account. If (as I believe) the poet gave a perfunctory account of the planets' natures in the great lacuna, he might well have considered his obligations duly discharged.

Even under the most favourable circumstances the composition of a didactic poem is likely to be a hazardous enterprise : when as in the case of astrology the theme involves numbers and tables, procedures and calculations, diagrams and catalogues, rhetorical virtuosity may with difficulty be attainable, but poetic distinction is likely to be beyond reach. At any rate the young Goethe thought so (*Ephemerides*, 1770): "I began to read Manilius's Astronomicon and soon had to put it down: no matter how much this philosophical poet festoons his work with lofty thoughts, he cannot redeem the barrenness of his subject . . . I consider that one has to debit the poet's account with the ill consequences of a subject. After all, he is the one who chose it." Yet those who have studied him most have on the whole rendered him generous applause, as Fletcher's collection of tributes shows. For Scaliger he was Ovid's equal in sweetness and his superior in majesty, a judgement endorsed by Bentley. On the other hand, Sikes has some justification for his disappointment with Manilius as a Stoic controversialist, eager as he is to refute the Epicurean views of Lucretius whenever opportunity offers, for in this match the later poet assuredly comes off *impar congressus Achilli.* Nevertheless here and there Manilius so far abandons his customary rhetorical manner as to give voice to his soul's passion for certain themes: the beauty of the skies, the eternal movement of the great celestial clock, the immanence and supremacy of reason, and heaven's call to man to elevate himself to godhead. And for all his earlier distaste Goethe showed an unerring appreciation of two of the poet's finest lines (2. 115 f.) when on September 4th, 1784 he climbed the Brocken and, his spirit stirred mightily, inscribed in the visitors' book kept at the summit:

Quis coelum posset nisi coeli munere nosse,
Et reperire Deum, nisi qui pars ipse Deorum est?

A GUIDE TO THE POEM

The science of astrology forms the principal theme of Manilius's work, but he postpones all treatment of it until he has set forth a descriptive account of the heavens, a "Sphaera," such as Aratus had composed nearly three centuries earlier : thus the first book is entirely astronomical.

[*History of astronomy* : Dicks ; Dreyer ; Heath [1] ; Heath [2] ; Neugebauer. *Sphaera* : Aratus ; *Comm. Arat.* ; Hipparchus ; Geminus ; Cicero ; Germanicus ; Avienius. *Catalogue* : Ptolemy, *Syntaxis* 7 and 8. *Mythology* : Eratosthenes ; Hyginus. *Names* : Allen ; Webb. *Figures* : Thiele ; Moeller. *Atlas* : Norton]

Astronomical symbols used in text and figures

ZODIAC

♈	Aries	♎	Libra
♉	Taurus	♏	Scorpio
♊	Gemini	♐	Sagittarius
♋	Cancer	♑	Capricorn
♌	Leo	♒	Aquarius
♍	Virgo	♓	Pisces

PLANETS

☉	Sun	♂	Mars
☾	Moon	♀	Venus
♄	Saturn	☿	Mercury
♃	Jupiter	⊕	Lot of Fortune

INTRODUCTION

BOOK ONE

1. 1-117 : Prooemium

Unlike the composer of epic or drama, who most effectively plunges *in medias res*, the didactic poet needs to prepare his ground. And in fact Manilius takes some time to get into his stride. We learn that he is the first to sing of astrology (1-24), and that the god Mercury stimulated interest in the sky and paved the way for the ultimate discovery that destiny is written in the stars (25-66). Before that time man led a primitive, if untroubled, existence, and it was the awakening of his intellect which brought about the progress that has led to his amazing knowledge of the heavens (67-112). The formal exordium closes (113-117) with a reiteration of his originality and a prayer for a comfortable old age to enable him to complete his design.

In traversing the early history of man our poet, with subtle allusions to the Golden Age, combats at every step the unromantic account given by his great Epicurean rival. For Lucretius the first men were unconcerned with celestial phenomena ; Manilius improbably portrays them as terrified lest day fail to follow night. "In dealing with the next state of society," says Sikes (176), "the Stoic shews to more advantage : while Lucretius is compelled to deny any altruism in social evolution, Manilius rightly sees that society cannot be formed by self-interest alone. Taking the actual words of Lucretius—*commune bonum*—he insists on the influence of society as a natural and not merely a conventional bond. The wild and selfish cave-man of Epicurean fancy is replaced by the 'noble savage,' who is led upwards to civilization through the providence of God."

His account of the rise of astrology, on the other hand, is both vague and slight ; and we are forced to conclude that he has nothing factual to impart. The kings of the orient referred to in verse 42 perhaps include Zoroaster (that is the

Persian sage Zarathustra, reputed in later tradition to have been a king of Bactria and the founder of astronomy) and Belus (for whom the Chaldeans make the same claim according to Achilles, *Isag.* 1, p. 27). Among the priests mentioned in 47 are presumably included Nechepso (habitually called ὁ βασιλεύς by Vettius Valens) and Petosiris, legendary names, once perhaps even borne by a Pharaoh and his high-priest, but owing their currency among ancient astrologers to a 2nd-century B.C. manual purporting to be written by them. Manilius seems to give precedence in astronomy to the Babylonians (41-45), in astrology to the Egyptians (46 ff.).

1. 118-254 : The origin and nature of the universe

Venturing into the field of cosmological enquiry the poet lists, as more than once elsewhere, a number of competing theories : Diels (229 ff.) sees in this feature a characteristic of Posidonius, Manilius's dependence on whom has been industriously if inconclusively argued by Mueller.

(1) 122-124 : Possibly the world has neither beginning nor end (Xenophanes) ;
(2) 125-127 : Or it was given birth by Chaos (Hesiod) ;
(3) 128-131 : Or it is composed of atoms (Leucippus) ;
(4) 132-134 : Or it is composed of fire (Heraclitus) ;
(5) 135-136 : Or it is composed of water (Thales) ;
(6) 137-144 : Or it is composed of four elements (Empedocles).

The poet professes to have an open mind on these matters and at once proceeds to expound (149-170) the Stoic view of the creation of the universe from the four elements (Chrysippus : Diels 465 f.).

The Earth's suspension in space (173 ff.) is to be argued from the apparent path of Sun and Moon and Venus and fixed stars beneath it. All these bodies are suspended ; they are spherical, too. And so is the Earth. This explains why we cannot see all the constellations in all parts of the world: the star Canopus is invisible north of Rhodes (so most authorities, *e.g.* Hipparchus, Geminus, Pliny, Cleomedes), and people who see Canopus above them

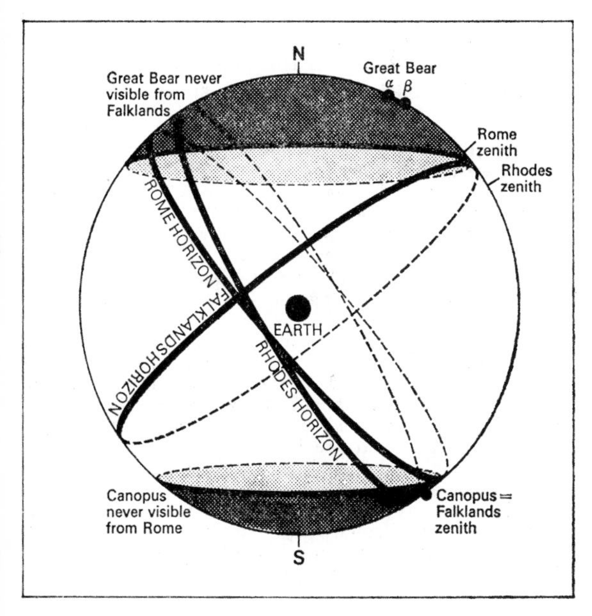

Fig. 1. The Invisibility of Certain Stars, 1. 215-220

—*e.g.* the inhabitants of the Falkland Islands—cannot see the Great Bear. Figure 1 is here introduced to clarify the reasoning. Lunar eclipses afford another proof of the Earth's sphericity: in different longitudes the eclipse occurs at different hours. Herein the poet may have misunderstood his source, for the relevance of lunar eclipses to this enquiry consists not in what he says at all, but rather in the striking fact that during them the shadow of the Earth's outline is cast on the Moon and is then seen to be irrefutably round (Aristotle, *De Caelo* 2. 297 b). Then (236 ff.) the poet speculates about the inhabited parts of the Earth's surface.

Unfortunately he has fallen into the error of confusing the western with the southern hemisphere, thus leaving us uncertain how exactly he envisaged the inhabited world whose geography he describes at 4. 585-817. In a fascinating chapter (*Comm. Somn. Scip.* 2. 9) Macrobius sets forth the theory (which may go back to Crates of Mallus in the 2nd century B.C.) that the land surfaces of the Earth are divided into four by two intersecting zones of Ocean, one encircling the equator (this presumably the *pontus* of 1. 246), the other circling the Earth longitudinally (as would the Atlantic and Pacific Oceans if connected at the poles). There are thus four land-masses cut off from each other: in the northern hemisphere our own (inhabited by σύνοικοι—to use the terminology of Geminus 16. 1—and comprising Libya, Asia, and Europe) and one 180 degrees of longitude away on the opposite side of the world (περίοικοι: the location of North America); and in the southern hemisphere that on our side of the world (ἄντοικοι: the location of southern Africa) and that diametrically opposite to us (ἀντίποδες). This view of the world was almost certainly known to Manilius and perhaps was held by him.

The climax (247-254) of the second section is signalled by a brief statement of the Stoic doctrine that a divine spirit governs the harmony of the four elements which make up the universe.

1. 255-531 : The stars of the sky

Manilius has now reached the central and most important part of his Sphaera, a description of the constellations. This he sets forth in the following order.

(1) 256-274 : the signs of the zodiac (put first by Geminus 1. 1 f. ; Vitruvius also gives the order zodiacal, northern, southern ; Aratus begins with the circumpolars and does not deal with the zodiac as a group at all).

(2) 275-370 : the northern signs are introduced by a paragraph on the polar axis, which leads naturally to the circumpolar stars and then to the rest of the constellations north of the ecliptic.

(3) 373-455 : finally the southern signs. The celestial antarctic lay beyond the observation of Manilius and his sources, but he infers by analogy the existence of stars in this region of the sky.

(4) 456-473 : we may wonder why the creator did not endow the constellations with fuller shapes of the figures he wished them to represent : the sky, we are told, could not have withstood such an extra quantity of fire.

(5) 474-531 : a poetical cadenza follows, in which Manilius marvels at the orderliness and immutability of celestial motions, a passage singled out by Jebb as exhibiting the author at his best.

History. The earliest Greek star-gazers first noted individual luminaries like Arcturus and then grouped neighbouring stars together in constellations like the Great Bear. They named their constellations, says Webb simply and dogmatically, after their supposed shapes. Of course, once the Great Bear was named, the consequent naming of a single star Arcturus with relation to it was a natural development.

In like manner, after Orion was identified, the bright star which ever follows at his heels was called his dog (Homer, *Il.* 22. 29). However, it is clear that many of the constellations in the Greek Sphaera are imports from abroad: this is particularly true of the zodiacal signs, for there is no evidence that the twelve signs as a group existed before the 4th century; indeed, what appear to be older appellations and figures have been incorporated in them, the Pleiades in Taurus and the Manger in Cancer, for example. It is significant that the only obviously Greek myth in the sky, that of Perseus and Andromeda and her parents (both Sophocles and Euripides wrote plays involving catasterisms of the principals, and perhaps this group-catasterism took place in the 5th century: certainly it was natural for Manilius to narrate this of all myths at length in Book 5), must antedate the zodiac, for the celestial tableau with Cetus is awkwardly broken by the dim stars of Aries and Pisces. Nor is this the only clue to the probability that the inventors of the zodiac were forced to make adjustments: the Scorpion once occupied so much more than thirty degrees that to get a duodecimal zodiac it had to be made into two signs; and the two legs of Ophiuchus protruded, and still do, right into the zodiac and even south of the ecliptic. Nowhere do we see more clearly the lateness of Greek preoccupation with the stars than in the mythology of the constellations. For the most part the stories are manifestly contrived, and conflicting versions are common. Manilius's general agreement with the astronomical mythographers (most conveniently presented in Robert's edition of Eratosthenes) makes it likely that he used a related source.

Pictorial representation. There is not and there cannot be a definitive Greek Sphaera. Eudoxus in the 4th century and Aratus in the 3rd laid the basis of a celestial picture-atlas, but Hipparchus was the first to define the outlines of constellations systematically in terms of stars identified by coordinates. His star-catalogue has not survived, though it seems pretty certain that we have it subsumed in the extant catalogue of Ptolemy, the value of which for a reconstruction of Manilius's globe will at once emerge from the first three entries of Ursa Minor (*Synt.* 7. 5 pp. 38 f.):

1. The star at the end of the tail:
   Gemini 0° 10′ North 66°; 3rd magnitude.

2. The next along the tail:
   Gemini 2° 30′ North 70° ; 4th magnitude.
3. The next, in front of the tail's issuance:
   Gemini 16° North 74° 20′ ; 4th magnitude.

The coordinates (given in respect of the ecliptic: modern astronomers use the celestial equator) readily enable the stars to be identified as αδε UMi. Ptolemy is no slave to imagined figures: if a star comes outside the perimeter of a constellation (even a bright star like Arcturus), he lists it separately as extra-territorial. What pictorial representations have survived directly or indirectly (or can be reconstructed from textual statements such as Vitruvius's descriptions in 9. 3-5) have been collected and handsomely edited by Thiele. Foremost among them stands the 2nd-century Farnese globe (carried by Atlas on his shoulders), which exhibits the figures of the constellations, the five parallels, the ecliptic (and the northern and southern limits of the zodiac), and the two colures; no stars are marked, nor the Milky Way. Most Manilian scholars hold that the poet had the use of just such a globe, Moeller (31) even going so far as to argue from such phrases as 1. 609 *ab excelso decurrens limes Olympo* and 1. 687 *inde per obliquum descendens* that it must have been a movable one.

The constellations today. The Greek Sphaera (to which no addition was made until the 17th century) has proved astonishingly long-lived. Only a few extra asterisms (and these all minor) have gained a place in the northern heavens, the chief innovations naturally being made in the antarctic region, which was soon apportioned among new groups. The last constellation was created in the 18th century. Today (as the result of an agreement made by the International Astronomical Union in 1930) the whole sky is distributed among these constellations, the boundaries between them being drawn along arcs of Right Ascension and Declination: three-letter abbreviations denote the constellations, and Bayer's system of Greek letters (devised in 1603) is for the brighter stars still universally used and will be used below.

Manilius's constellations. For the reader's reference I append a list of the constellations mentioned by the poet, giving (1) name (and the verse-number of its first occurrence in Manilius), IAU abbreviation, and alternative names occurring in the Latin text, (2) mythological and other information given in or relevant to the poem, and (3) data indicating the poet's pictorial conception. Sometimes Manilius describes different representations (as in the case of Libra) ; and to illustrate some of his errors (as with Draco) would involve a distortion of star positions. But for the most part the poet's views can be and have been incorporated in star-charts : these will be found at the end of the volume.

(a) Zodiacal

1 ARIES 263 (*Ari* : *Corniger*, *Laniger*) : The ram which swam to Colchis with Phrixus on its back after losing Helle and its left horn ; then stripped of its golden fleece and sacrificed to Jupiter : 2. 34, 532 ; 3. 304 ; 4. 515 ff., 746 ff. ; 5. 32. *Looks back* 1. 264 ; 2. 212 ; *runs* 2. 246.

2 TAURUS 264 (*Tau*) : The bull (Jupiter in disguise) which carried Europa over the sea to Crete : 2. 489 ff. ; 4. 681 ff. *Rises backwards* 1. 264 ; *limps and has one leg doubled under it* 2. 259.

PLEIADES [371] : A cluster in Taurus, regarded as a separate group representing seven sisters : 5. 142, 710.

HYADES [372] : αγδεθ *Tau*, regarded as an independent constellation : 5. 119 ff. : the name means "Rainers," mistranslated into Latin as *Suculae* "Piglets," and so interpreted by Manilius : 5. 125 f.

3 GEMINI 265 (*Gem*) : Brothers, not named by Manilius, though Eratosthenes and others identify them with the Dioscuri, Castor and Pollux : 2. 568 ; 4. 756. *Naked* 2. 163, 184 ; *arms about each other* 2. 164.

4 CANCER 266 (*Cnc*) : The crab sent by Juno to assail Hercules in combat with the Hydra ; it bit the hero, but was killed : rewarded with catasterism : 2. 33. At 4. 530 Manilius refers to the cluster *Praesaepe* "the Manger." *Has no eyes* 2. 259 f.

5 LEO 266 (*Leo* : *Nemeaeus*, *-eeius*) : The Nemean lion, whose skin Hercules wore as spoil : 2. 32, 531. *Runs* 2. 246.

6 VIRGO 266 (*Vir* : *Erigone*) : Erigone (the name occurs frequently), who on finding her father Icarius dead

hanged herself in grief and was raised to heaven for her piety. An alternative story (*cf.* Aratus, *Phaen.* 98 ff.) identified her as Astraea (or Justitia), daughter of Jupiter (or Astraeus), who at the advent of the Bronze Age fled to heaven: Manilius has inconsistently introduced her at 4. 542 f. *Has wings on her shoulders* (by implication: *cf.* Ptolemy, *Synt.* 7. 5 pp. 102 f. and Farnese globe) 2. 176, 662; *carries wheatsheaf* 2. 442.

7 LIBRA 267 (*Lib*: *Chelae* and so always named by Hipparchus): Originally the claws of the Scorpion, which were conceived as the pans of a balance: *Ζυγός* (first in Geminus) and *Libra* are not found before 1st cent. B.C. Manilius's picture of the sign varies: *the claws holding a balance* 1. 611, 4. 547 f.; *the scales collapsed* 2. 251; *the scales held by a male figure* 2. 528.

8 SCORPIUS 268 (*Sco*: *Nepa*): The scorpion which killed Orion at the behest of Diana (*cf.* Aratus, *Phaen.* 636 ff.: note that Orion sets at the rising of Scorpio): 2. 32. *Has lost its claws to Libra* 2. 258.

9 SAGITTARIUS 270 (*Sgr*: *Arcitenens*, *Centaurus*, *Sagittifer*): Originally the Archer seems to have been a satyr (Crotus son of Eupheme, according to Eratosthenes); and often he is pictured (*e.g.* on the Farnese globe) as having two legs only. For others, however, including Hipparchus, he is a four-footed centaur, and Manilius's frequent use of the name Centaurus shows that he also takes this view. *Aims arrow* 1. 269 f.; *runs* 2. 246; *has one eye only* (=profile) 2. 260; *wears cloak* 4. 560.

10 CAPRICORNUS 271 (*Cap*: *Caper*): An import from abroad. Here and elsewhere (2. 252, 445) termed a cramped constellation, because it does not fill up thirty degrees. *A goat-fish* 2. 231, 4. 795.

11 AQUARIUS 272 (*Aqr*: *Iuvenis*, *Urna*): Ganymede, carried off by the Eagle for Jupiter: 5. 487. The water poured out from his urn is considered by Manilius as joining *Eridanus* to form a composite sign, see 44 *Flumina.* *Naked* 2. 511.

12 PISCES 273 (*Psc*): On the banks of the Euphrates Venus and Cupid (Manilius, however, does not mention Cupid), suddenly confronted by the giant Typhon, dived into the river and eluded pursuit by changing themselves into two fishes (Hyginus, *Poet. Astr.* 2. 30): 2. 33; 4. 579 ff., 800 f. Evidently of Babylonian origin. *The fishes face different directions* 2. 165.

(b) Northern

13 HELICE 296 (*UMa = Ursa Major*: *Arctos*, *Ursa*): Helice (so Aratus) is doubtless an old appellation, given because of the sign's winding revolution round the pole. Callisto, daughter of Lycaon, ravished by Jupiter and turned into a bear by Diana; Jupiter, to save her when she was pursued by Arcadians, set her among the stars: 2. 29; 3. 359. *The Bears follow each other* 1. 303 f.

14 CYNOSURA 299 (*UMi = Ursa Minor*): The name Dog's Tail reflects an old conception of the constellation (Schol. Aratus, *Phaen.* 27: Callisto's dog). A nymph of Cretan Ida who nursed Jupiter: 2. 30.

15 DRACO 306 (*Dra*: *Anguis*, *Serpens*): Usually said to be the snake set by Juno to guard the golden apples of the Hesperides, with which Manilius (5. 16) has identified Hydra, perhaps confusing the two. *Separates and surrounds the Bears* 1. 305 ff.: this, however, is an error (not confined to Manilius) arising out of a misinterpretation of Aratus, *Phaen.* 45 f. (Malchin 49).

16 ENGONASIN 315 (*Her = Hercules*): A man on his knees, not identified by Manilius, who emphasizes the mysterious nature of this sign (generally taken by the mythographers to be Hercules exhausted after his labours).

17 BOOTES 316 (*Boo*: *Arctophylax*): A drover, likewise not identified (others make him Arcas, son of Callisto, or Lycaon, Callisto's father). *Presses forward like a drover* 1. 317 (but this need be no more than an etymology).

18 CORONA 319 (*CrB = Corona Borealis*): The crown worn by Ariadne (the mythographers represent it

as golden, but Manilius may have thought of it as a wreath, *cf.* 5. 263).

19 LYRA 324 (*Lyr*: see also 46 *Fides*): The lyre which Orpheus received from its inventor Mercury and with it overcame the barrier of the underworld: 5. 325 ff. *Inverted* (as seen on globe) 1. 627.

20 OPHIUCHUS 331 (today classified as two constellations, Snake-holder and Snake: *Oph* and *Ser* = *Serpens*: *Anguitenens*): No identification (others see in it Aesculapius). *The snake's face turns back to regard the man* 1. 334.

21 CYCNUS 337 (*Cyg* = *Cygnus*: *Olor*): The swan which served as Jupiter's disguise when he seduced Leda: 1. 337 ff.; 2. 31; 5. 25, 381. *Wings outspread* 1. 341; *slanting* (as seen on globe) 1. 687.

22 SAGITTA 342 (*Sge*): An arrow not further described by Manilius: its occasional figuration in the Eagle's talons arises from a mistaken interpretation of Germanicus.

23 AQUILA 343 (*Aql*): The eagle which carried off Ganymede: 5. 487. *Supine* (as seen on globe) 1. 688.

24 DELPHINUS 346 (*Del*): Manilius seems to accept Eratosthenes' identification with the Dolphin who persuaded Amphitrite to marry Neptune, for which he honoured it in sea and heaven alike: 1. 347.

25 EQUUS 348 (*Peg* = *Pegasus*): Originally just a horse (so 1. 348 f.), but in Book 5 Manilius seems to accept the later identification of it with the winged horse Pegasus. *Tries to catch Dolphin* 1. 348; *cut off in the constellation Andromeda* 1. 350; *flies* 5. 24, 633.

26 DELTOTON 353 (*Tri* = *Triangulum*: *Trigonum*): simply a triangle. *Isosceles* 1. 351 f.

27 CEPHEUS 354 (*Cep*): Cassiepia, wife of Cepheus, boasted that her beauty (so Aratus, *Phaen.* 657 f.: Andromeda's, Hyginus, *Fab.* 64) excelled that of the Nereids; Neptune in anger sent a flood and a monster (Cetus) to invade the land, and an oracle ordered that their daughter Andromeda be sacrificed to it: her ordeal and rescue by Perseus are told at length 5. 540-618. *Wears the garb of a tragic actor* 5. 470 (Moeller 29; Farnese globe).

28 CASSIEPIA 354 (*Cas* = *Cassiopeia*: *Cassiope*): See on 27. Ancient authors consistently give her name as *Cassiepia* (*Κασσιέπεια*), and this form I have used in preference to the modern astronomical *Cassiopeia*. This is a hybrid influenced by the variant *Cassiope* (see note on 5. 504). *Upside down* (*cf.* Aratus, *Phaen.* 654 ff.) 1. 686.

29 ANDROMEDA 356 (*And*: *Cepheis*): See on 27.

30 PERSEUS 358 (*Per*): See on 27, but catasterized for his victory over Medusa: 5. 22. I have wondered, reflecting that the story of Perseus and Andromeda is not known to Greek literature before Herodotus 7. 150, whether the hero's association with the heroine may not even have been created to explain their contiguity in the skies (*cf. Proc. Afr. Class. Ass.* 2 [1959] 10-15 and also above p. xxii). *Sickle in hand*: 5. 22.

31 HENIOCHUS 362 (*Aur* = *Auriga*): Erichthonius, first to drive a four-horse chariot. Manilius presents an inconsistent picture: *steps on ground* 1. 361; *continues to drive his chariot in heaven* 5. 20, 68 ff.

HAEDI 365 (*ζη Aur*: see also 45 *Haedus*): These, though stars in Auriga, undoubtedly belong to a previous independent conception: twin kids of Capella according to Hyginus. *On the Charioteer's left wrist* (Ptolemy, *Synt.* 7. 5 pp. 66 f.).

CAPELLA 366 (*α Aur*: *Olenie*): A star in Auriga, but like the Kids, whose mother it is said to be, it seems to have possessed an earlier, independent identity as the goat that nourished Jupiter in Crete: 1. 366 ff.; 2. 30; 5. 130 ff. *On the Charioteer's left shoulder* (Ptolemy, *Synt.* 7. 5 pp. 66 f.).

(c) Southern

32 ORION 387 (*Ori*): Apart from his being killed by a scorpion (see 8 *Scorpius*) the poet tells us nothing of Orion's story. The largeness of the constellation is mentioned 1. 388; 5. 12, 58. Individual stars referred to: *shoulders* 1. 390; *sword* 1. 391; *head* 1. 393; *belt* (this must be the reference of *Iugulae*) 5. 175. *Near the Twins* 1. 387.

33 CANICULA 396 (*CMa = Canis Major*: *Canis*): Originally Orion's dog (and probably only the star Sirius); known to Homer (*Il.* 22. 29). Manilius attempts no explanation. *With flashing face* (*i.e.* Sirius = α *CMa*, the brightest star in the sky) 1. 407; *tries to seize the Hare* 5. 233.

34 PROCYON 412 (*CMi = Canis Minor*): The precursor of the dog *Canicula*: no further information given by Manilius.

35 LEPUS 412 (*Lep*): A hare placed in heaven, says Eratosthenes, on account of its swiftness; which Manilius seems to have in mind at 5. 159 ff. *Perpetually chased by Canicula* 5. 233.

36 ARGO 412 (today divided into four constellations: *Car = Carina*; *Pup = Puppis*; *Pyx = Pyxis*; *Vel = Vela*: *Ratis*): The first ship, given apotheosis for protecting gods (*i.e.* the heroes who sailed in quest of the Golden Fleece): 1. 413 ff. *Rudder and stern-top located* 1. 623, 694; *sails among the stars* 5. 13, 36; *drawn by Aries to its side* (but this is an error) 5. 36. Ancient authorities concur in figuring only the sternward part of the ship.

37 HYDRA 415 (*Hya*: *Anguis*): The wakeful warder of the Hesperides (though others make this *Draco*, and in this Manilius may have erred): 5. 16. The mythographers generally link Hydra, Corvus, and Crater together, evolving a droll story to account for their juxtaposition (*cf.* Ovid, *Fasti* 2. 243 ff.).

38 CORVUS 417 (*Crv*): The raven sacred to Apollo, who adopted its shape to elude Typhon: 1. 417, 783. See also on 37.

39 CRATER 418 (*Crt*): A bowl, not further described. See also on 37.

40 CENTAURUS 418 (*Cen*): The identification with Chiron made by Eratosthenes and others is reflected in the endowments the sign bestows at 5. 348 ff. Many authorities depict the Centaur as holding up a beast: on this see 50 *Bestia*.

41 ARA 421 (*Ara*): The altar set up by Jupiter before the battle with the Giants (rather the Titans): 5. 341 ff.

42 CETUS 433 (*Cet* : *Pristis*) : See on 27. *Has jaws wide open* 1. 356 ; *scaly back and coils* 1. 433 f. ; *amphibious* 5. 15 ; *makes for Andromeda* 5. 657.

43 PISCIS NOTIUS 438 (*PsA* = *Piscis Austrinus*) : Manilius knows nothing about this sign save its place and name.

44 FLUMINA 440 (*Eri* = *Eridanus*) : This constellation (here and at 5. 14) is peculiar to Manilius. The chief member is the star-group generally called Eridanus (no doubt originally just a river), which flows from Orion's left foot into Cetus. For other authorities it then flows out again, almost reversing its direction before turning south. Manilius, however, seems to have imagined it flowing right through Cetus and on to the mouth of the Southern Fish, before reaching which it had necessarily to be joined by the waters from Aquarius's stream (sometimes reckoned as a separate constellation). The two streams constitute the poet's *Flumina.*

Ghost constellations

45 HAEDUS 5. 311 : a mistaken duplication of *Haedi* (part of 31 *Heniochus*) : see below pp. xciv f.

46 FIDES 5. 409 : a mistaken duplication of 19 *Lyra* : see below p. xcvi.

Omitted constellations

47 EQUULEUS (*Equ*) : *Προτομὴ Ἵππου* Geminus 3. 8 ; Ptolemy, *Synt.* 7. 5 pp. 76 f. ; Allen 212 ff.

48 COMA BERENICES (*Com*) : Geminus 3. 8 ; Allen 168 ff.

49 CORONA AUSTRALIS (*CrA*) : Geminus 3. 13 ; Ptolemy, *Synt.* 8. 1 pp. 164 ff. ; Allen 172 ff.

50 BESTIA (*Lup* = *Lupus*, a mistranslation from the Arabic, *cf.* Allen 278) : The beast held by Centaurus (and perhaps for Manilius contained within that constellation), in Greek *Θηρίον* : Aratus, *Phaen.* 442 ; Vitruvius 9. 5 ; Ptolemy, *Synt.* 8. 1 pp. 162 ff. The asterism is of interest as explaining the error at 1. 433, where next to Ara the poet places Cetus (which is on the other side of the sky) : G. R. Mair (Loeb Aratus, *Phaen.* 442 n.) has seen that

Manilius, noting that a beast was catasterized next to Ara, has in his haste announced it to his readers as Cetus; then (as I think, comparing his mistakes over Lyra and Orion in Book 5: see below p. xcvi) he decided to let his error stand when he discovered it, as he was bound to do on reaching the true Cetus. I formerly accepted the transposition of 1. 433-437 after 442, independently suggested by Garrod[1] and Naiden-Householder (the latter make the further necessary change of *quam* to *quae*), but now recognize that this corrects the poet rather than his scribes.

1. 532-538, 805-808: The planets

Now that Housman has definitively transposed verses 805-808 to their proper place here, one can appreciate that this little paragraph neatly serves to conclude, with a mention of the planets, Manilius's account of the heavenly bodies before he moves on to describe the measurements and boundaries of heaven. Indeed, Aratus had followed just this order (*Phaen.* 19-453 constellations; 454-461 planets; 462-558 circles).

The curt treatment accorded the planets may to some seem strange in an astrological writer. It needs to be remembered, however, that it is difficult to give simple instructions for discriminating between them or to describe for the layman their celestial vagaries: οὐδ' ἔτι θαρσαλέος κείνων ἐγώ "when it comes to them my daring fails," says Aratus, *Phaen.* 460, understandably. Nevertheless, we may perhaps infer that Manilius, who is always ready to turn numbers into hexameters, is following some commentator on Aratus similar to the source of Achilles (*Isag.* 16 p. 42, 25 ff.), who simply lists the planets, rather than Geminus (1. 24-30), who would have provided him with the periods of their zodiacal orbits. These are (with their sidereal periods in parentheses): Saturn: about 30 years (29.46); Jupiter 12 years (11.86); Mars: 2½ years (1.88); Sun: 1 year; Venus: about the same (.62); Mercury: about the same

(.24) ; Moon : 1 month. The poet similarly arranges the planets at 5. 6 f., and for the first of those verses repeats 1. 807 verbatim.

1. 539-804 : The circles of the sky

Preparatory to an exposition of the celestial circles Manilius treats us to a little geometry of the circle itself (Euclid, *Elem.* 4. 15 : the radius of a circle is equal to the side of an inscribed regular hexagon). He cannot resist the temptation to put π into Latin verse, although it hardly advances his purpose. But at 561 he gets down to the business in hand. A lacuna of some six verses has robbed us of the beginning of the parallel circles, in particular of the Eudoxean division of a circle into 60 degrees (a scale adopted by Geminus 5. 46 in the same context). Since Manilius in his astrological books uses the 360° circle, it would seem that, when he uses different sources, he does not trouble to harmonize them. Figure 2 represents diagrammatically the fixed circles of the sky (1. 561-630).

(1) 563A-602 : He deals first with the five **parallels.** Originally the celestial arctic circle was defined as bounding that part of the heavens which never sets, the antarctic that which is never visible (Diels 340b 15 ff.). But it would then follow that the circle would vary for different latitudes : Posidonius seems to have been the first to fix an immovable arctic circle by re-defining it as the most southerly latitude in which at the summer solstice the Sun remains above the horizon all day. Manilius's location does not, however, con-form to this definition, being rather the original variable arctic as determined for the latitude of Rhodes (6° Eud. from pole = 54° lat. N). The celestial tropic

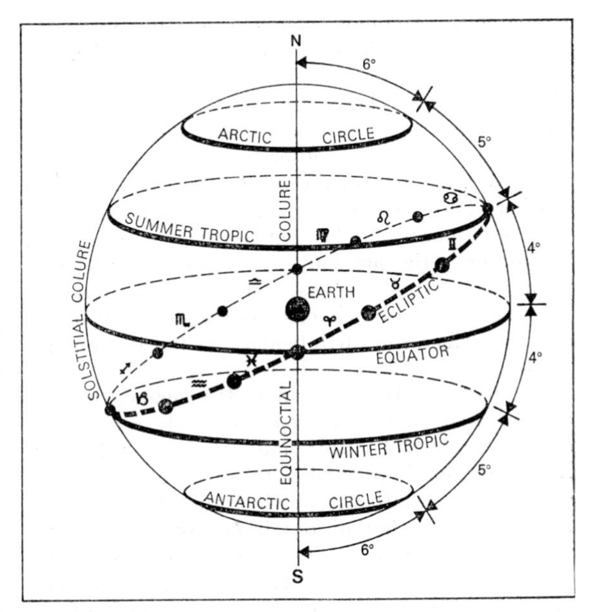

Fig. 2. The Fixed Circles of the Sky, 1. 561-630

of Cancer (given as 24° N), equator, and tropic of Capricorn are then briefly defined and located, followed by the antarctic circle.

(2) 603-630 : The **colures** are two in number, (a) the meridian (or hour-circle) which passes through the equinoctial points and the poles, and (b) the meridian which passes through the solstitial points and the poles. Both are marked on the star-charts at the end of this volume, where Manilius's description may conveniently be traced. Like Aratus and

Hipparchus, Manilius places the equinoxes and solstices at the beginnings of the tropic signs (but see further p. lxxxi).

(3) 631-662 : There are two circles, however, which, unlike the previous ones, have no fixed abode : they move with the observer. The first is the **meridian** : it runs down from the north pole (to both southern and northern horizons) ; it cuts the sky in two, and marks midday when the Sun crosses it. The other is the **horizon**, *i.e.* the observer's horizon projected on to the celestial sphere ; this also cuts the sky in two—that hemisphere which we can see, and that which the earth beneath us blocks from our vision.

(4) 666-680 : We proceed to the **zodiac**, the first of two great circles which are specially remarkable as being real and visible and possessing thickness. It may be simply defined as the ecliptic given width (properly the width within which the planets move). Manilius's omission of detail has prompted some versifying editor, whose technique can more clearly be seen at 2. 732-734 and 2. 968-970, to interpolate (681-683) the conventional measurement of 12 degrees (*cf.* Geminus 5. 53), which he has been careful to point out are non-Eudoxean ones.

(5) 684-804 : The second of the visible circles is the **Milky Way**. This, we learn from an anonymous commentator on Aratus (p. 95, 30 f. Maass), was marked on some ancient star-globes (though not on the Farnese), and it is very likely that our poet was using one such.

Manilius (as also Manetho 2. 118 ff.) starts from its northern intersection with the equinoctial colure and traces the following circuit : Cassiepia, Cycnus, summer tropic, Aquila, equator, Scorpio-Sagittarius, Southern Centaur, Argo, equator, Gemini, Heniochus, Perseus, and back to Cassiepia.

We can hardly wonder at the failure of the ancients to discover that the Milky Way is a huge disc of stars, of which the solar system is an undistinguished component. Manilius assembles various theories about its origin, with which may be compared Achilles (55, 7 ff. Maass), Macrobius, *Comm. Somn. Scip.* 1. 15. 3 ff., and Diels 229 ff.

(1) 718-728 : The juncture of two hemispheres coming apart (Theophrastus) or, it may be, coming together (Diodorus of Alexandria) ;
(2) 729-734 : The former path of the Sun (Oenopides of Chios) ;
(3) 735-749 : The disastrous route of Phaethon (Pythagoreans) ;
(4) 750-754 : Milk spilled from Juno's breast (cf. Eratosthenes, *Catast.* 44, pp. 198 f.) ;
(5) 755-757 : A concentration of countless small stars (Democritus) ;
(6) 758-761 ff. : The abode of the souls of heroes (Cicero, *De Rep.* 6. 16).

Here the poet takes over from the doxographer and surveys the glorious dead in paradise, inspired perhaps by the pageant at the end of *Aeneid* 6, though he does not repeat Virgil's success.

1. 809-926 : Comets

Having now reached the end of the astronomical portion of his poem, Manilius subjoins a chapter on comets (and meteors, which, except for the special mention of the latter in 847-851, he seems not to have distinguished) to enable him to introduce and develop his peroration.

He advances three explanations.

(1) 817-866 : Inflammable earth-vapours ignited by dry air (Aristotle, *Meteor.* 1. 341 b) ;
(2) 867-873 : Stars attracted to and then released by the Sun (Diogenes Apolloniates : Diels 366) ;
(3) 874 ff. : God's means of warning mankind against impending calamity.

Says Breiter perceptively : "As in his account of the Milky Way, the poet closes with the view that is scientifically least plausible ; but it leads into the splendid digression which concludes the book."

Manilius's types of comet (835-851) are readily identifiable with the classifications we meet with elsewhere (*e.g.* [Aristotle], *De Mundo* 4 ; Seneca, *N.Q.* 7 *passim* and 1. 15. 4 ; Pliny, *N.H.* 2. 89 ff. ; also among the astrologers, *cf.* Ptolemy, *Tetr.* 2. 9 ; Hephaestion 1. 24).

(1) 835-839 : Hairy :
crinitae, κομῆται, long-haired ;
pogoniae, πωγωνίαι, beards ;
(2) 840 f. : Geometrical :
trabes, δοκίδες, beams ;
columnae, κίονες, columns ;
(3) 842-846 : Various :
pithei, πίθοι, casks ;
hirci, αἶγες, goats ;
faces, λαμπάδες, torches ;
(4) 847-851 : Shooting-stars :
faces (or *bolides*), διᾴττοντες, meteors.

These by no means exhaust comet-types, over thirty kinds of which Isidore (*Orig.* 3. 71. 17) tells us the Stoics recognized.

For his peroration the poet has clearly taken as a model the ending of the first book of the *Georgics*.

BOOK TWO

With this book Manilius embarks upon astrology proper, so that this is perhaps the most suitable place to assemble a select bibliography of the subject. In his introductions to Books 2 to 5 Housman admirably elucidates, with copious quotation from ancient sources, the technical passages in the text, and in his review of van Wageningen (ed. 5² p. xxviii) expresses the opinion that Sextus Empiricus's polemical treatise is the best introduction to Greek astrology.

[*General*: Bouché-Leclercq; Cramer; Lindsay. *Ancient sources*: (*survey*) W. and H. G. Gundel: (*authors*) Geminus; Dorotheus; Ptolemy, *Tetrabiblos*; Vettius Valens; Sextus Empiricus; Porphyry; Firmicus Maternus; Manetho; Paulus Alexandrinus; Hephaestio Thebanus; Heliodorus; *CCAG*. *Barbaric sphere*: Boll. *Horoscopes*: Neugebauer and Van Hoesen]

2. 1-149: Prooemium

In a studied exordium, a feature of all his books, the poet surveys the genres of hexameter poetry from Homer, who is acknowledged to be the fountainhead of all, down to his own time. By making special mention of the various themes of didactic poets he stakes his own claim to a place on the roll of innovators. Naturally, the poets named or alluded to are Greek, but it is surprising that Manilius refrains from commenting on the fact and from seeking applause as a Latin poet. He proceeds to treat generally of Stoic beliefs and the lonely majesty of Stoicism; and here and there succeeds, as rarely in his work, in striking a note of sublimity.

2. 150-269 : The signs of the zodiac

For our first lesson in astrology we are to learn the characteristics of the signs of the zodiac : these the poet expounds under ten heads.

(1) 150-154 : Masculine (Aries, Gemini, Leo, Libra, Sagittarius, Aquarius) and feminine (Taurus, Cancer, Virgo, Scorpio, Capricorn, Pisces). The criterion is not so much sex as the Pythagorean notion that odd numbers are male, even female. Of the male signs Libra is occasionally conceived (so apparently at 2. 528) as a male figure holding the balance. Taurus seems strange in group of females, though the absence of its hind quarters and its backward rising make the oddest speculations possible ; and Capricorn's great beard is yet a further challenge to one's credulity.

(2) 155-157 : Gemini, Virgo, and Aquarius (to which Libra is possibly to be added : see 2. 528) are human ; Aries, Taurus, Cancer, Leo, Scorpio, Capricorn, and Pisces are bestial (the first two cattle, the rest ferine). Sagittarius combines both natures.

(3) 157-196 : Aries, Taurus, Cancer, Leo, Libra, Scorpio, and Aquarius are single. The remainder are double. Of these Gemini and Pisces are pairs, Sagittarius and Capricorn are composites, man-horse and goat-fish respectively, while Virgo (who has a pair of wings) demands classification as a double by belonging to the square ♊♍♐♓, each of whose members effects a juncture of two seasons.

(4) 197-202 : Unlike the other signs, which rise in upright posture, Taurus, Gemini, and Cancer do so upside down. In Gemini the Sun is (or was during the Hellenistic age) in apogee, that is to say at its

farthest from the Earth (for in fact its distance varies during the year), and its motion is consequently slowest in this and the adjacent signs Taurus and Cancer ; and this tardiness Manilius fancifully ascribes to the reverse position of these signs. It is improbable that he understood the eccentricity of the Sun's orbit, which was discovered by Hipparchus. Geminus, not quite agreeing with Manilius, gives the length of the Sun's passage between tropic points as follows : in Aries, Taurus, and Gemini 94½ days ; in Cancer, Leo, and Virgo 92½ days ; in Libra, Scorpio, and Sagittarius 88⅛ days ; and in Capricorn, Aquarius, and Pisces 90⅛ days (*cf. Isag.* 1. 13 ff. and Manitius's note p. 253).

(5) 203-222 : Diurnal and nocturnal : the poet expounds three classifications, of which he evidently prefers the first (211-217). This separates the signs into alternate pairs : Pisces and Aries, Cancer and Leo, and Scorpio and Sagittarius are diurnal (*i.e.* the trigons of Aries and Cancer, the signs in which respectively the day becomes longer than the night and the day attains its greatest length) ; the remainder are nocturnal (the trigons of Libra and Capricorn, the signs in which respectively the night becomes longer than the day and the night attains its greatest length).

Some (218-220) hold that the signs from Aries to Virgo (*i.e.* in which the day is longer than the night) are diurnal, and the remainder nocturnal (by the same principle).

The third classification (221 f.) repeats the one we learnt in 150-154 : masculine signs are diurnal, feminine nocturnal. Here too the basis for it is Pythagorean and numerical, light being odd and darkness being even (Aristotle, *Metaph.* 1. 5 986a).

(6) 223-233 : Cancer and Pisces are aquatic ; Aries, Taurus, Leo, and Scorpio terrene ; Capricorn and Aquarius amphibious. Manilius has omitted Gemini, Virgo, Libra, and Sagittarius, which are nothing if not terrene : possibly their human nature excluded them from classification, Aquarius being exceptionally mentioned on account of his association with water, which provided too good an opportunity to miss.

(7) 234-243 : The signs are now assessed in terms of fertility : Cancer, Scorpio, and Pisces are prolific, like the creatures their names designate ; Virgo is naturally unfruitful, and thus too are reckoned Leo (the lioness was thought to bear cubs once only) and Aquarius (who, if not rendered sterile by being Ganymede, is likely to suffer from his urn's incapacity to hold what it has received) ; the remainder are intermediate.

(8) 244-255 : Posture (a criterion peculiar to Manilius) determines the next grouping. Three are running signs, Aries, Leo, and Sagittarius ; three standing, Gemini, Virgo, and Aquarius ; three sitting, Taurus, Libra (*i.e.* with its scales collapsed), and Capricorn ; three lying, Cancer, Scorpio, and Pisces.

(9) 256-264 : Disfigurement (another criterion peculiar to Manilius) specially characterizes four signs : the Scorpion's claws have been appropriated by Libra, Taurus limps with a leg doubled under it, Cancer has no eyes, and Sagittarius (who is figured in profile) only one.

(10) 265-269 : The four seasons are each allotted three signs. Spring gets Pisces, Aries, and Taurus ;

summer Gemini, Cancer, and Leo ; autumn Virgo, Libra, and Scorpio ; and winter Sagittarius, Capricorn, and Aquarius. What Manilius has said at 2. 176-196 confirms this division, although other astrologers begin the seasons with the tropic signs.

2. 270-432 : The conjunctions of the signs

We proceed to the relationships of the signs with each other : geometrical figures inscribed within the circle of the zodiac indicate trine, quartile, and sextile aspects, and diameters indicate opposition.

(1) 273-286 : The construction of four equilateral triangles produces the following scheme of trigons (figure 3) :

1. Aries, Leo, Sagittarius.
2. Taurus, Virgo, Capricorn.
3. Gemini, Libra, Aquarius.
4. Cancer, Scorpio, Pisces.

(2) 287-357 : The construction of three squares produces the following scheme of tetragons (figure 4) :

1. Aries, Cancer, Libra, Capricorn.
2. Taurus, Leo, Scorpio, Aquarius.
3. Gemini, Virgo, Sagittarius, Pisces.

But we must be careful to calculate aright : we may not connect any degree of a sign we please with any degree of its trigonal or quadrate neighbours ; the side of a trigon requires exactly 120, that of a square exactly 90 degrees. Such kindergarten instruction takes the poet many lines, but, as Housman with unwonted geniality happily puts it : "The inordinate

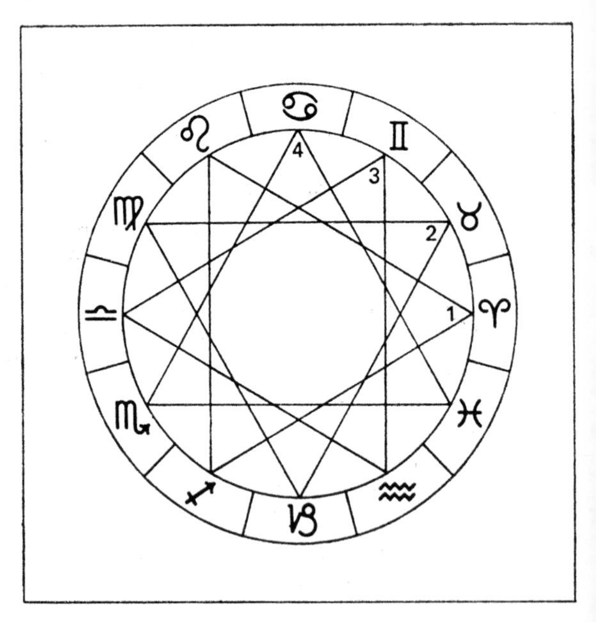

Fig. 3. Trigona, 2. 273-286

length of Manilius's exposition is perhaps after all less due to a low estimate of his reader's knowledge of ciphering than to the pleasure he takes in exercising that eminent aptitude for doing sums in verse which is the brightest facet of his genius." In 352-357 we are told that the trigon's power is far greater than the square's, though this is not orthodox doctrine : for most astrologers the trigon's power is for good, the square's for ill.

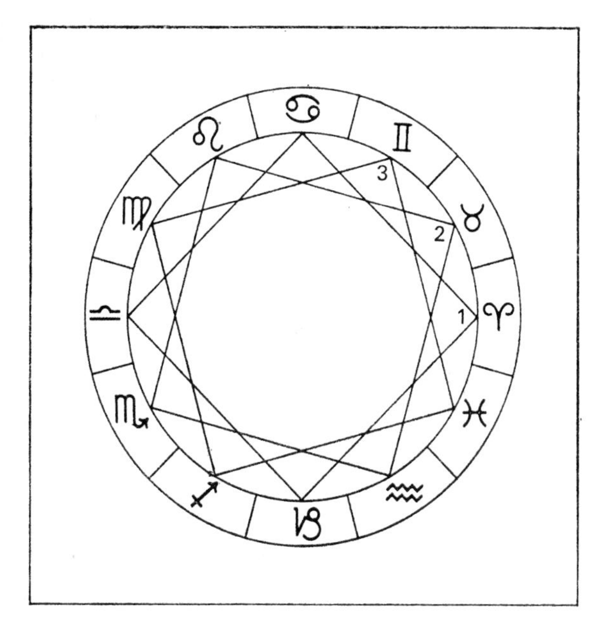

Fig. 4. Quadrata, 2. 287-296

(3) 358-384 : The construction of two hexagons produces the alliance of masculine and feminine signs (figure 5) :

1. Aries, Gemini, Leo, Libra, Sagittarius, Aquarius.
2. Taurus, Cancer, Virgo, Scorpio, Capricorn, Pisces.

(4) 385-432 : Before enumerating the diametric

signs Manilius considers the relationship between neighbouring signs as members of an inscribed dodecagon, but judges this aspect to be powerless,

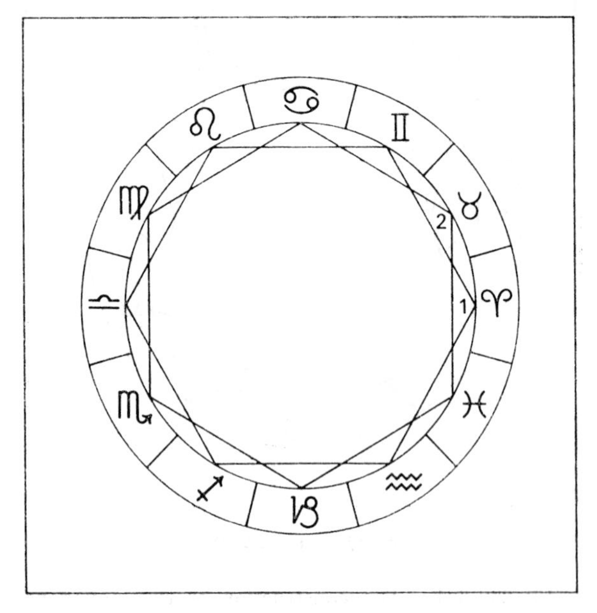

Fig. 5. Hexagona, 2. 358-384

as he does that between signs separated from each other by four intervening signs ; in the latter case the line of relationship does not form the side of an equilateral figure. The aspect of direct opposition, which is strong though for the most part discordant, produces the following scheme (figure 6) :

1. Aries : Libra.
2. Taurus : Scorpio.
3. Gemini : Sagittarius.
4. Cancer : Capricorn.
5. Leo : Aquarius.
6. Virgo : Pisces.

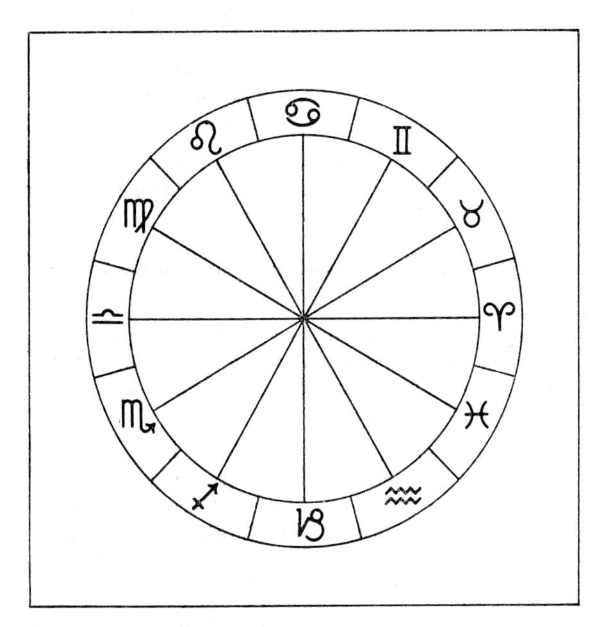

Fig. 6. Diametra, 2. 395-432

2. 433-452 : The guardians of the signs

Here Manilius digresses. Mention of the diametric signs has reminded him that the gods and goddesses in pairs have been allotted the guardianship of the diametric signs in pairs, and he now sets forth the following scheme :

1. Aries → Minerva	:	Vulcan ← Libra.
2. Taurus → Venus	:	Mars ← Scorpio.
3. Gemini → Apollo	:	Diana ← Sagittarius.
4. Cancer → Mercury	:	Vesta ← Capricorn.
5. Leo → Jupiter	:	Juno ← Aquarius.
6. Virgo → Ceres	:	Neptune ← Pisces.

This arrangement has nothing to do with the plan by which the signs of the zodiac become planetary houses (for which see below p. xcix).

2. 453-465 : The parts of the body allotted to the signs

The digression affords our teacher an opportunity to tell us of the dominion exercised over the several parts of the body by the individual signs. The allotment is as follows :

1. Aries : head.	7. Libra : loins.
2. Taurus : neck.	8. Scorpio : groin.
3. Gemini : arms.	9. Sagittarius : thighs.
4. Cancer : breast.	10. Capricorn : knees.
5. Leo : sides.	11. Aquarius : shanks.
6. Virgo : belly.	12. Pisces : feet.

Herein for once astrologers find themselves in complete agreement. Manilius, remarkably, repeats the list at 4. 701-709.

2. 466-692 : More relationships of the signs

(1) 485-519 : The poet now propounds a further set of relationships determined by parallel lines

drawn within the circle. (*a*) The diameter connecting Aries and Libra and two parallels on either side connecting the other signs (save Cancer and Capricorn, which are thus excluded from the scheme)

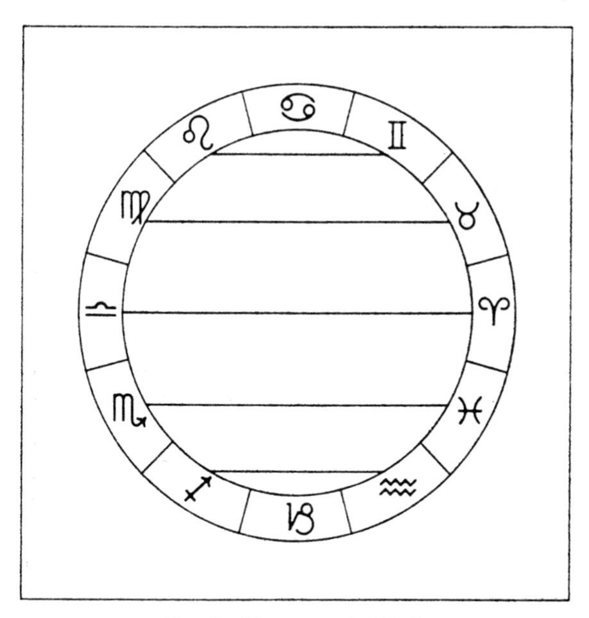

Fig. 7. Videntia, 2. 466 ff.

identify those which see each other (*videntia*, *βλέποντα* : see figure 7). Some authorities, however, connect the *videntia* by points (0°♈ with 0°♎) and not signs as a whole ; and as a consequence all the signs are paired (*i.e.* Gemini and Cancer ; Taurus and

Leo ; Aries and Virgo, etc.). (*b*) The diameter connecting Cancer and Capricorn and two parallels on either side connecting the other signs (save Aries and Libra, which are thus excluded from the scheme)

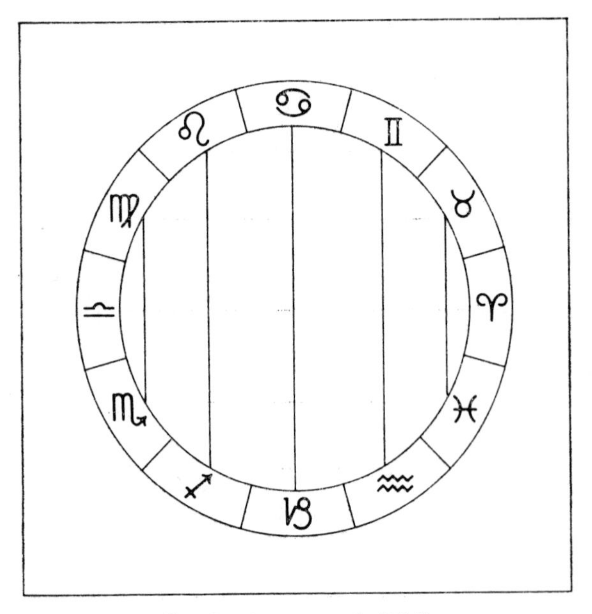

Fig. 8. Audientia, 2. 466 ff.

identify those which hear each other (*audientia* ἀκούοντα : see figure 8). Here again some authorities connect the *audientia* by points (0°♋ with 0°♑) and not signs as a whole, so that all the signs are paired (*i.e.* Pisces and Aries ; Aquarius and Taurus ;

Capricorn and Gemini, etc.). (*c*) The diameter drawn from the point dividing the first masculine sign Aries and the first feminine sign Taurus acts as a wall at which each masculine sign pays court to the

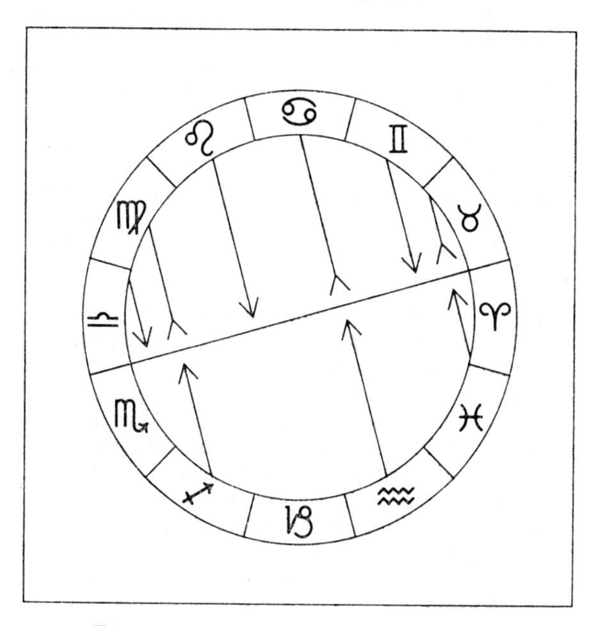

Fig. 9. Amantia and Insidiantia, 2. 466 ff.

feminine sign equidistant with himself from it; and she, if a northern sign, responds treacherously, if a southern, apparently not at all. The *amantia* and *insidiantia* (figure 9) are exclusive to Manilius.

(2) 520-607: We proceed to enmities. To begin

with, the first and third trigons (♈♌♐ and ♊♎♒ respectively) are at daggers drawn : not only is trigon opposite to trigon, but the signs of the former are all bestial, of the latter all human (Libra, as was indicated above, being here at any rate conceived as a male balance-holder). Then each individual sign has to suffer various assaults which Manilius arranges in three classes : (*a* : 572-578) from opposite signs and members of that opposite sign's trigon ; (*b* : 536-538) from bestial signs (if a human) and from human signs (if a beast) ; and (*c* : 539 f.) from signs motivated by sheer caprice. After enunciating the principles the poet works out their application (541-569), specifying each sign's individual assailants thus :

ARIES : *ab* Libra, Gemini, Aquarius ; *b* Virgo.
TAURUS : *a* Scorpio, Cancer, Pisces ; *b* Libra.
GEMINI : *ab* Sagittarius, Aries, Leo.
CANCER : *ab* Virgo ; *a* Capricorn, Taurus ; *b* Libra.
LEO : *ab* Aquarius, Libra, Gemini ; *b* Virgo.
VIRGO : *ab* Pisces, Cancer, Scorpio ; *b* Capricorn.
LIBRA : *ab* Aries, Leo, Sagittarius ; *b* Taurus, Cancer, Scorpio, Capricorn.
SCORPIO : *ab* Virgo ; *a* Taurus ; *b* Gemini, Libra, Aquarius ; *c* Leo, Sagittarius.
SAGITTARIUS : *ab* Gemini, Libra, Aquarius ; *b* Virgo.
CAPRICORN : *b* Gemini, Libra, Aquarius, Virgo.
AQUARIUS : *ab* Leo, Aries, Sagittarius.
PISCES : *ab* Virgo ; *b* Gemini, Aquarius ; *c* Sagittarius.

(3) 608-641 : Heaven's friendships are restricted to signs enjoying trigonal association, and even this

is no guarantee of good fellowship : in fact only Gemini's trigon leads a peaceful life. In his trigon Aries has to keep a sharp eye on his unpleasant partners ; Taurus finds Capricorn incompatible and Virgo, though attractive, infuriating ; and Cancer has in Scorpio a false, in Pisces a fickle friend.

(4) 643-692 : The innate relationships of the signs which the poet has expounded undergo modification as a result of the perpetual revolution of the zodiac. If a line is rising, its effects will be enhanced ; if setting, diminished. One modification is especially stressed : when quadrate signs pass through the framework of cardinal points (to be described in 2. 788 ff.), the cardinal takes precedence over the quadrate association and significantly affects marriage, children, and the near degrees of kinship. Similarly, the powers of trigons, hexagons, and even dodecagons will be intensified or depressed (though not altered) by passage through the cardinal points. Finally (687 ff.), there are modifications sustained not by the whole of a sign but only by part, and this brings us to a completely novel subject.

2. 693-737 : Zodiacal dodecatemories

Each zodiacal sign is divided into twelve equal parts or dodecatemories ; these are allotted to the signs in order, the first to the sign in which it is situated, the second to the next, and so on. Take for instance the first two signs : Aries gives its first dodecatemory (*i.e.* the first two and a half of its thirty degrees) to itself, the next to Taurus, and so on till the last, which it gives to Pisces ; Taurus gives its first dodecatemory to itself, its second to Gemini, and its twelfth to Aries (*cf.* figure 10). The signifi-

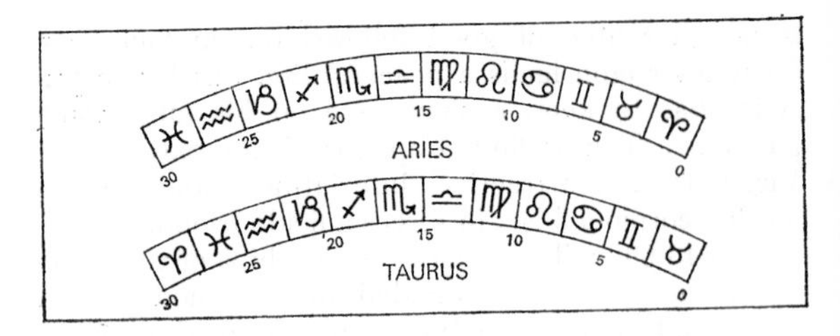

FIG. 10. ZODIACAL DODECATEMORIES, 2. 693-721

cance of the dodecatemories is this : not only is a planet's influence affected by the sign in which it stands, but it is further modified by the sign of the particular dodecatemory which it occupies.

No more than this need the poet have said. But operating with multiples of $2\frac{1}{2}$ in Latin calls for mental dexterity, and our teacher cannot resist helping with our arithmetical calculations. We should therefore note the degree of a sign that the Moon (given as the most obvious example of a planet) occupies at the moment of a nativity (let us suppose it to be the 9th degree of Aries) : it is desired to ascertain in what sign's dodecatemory that degree falls. Multiply the number by twelve ($9\times12 = 108$) ; dole out thirty apiece to the signs in order beginning with itself (30 to Aries, 30 to Taurus, 30 to Gemini), and note the sign in which the multiples of thirty give out (here Cancer, for $108-90 = 18$) : that will be the sign of the Moon's dodecatemory. This is merely a variation, of course, on what the poet has previously told us, and the whole point of multiplying by twelve is that we may operate with the easier number 30 instead of $2\frac{1}{2}$.

[2. 732-734 : The dodecatemory of the Moon]

So much is easy. But the explanation of the lines 2. 732-734 is complicated : "*When the number fails, then let the remainder be divided into portions of two and a half, so that these may be distributed in order among the remaining signs.*"

The term "dodecatemory," which is simply the Greek for "a twelfth," is frequently used as a synonym for "a sign of the zodiac," but occasionally we encounter it in a special conception, the dodecatemory of the Moon. This significant place in the zodiac (somewhat like the Lot of Fortune, to which we shall come at 3. 160-202) one found by a procedure described by Porphyry, *Isag.* 194 Wolf (= *CCAG* V. 4. 211) and repeated in essentially the same form at *CCAG* VIII. 2. 50 : "Count the number of degrees from the Sun to the Moon, and from these take away all multiples of thirty ; *as for the remainder, divide it into portions of two and a half, and distribute them in order from the zodiacal sign in which the Moon is situated.*" Let the Sun be situated in the 15th degree of Gemini, the Moon in the 23rd of Scorpio : the distance is 158 degrees. Take away 5×30 : that leaves 8. Divide this into portions of $2\frac{1}{2}$: one will go to Scorpio, one to Sagittarius, one to Capricorn, and the remainder comes to an end in Aquarius. The dodecatemory of the Moon is thus Aquarius.

Now this has nothing to do with Manilius, whose rule when the Moon stood in the 23rd degree of Scorpio would lead to the following computation : (726-728) $23 \times 12 = 276$; (729 f.) of this 30 go to Scorpio, leaving 246 ; (731) 8×30 go to Sagittarius, Capricorn, Aquarius, Pisces, Aries, Taurus, Gemini, and Cancer, with 6 left over ; (735 f.) so the thirties give out in Leo, and therefore when the Moon (or any other planet) occupies the 23rd degree of Scorpio it stands in the dodecatemory of Leo.

We can now see what has happened. At the point in Manilius's procedure where the casting out of thirties left a remainder, some astrological interpolator, believing this casting out of thirties to be part of the Porphyrian procedure, has fabricated three verses to express the rest of the operation (italicized above). His composition, which shrewdly included several echoes of Manilian language, was good enough to deceive Housman, though it is exposed as a fake by the anapaest *reliquis*, *cf.* Lachmann on Lucretius 5. 679. The interpolator assuredly meant to delete the lines 735-737, which he intended his own to supplant; and it is fortunate that his purpose was frustrated, for the survival of these lines confirms that Manilius did not (as Housman unaccountably supposes) write those, namely 732-734.

We have further evidence of this interpolator's existence and activity at the end of this book, where we encounter precisely the same phenomenon: holding again a different view from Manilius, he has composed three lines of his own (968-970) to take the place of three of the original (965-967); and by some kindly providence the latter have once more survived. Indeed, it is tempting to ascribe to the same individual much other of the interpolation detected in our author.

2. 738-748: Planetary dodecatemories

There are other subdivisions, also called dodecatemories, allotted not to the signs but to the planets. In size they each occupy half a degree, and there are five of them to every zodiacal dodecatemory. Manilius has inconsiderately forgotten to tell us the order in which the planets claim these planetary dodecatemories, but Housman refers us to 1. 807 and 5. 6 and suggests Saturn, Jupiter, Mars, Venus, and Mercury: this is illustrated in figure 11.

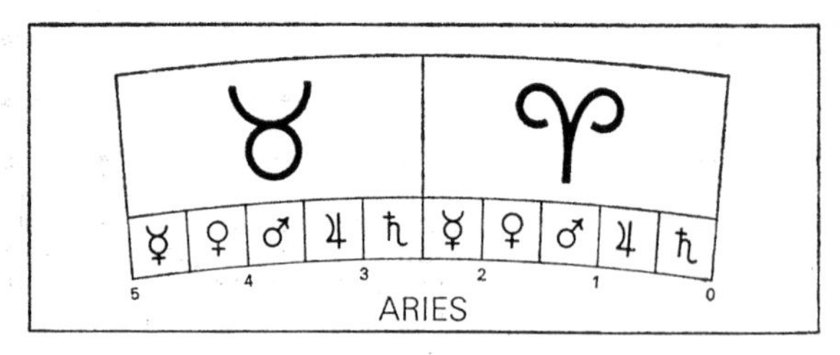

Fig. 11. Planetary Dodecatemories, 2. 738-748

2. 749-787 : Intermezzo

No paragraph actually begins at this point, but in broaching the subject of the dodecatemories and treading, as it were, on planetary ground, Manilius seems to have gone as far as his resources for the moment permitted ; and he draws back, excusing himself with the plea that students must not proceed to the advanced stages of a course before they have mastered the elementary. Suddenly, after a number of trite remarks, he launches upon an extended simile (praised by Scaliger and anthologized by Garrod) in which is portrayed the building of a city.

2. 788-967 : The fixed circle of the observer

It is no use taking bearings of the zodiac and planets, however accurate, until we know what to do with the information. We first need to be able to construct the fixed frame upon which every horoscope is cast. This subject will occupy Manilius for the rest of Book 2.

The casting of a horoscope is based on the belief that the zodiac and planets, as they ceaselessly revolve about the earth, exert different influences according to their different positions relative to the native. The astrologer, whom let us call the observer, regards the heavens as presenting the same frame to everyone. And we can draw this frame quite easily, because it is simply a circle, quartered where the horizontal and vertical diameters cut it (*i.e.* at the points corresponding to 9 and 3 and 12 and 6 on a clockface). What these points represent Manilius proceeds at once to tell us.

(1) 788-840 : They are the four cardinals, the

points at which the zodiac intersects (a) the eastern horizon (*horoscopos* or *Hor.*), (b) the western horizon (*occasus* or *Occ.*), (c) the meridian overhead (*medium caelum* or *MC*), and (d) the meridian underfoot (*imum caelum* or *IMC*)—see figure 12. The points, like the quadrants and the temples to which we shall presently come, possess no power themselves but direct the power of the zodiacal degrees occupying them towards certain areas of human experience. Manilius arranges the cardinals in this order: (c) the MC, affecting distinction, honours, and success generally; (d) the IMC, affecting wealth and its sources; (a) the Horoscope, affecting life generally, character, and station; and (b) the Occident, affecting the consummation of things (including marriages and the banquet which comes at the end of the day).

(2) 841-855: The quadrants of the circle linking the cardinal points also impart special direction to the forces of the zodiac whilst revolving through them: they hold dominion over the quarter-periods of human life, as follows: from the Ascendant to the MC over infancy; thence to the Occident over youth; thence to the IMC over adulthood; and thence back to the Ascendant over old age: this is illustrated in figure 12. Housman accuses Manilius of inconsistency in here attributing infancy to the first quadrant when at 833 he had given the early years to the Horoscope; and less easily resolved are the conflicting claims on old age of the Occident (839) and now the fourth quadrant (854 f.). Paul of Alexandria alone retails a like doctrine, and is no less the object of Housman's castigation.

(3) 856-967: We now come to the commonest

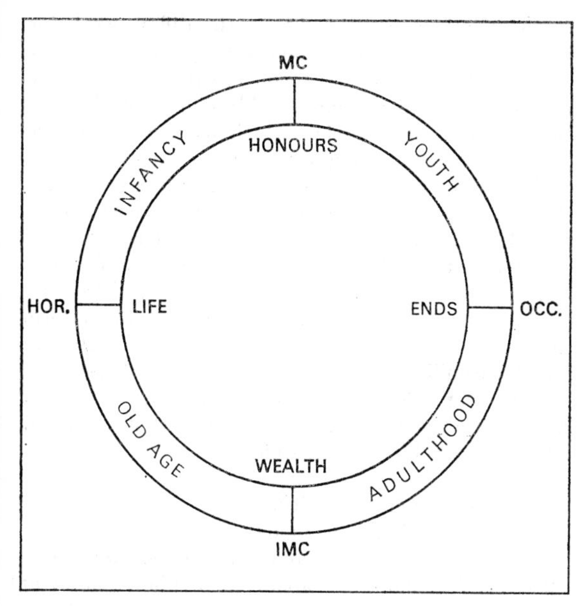

Fig. 12. Cardines and Intervalla, 2. 788-855

method of dividing the observer's fixed circle: the *dodecatropos* or division into twelve equal parts or *temples* (modern astrologers call the divisions houses, a term best avoided here since it means something else in Greek astrology, see p. xcix). Four of these temples embrace the cardinal points, and the intervening segments of the circle are bisected to provide a further eight. In fact the numbers denoting the twelve temples could be placed in the same places as the numbers on a clockface, the difference being that astrologers conventionally assign first place to the temple at 9 o'clock, second to that at 8 o'clock, and so on round the circle. Manilius himself gives no numbers, an omission which leads to complications; nor does he begin with the quarter temples, which indeed he leaves till last. His order is 12, 6, 8, 2; 11, 5, 9, 3; 10, 4, 1, 7: in the case of each temple Manilius gives its name, the planet (if any) which has special honour in it, and the area of human experience over which it presides. It will be convenient to arrange this information in tabular form as well as in a diagram (see figure 13): items on which Manilius agrees with other astrologers are given in capitals, items on which they agree against him, in italics (*cf.* Sextus Empiricus 12 ff.; Paulus 14 pp. 53 ff.; Firmicus 2. 15 ff.).

1 (939-947)	name	=	Stilbon
	abode	=	MERCURY
	area	=	children (*life and character*)
2 (866-870)	name	=	abode of Typhon (*portal of Pluto*)
	area	=	(*estate and fortune*)
3 (910-917)	name	=	DEA
	abode	=	MOON
	area	=	BROTHERS
4 (929-938)	name	=	Daemonium
	abode	=	Saturn
	area	=	PARENTS
5 (891-904)	name	=	Daemonie (*Bona Fortuna*)
	abode	=	(*Venus*)
	area	=	health (*children*)
6 (877-879)	name	=	portal of toil (*Mala Fortuna*)

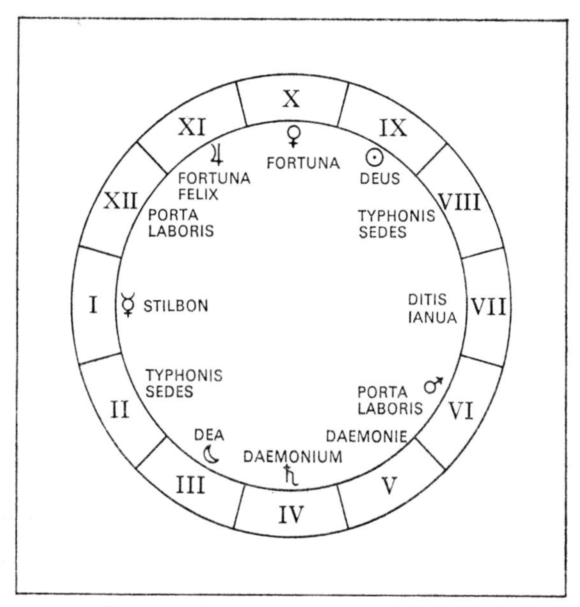

Fig. 13. The Dodecatropos, 2. 856-967

	abode	=	(*MARS, perhaps omitted in error*)
	area	=	(*health*)
7 (948-958)	name	=	portal of Pluto
	area	=	kind of death (*marriage*)
8 (871-879)	name	=	abode of Typhon
	area	=	(*kind of death*)
9 (905-909)	name	=	DEUS
	abode	=	SUN
	area	=	bodily vicissitudes (*travels*)

10 (918-927) name = Fortuna
abode = Venus
area = marriage (*honours and preferments*)

11 (881-890) name = Felix Fortuna (*Bonus Daemon*)
abode = JUPITER
area = (*friendships*)

12 (864-866) name = portal of toil (*Malus Daemon*)
abode = (*Saturn*)
area = (*enemies and misfortunes*)

Housman sternly observes that the areas governed by the cardinal temples are very different from those governed by the cardinal points contained in them (808-840). But more serious awkwardnesses lie concealed beneath the foregoing account.

For one thing the zodiacal wheel never outside the tropics revolves so as to pass through the zenith overhead or the nadir underneath. A more unsettling discrepancy occurs as a result of the obliquity of the ecliptic. If it were the rotating circle of the celestial equator we were minded to superimpose on the circle of the twelve temples, all would be well: the celestial equator always rises due east and sets due west, and its arc is always bisected by the meridian into two halves of 90 degrees. But the ecliptic is a great circle tilted 23½ degrees to the celestial equator: when in northern latitudes the equinoctial point of Aries crosses the meridian, the degree of the ecliptic rising above the eastern horizon is not the solstitial point of Cancer 90 degrees away, but a point more distant (in the latitude of Reykjavik, to take an extreme case, over 120 degrees of the ecliptic will be contained in the quadrant Hor. to MC); and conversely less than 90 (at Reykjavik less than 60) degrees will at that time be contained in the quadrant MC to Occ. To make computation more difficult, the proportions vary throughout the day, for when the equinoctial points coincide with the horizon, then a solstitial point will in fact be crossing the meridian, and the ecliptic will be exactly quartered. Manilius shows no awareness of these awkward facts, but some ancient authorities (*e.g.* Porphyry, *Isag.* 197 f. Wolf [= *CCAG* V. 4. 215 f.]) do face up to the difficulty and with varying success attempt to rectify it. Modern astrologers (who like Firmicus 2. 19. 2 locate the initial point of the first temple precisely at the cardinal, so that its domain extends 30 degrees downward from the quarter-point, and so on round the circle) generally adjust the zodiacal arcs to fit the twelve equal temples (using tables drawn up by the 17th-century astronomer Placidus), but Regiomontanus has persuaded some to retain the signs of the zodiac evenly segmented and manipulate the dodecatropos. House-division has generated violent schisms among professional star-prophets, and F. Wiesel discusses no less than fourteen different systems in *Das astrologische Häuserproblem* (Munich 1930).

[2. 968-970 : The octatropos]

Manilius's omission to specify the number of temples in the fixed circle of the observer has led to the following complication. Some astrological interpolator, either believing (like Bouché-Leclercq 276 ff.) or wishing it believed that the poet divided the circle into eight temples only (864-917) and in 918 ff. referred merely to the cardinal *points*, has attempted to enforce this belief on readers by composing three lines labelling this chapter *octotropos*. Fortunately the nature of his interpolation is perfectly clear : 968-970 were intended to supplant 965-967, as is proved by the otherwise inexplicable duplication of the promise to deal with the planets later. This interpolator we have already encountered at 2. 732-734 (*q.v.*).

Salmasius (187) and Bentley between them perfect the metre and Latinity of 2. 969 by reading *octatropon ; per quam stellae in diversa volantes*, but Housman is probably right in leaving the verse as we find it in the MSS. The terms δωδεκάτροπος and ὀκτάτροπος (feminine adjectives with an understood ἀγωγή or τοποθεσία according to Salmasius 189) occur too rarely (see Housman *ad loc.*) for us to be sure that the interpolator spelled the latter correctly.

Moreover, the octatropos is not, as the interpolator would have it, a division of the circle into eight. It is simply the first eight temples of the dodecatropos, considered as a sequence from birth to death : 1 life ; 2 estate ; 3 brothers ; 4 parents ; 5 children ; 6 health ; 7 marriage ; 8 death (so Antiochus, *CCAG* VIII. 3. 117 ; Firmicus, *Math.* 2. 14). Manifestly this system is irreconcilable with that of Manilius, in which death occurs in the seventh temple.

BOOK THREE

3. 1-42 : Prooemium

The poet again asserts his originality. He enumerates themes of others which he will not pursue and, with an eye to the technical nature of the book, warns readers not to expect a glamorous composition. The list of works alluded to is remarkable as containing in a reference (3. 23 ff.) to the *Annals* of Ennius Manilius's one solitary notice of Latin literature.

3. 43-159 : The circle of the twelve athla

It seems that here, as at the beginning of his fourth and fifth books, our instructor has turned to a new source, for certainly his initial chapter, in which he describes the circle of the twelve athla (or lots), conflicts in principle as well as in detail with the doctrine of the dodecatropos or twelve temples so recently propounded (at the end of Book 2). These two circles, for all their differences, have precisely the same function, which is to provide a spectrum of human experience against which the zodiac with its

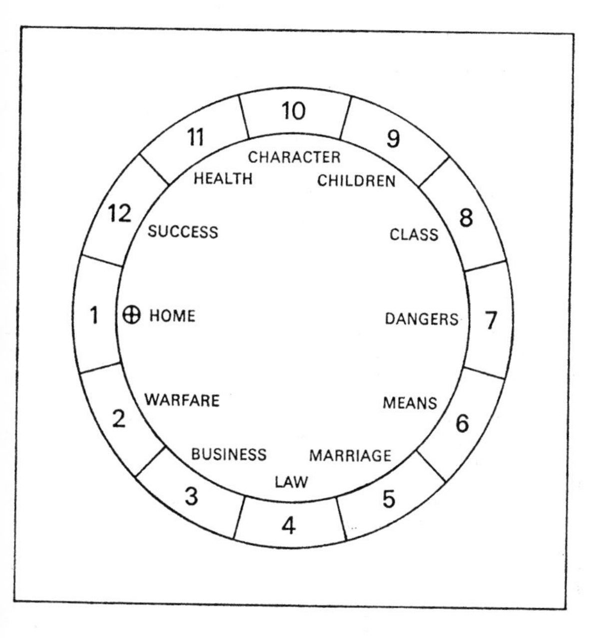

Fig. 14. The Circle of the Athla, 3. 43-159

ever-varying planetary pattern can form a kaleidoscope reflecting the infinite variety of man. Figure 14 compendiously illustrates the circle of the twelve athla, which, together with a reminder of the activities already assigned in Book 2, may be tabulated as follows.

1	Home, household, and property	
2	Warfare and foreign travel	
3	Civil life and business relationships	
4	Law : litigation and oratory	
5	Marriage ; friendships	(Temple 10 : 2. 295)
6	Means and prosperity	
7	Dangers	
8	Class : social rank and repute	
9	Children	(Temple 1 : 2. 946)
10	Way of life ; character	(Horoscope : 2. 831)
11	Health and sickness	(Temple 5 : 2. 901)
12	Success in attaining aims	

The exact astrological doctrine which lies buried beneath the confusion of Manilius's exposition in 3. 43-159 is probably no longer recoverable, but two clues survive to suggest that it was not simply an alternative to the dodecatropos.

For one thing the Lot of Fortune, as Manilius calls his first athlum or lot, is in Greek astrology a term (*κλῆρος Τύχης*) so frequent and so important that it has the privilege of being denoted by a special symbol (⊕), like a planet. Indeed, like a planet it truly is, for Ptolemy (*Tetr.* 3. 10) attributes to it an authority commensurate with that of the Sun, the Moon, and the Horoscope. And the system of lots, of which Fortune is one, usually comprises seven, one for each of the planets ; they are not segments which together make up a complete circle, but are rather specially endowed points in the chart of a nativity.

But even more significant are two phrases which Manilius has let escape his lips in expatiating on the 11th and 12th lots : (142) "no other seat is there which claims . . . *the moment* for administering (the medicine)" and (154) "this is the portion in which *day and hour* for decision shall be given." This is language quite alien to genethliacal astrology, that is the determination of a person's future from the posture of the heavens at his birth. A strict determinist, such as in several of his utterances (like the proem to Book 4) Manilius reveals himself as being, would argue that everything was predestined at the moment of nativity and that the complete record of a native's predestination could be read in his horoscope, when properly cast. Many astrologers, however, took a less doctrinaire view, holding that the outcome of actions is settled or at any rate affected by the celestial disposition obtaining at the time they are performed. This led to a different application of astrology, whereby a more enlightened Oedipus might confidently expect to evade disaster written in the stars by engaging in (or refraining from) activity at times of favourable (or unfavourable) conjunction ; and ancient documents frequently tell us of astrologers being consulted about the prospect of embarking on a journey or a marriage or initiating some other kind of enterprise. The Greeks called this branch of astrology *καταρχαί* or "initiatives." Maximus wrote a poem (still extant) so titled, and Dorotheus's fifth book (on which Hephaestion largely drew for his third) was exclusively devoted to the business. This then is the type of astrological source which Manilius may have been following.

3. 160-202 : The Lot of Fortune

Unlike the dodecatropos, which is fixed in relation to the observer's framework of cardinal points (Hor., MC, Occ., IMC), the circle of the twelve athla constantly varies its position as the Sun and the Moon constantly vary their positions. To discover how it should be adjusted to the observer's framework, we calculate the position of the Lot of Fortune (the first athlum), which will as a consequence determine the position of the circle as a whole. The calculation differs, however, according as the nativity occurs by day or by night.

If it occurred by day, count the degrees from the Sun to the Moon, and then measure off this amount along the zodiac from the first cardinal point or Horoscope : where the measurement stops, there will be found the Lot of Fortune. For example (and reckoning for simplicity's sake by whole signs), let the Sun be in Cancer, the Moon in Virgo, the Horoscope

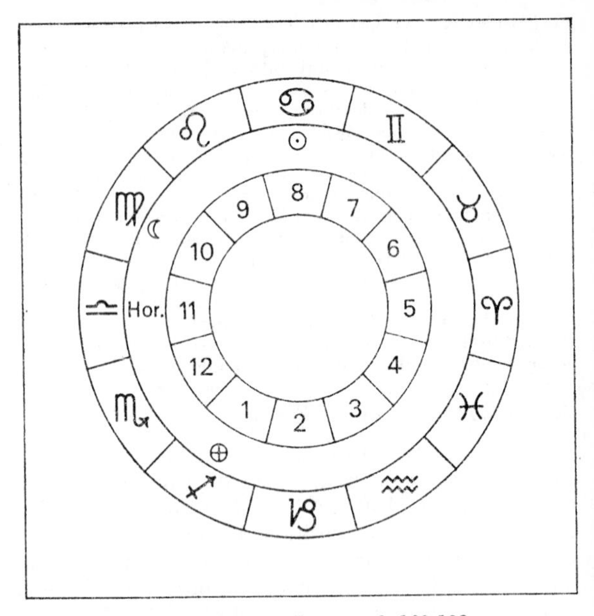

Fig. 15. Ratio Diurna, 3. 160-193

in Libra : from the Sun to the Moon is two signs : measure off two signs from the Horoscope and, as will be seen from figure 15, the Lot of Fortune will be found in Sagittarius ; the second lot (of war) will be found in Capricorn, and so on. Actually this is the procedure which Ptolemy (*loc. cit.* above) would have us follow at all hours, day or night.

For Manilius, however, there is a big difference. If the nativity occurred by night, reverse the pro-

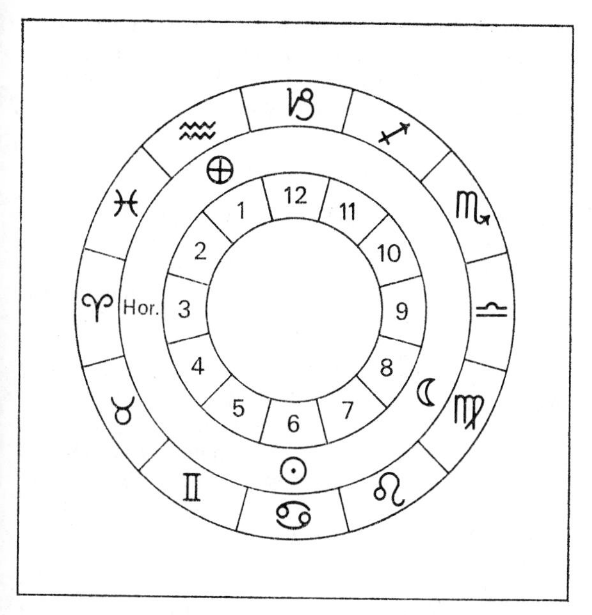

Fig. 16. Ratio Nocturna, 3. 194-202

cedure. This time count the degrees, not from the Sun to the Moon, but from the Moon to the Sun, and then proceed as before, measuring off the amount from the Horoscope to find the Lot of Fortune. For example, let the Sun and Moon be, as before, in Cancer and Virgo respectively, and let the Horoscope be in Aries (we cannot keep Libra for our illustration, since this is a nocturnal geniture and the Sun must *ex hypothesi* stand below the horizon : see figure 16) :

from the Moon to the Sun is ten signs ; measure off ten signs from the Horoscope, and the Lot of Fortune will be found in Aquarius ; the second lot (of war) will be found in Pisces, and so on.

3. 203-509 : How to find the Horoscope

Before we can perform the operations just described, we need some means of accurately determining the Horoscope, that degree of the ecliptic rising above the horizon at the moment of a nativity. If we get this wrong, we shall get everything else wrong.

218-246 : We must first beware of a common method of calculation. This attributes two hours and thirty degrees of the ecliptic to the rising of each sign : having noted (by the sundial) that at the moment of nativity the Sun was risen (let us say) six hours, and having noted (from our ephemerides) that the Sun is (let us say) at the first point of Aries, we find that this calculation gives us a distance of three signs or 90 degrees of the ecliptic to the Horoscope, which would thus be at the first point of Cancer. But this is not likely to be the case (at Reykjavik—see above p. lx—the Horoscope would be at least a further thirty degrees away, in Leo). To begin with, the signs of the zodiac make different angles with the horizon and take different times to rise. Furthermore the length of a day varies throughout the year, and so also must vary the length of an hour, a natural hour, that is, which by definition is a twelfth of the natural day ; and even if a sign takes two hours to rise one day, it will take a different period of time to rise on another.

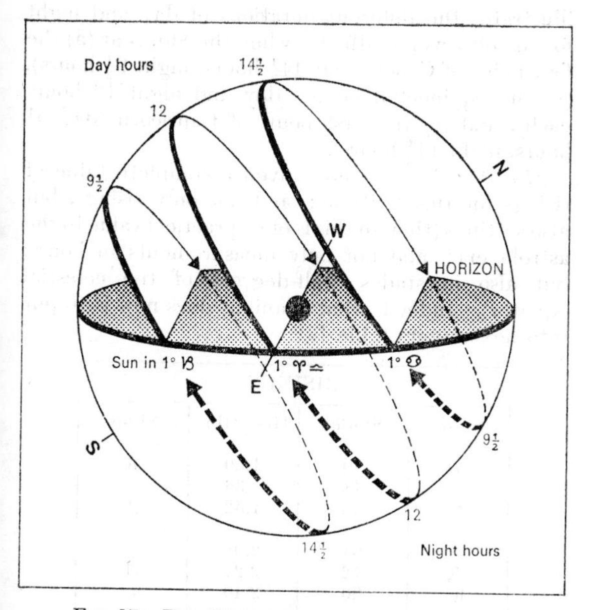

FIG. 17. THE OBSERVER AT RHODES, 3. 203-274

247-274 : We must therefore establish a standard hour. And such an hour we have at the equinoxes, when day and night are equal and may conveniently be each divided into twelve equal hours : let the astrologer work with these standard hours the whole year round. If he does so, he will find that at Alexandria (an error : the poet should have said Rhodes [Ptolemy, *Synt.* 2. 6 p. 108 f.]) the ratio of day and night at the solstices is $14\frac{1}{2}$ to $9\frac{1}{2}$. Figure 17 illustrates the different durations of day and night for the observer at Rhodes when the Sun is at (a) the first point of Cancer (day 14½ hours, night 9½ hours), (b) the equinoctial points (day and night 12 hours each), and (c) the first point of Capricorn (day 9½ hours, night 14½ hours).

275-300 : We are now given a complete table of risings for this latitude ; and not only risings, but also settings (though this is of no practical value to the astrologer) ; and not only measurements in hours, but also in stades (half-degrees of the celestial equator). Indeed what Manilius does now is to put into hexameters table 1.

RISINGS			
Signs	Stades	Hrs. Mins	Signs
♈	40	1.20	♓
♉	48	1.36	♒
♊	56	1.52	♑
♋	64	2.08	♐
♌	72	2.24	♏
♍	80	2.40	♎

SETTINGS			
Signs	Stades	Hrs. Mins	Signs
♈	80	2.40	♓
♉	72	2.24	♒
♊	64	2.08	♑
♋	56	1.52	♐
♌	48	1.36	♏
♍	40	1.20	♎

TABLE 1. RISINGS AND SETTINGS OF ZODIAC (1), 3. 275-300

301-384 : This table of ascensions, however, is only valid in one latitude. But days and nights (like the inclination of the ecliptic to the horizon) are not the same the whole world over. At the terrestrial equator each sign of the zodiac takes two hours to rise . . .

Here we may pause to notice a slight inaccuracy, as may be seen from figure 18 (wherein the straight line SN represents not only the celestial axis but the eastern horizon, the

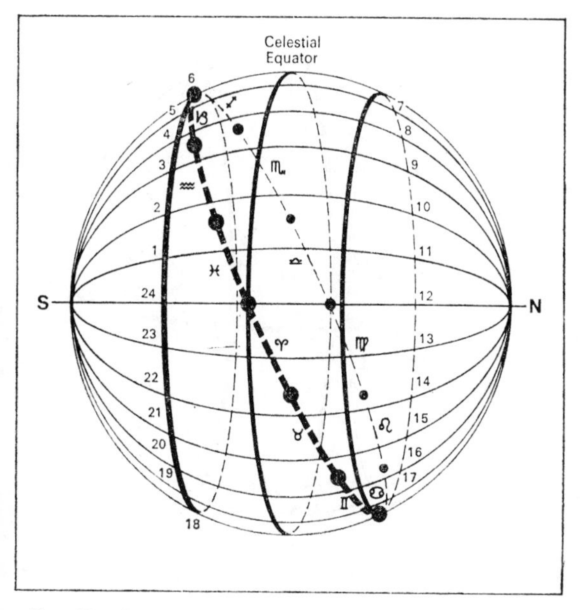

FIG. 18. ZODIACAL RISINGS AT THE EQUATOR, 3. 304-308

first curved line SN above it the first hour-circle, and so on): the ecliptic does not rise at right angles to the horizon, and as a consequence the signs closest to the equator (Virgo, Libra, Pisces, Aries) rise in less than two hours (for clarity the proportions in the figure are exaggerated), whilst those farthest from it (Gemini, Cancer, Sagittarius, Capricorn) take longer.

... and all the year round night is equal to day, as may be seen from figure 19, in which the shaded area

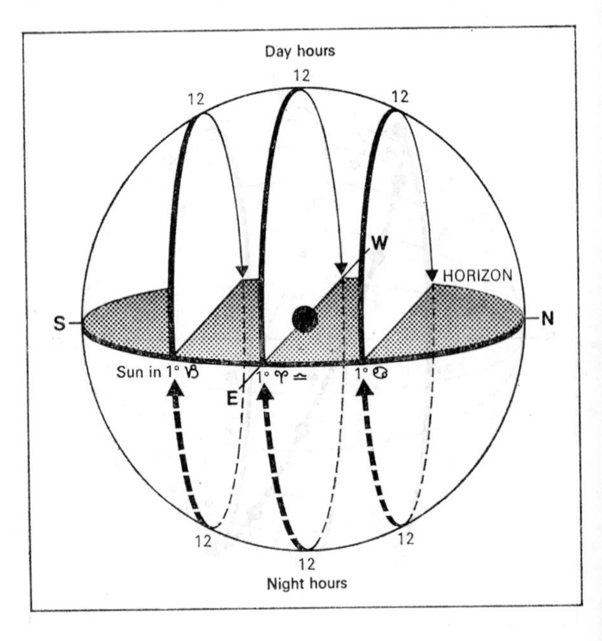

Fig. 19. The Observer at the Equator, 3. 301-322

represents the plane of the observer's horizon: obviously, no matter how far towards the south or north the Sun rises on the eastern horizon, it will always describe exactly a semi-circle above the horizon and a semi-circle below. But (323) as you move north towards the pole, the zodiac is tilted more and more to the south until the southern signs gradually cease to appear at all above the horizon and eventually disappear completely beneath it;

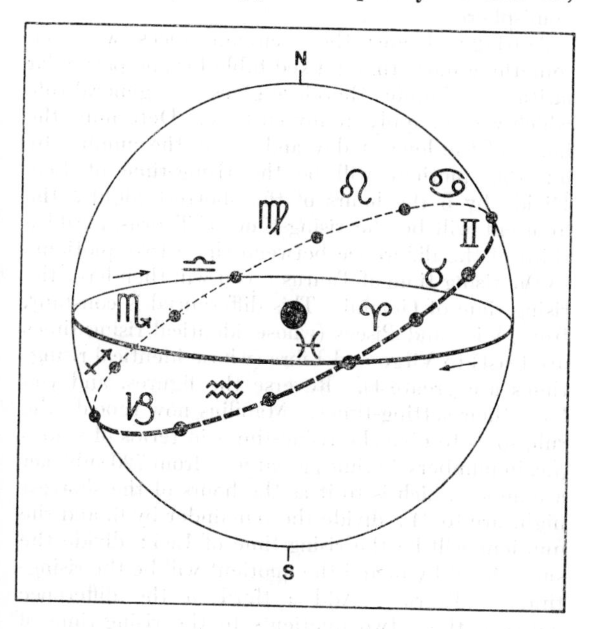

Fig. 20. The Observer at the North Pole, 3. 356-384

and if nature were to permit one to stand at the north pole, one would see six signs only. Figure 20 depicts the observer at the north pole : it is clear that since the celestial sphere spins round his zenith, he can never see more than that half of it which is above his horizon ; every day (from the vernal equinox on) the Sun makes one complete revolution, gradually rising to the tropic and then at the end of six months setting into the ever-invisible southern hemisphere.

385-442 : Under these circumstances we need something more than a fixed table for one particular latitude. Manilius therefore gives us a general rule which we can apply in any latitude. Determine the hours of the longest day and divide the number by 6 : the quotient will be the rising-time of Leo. Divide by 6 the hours of the shortest night : the quotient will be the rising-time of Taurus. Add a third of the difference between these two quotients to the rising-time of Taurus : you will then have the rising-time of Gemini. This differential is constant, from Aries and Pisces (whose identical rising-times are least) to Virgo and Libra (whose identical rising-times are greatest). Reverse the figures, and you have their setting-times. Manilius now repeats the rule so as to give the rising-times in terms of stades (*i.e.* in numbers 30 times greater) : from 720 subtract a number which is to it as the hours of the shortest night are to 24 : divide the remainder by 6, and the quotient will be the rising-time of Leo : divide the subtrahend by 6, and the quotient will be the rising-time of Taurus. Add a third of the difference between these two quotients to the rising-time of Taurus, and you will have Gemini's. Similarly with the rest of the procedure set forth for risings and settings in hours. The arithmetic is illustrated in table 2 (with 15 and 9 hours as the lengths of the longest days and shortest nights serving as an example).

RISINGS			
Signs	Stades	Hours	Signs
♈	35	$1\frac{1}{6}$	♓
♉	45	$1\frac{1}{2}$	♒
♊	55	$1\frac{5}{6}$	♑
♋	65	$2\frac{1}{6}$	♐
♌	75	$2\frac{1}{2}$	♏
♍	85	$2\frac{5}{6}$	♎

SETTINGS			
Signs	Stades	Hours	Signs
♈	85	$2\frac{5}{6}$	♓
♉	75	$2\frac{1}{2}$	♒
♊	65	$2\frac{1}{6}$	♑
♋	55	$1\frac{5}{6}$	♐
♌	45	$1\frac{1}{2}$	♏
♍	35	$1\frac{1}{6}$	♎

Table 2. Risings and Settings of Zodiac (2), 3. 385-442

443-482 : Here Manilius digresses to tell us at what rate daylight in any latitude increases between midwinter and midsummer. Take a sixth of the difference between the shortest day and the longest night (here as elsewhere the solstice is assumed to occur at the beginning of the sign), and give it to

Aquarius ; a half of this amount is given to Capricorn ; and one and a half times the amount is given to Pisces. These increments, however, must be compounded, so that Aquarius's growth progresses from Capricorn's final amount, and Pisces' growth from Aquarius's final amount. Although an increase in the number of hours continues to occur from Aries to Gemini inclusive, it does so at a reverse rate, Aries' increment being the large one of Pisces, Gemini's the small one of Capricorn. From the first point of Cancer a corresponding diminution occurs until the last of Sagittarius. The poet takes as an example the latitude of Rome, where the shortest day has 9 hours, and this illustration is fully worked out in table 3.

Signs	Increase in hours	Individual	Accumulated
♑	9 to 9½	½	½
♒	9½ to 10½	1	1½
♓	10½ to 12	1½	3
♈	12 to 13½	1½	4½
♉	13½ to 14½	1	5½
♊	14½ to 15	½	6
Signs	**Decrease in hours**	**Individual**	**Accumulated**
♋	15 to 14½	½	½
♌	14½ to 13½	1	1½
♍	13½ to 12	1½	3
♎	12 to 10½	1½	4½
♏	10½ to 9½	1	5½
♐	9½ to 9	½	6

Table 3. Variation of Days at Rome, 3. 443-482

483-509 : Our instructor resumes his task of teaching us how to find the Horoscope by setting forth another method. If the nativity occurs by day, note the natural hour and multiply it by 15 ; add the number of degrees the Sun has completed in his sign : from the total thus obtained distribute 30 to each sign (beginning with the Sun's) and in the degree where the number gives out will be found the Horoscope. If the nativity occurs by night, reckon the hour from sunset and multiply it by 15 ; add the number of degrees the Sun has completed in his sign, and add 180 : from the total thus obtained distribute 30 to each sign and, once again, in the degree where the number gives out will be found the Horoscope.

"Alas, alas ! This alternative method of yours, my poor Marcus, is none other than the vulgar method which in 218-24 you said you knew, and which in 225-46 you exposed as false. The wolf, to whom in his proper shape you denied admittance, has come back disguised as your mother the goose, and her gosling has opened the door to him" (Housman). Of course, to assume that 15 degrees of the zodiac rise in any hour of the day (488 f.) is the same as assuming that each sign (or 30 degrees) of the zodiac rises in two natural hours.

3. 510-559 : Chronocrators

The poet abruptly introduces us to *Chronocrators*, those celestial influences that govern the various divisions of a man's life. He expounds two systems, which, since they are different, cannot both be true.

According to the one (514-536) the first year of a

person's life is governed by the sign the Sun occupied at his birth (Aries, let us say); the second by the next (Taurus); and so on until the thirteenth year, when we begin all over again (with Aries). The governors of the months are determined by the sign the Moon occupied at the nativity. Suppose it was Taurus: then the first month (like the 13th, 25th, etc.) of the native's life goes to Taurus, the second to Gemini, and so on. The first day of his life is governed by the sign of the Horoscope, the second by the next, and so on. The hours, like the days, also have the Horoscope as their ἀφέτης or starter.

This tendency towards simplification may have led to the other system (537-559), which has standardized the principle of beginning with the sign of the Horoscope: years, months, days, and hours all begin their cycle therefrom, and the 13th in the series starts the cycle again.

Manilius is alone in making the signs of the zodiac his chronocrators; the other astrologers regard the planets as performing this function. Notice that, if one takes the first hour of the first day, assigning that to Saturn, and continues to assign an hour to each planet in the supposed order of distance from the Earth (*i.e.* Saturn, Jupiter, Mars, Sun, Venus, Mercury, Moon: Bouché-Leclercq 107 ff.), the first hour of the second and subsequent days will fall respectively to the Sun (Sunday), the Moon (Monday), Mars (*mardi*, Tuesday), Mercury (*mercredi*, Wednesday), Jupiter (*jeudi*, Thursday), and Venus (*vendredi*, Friday); on the eighth day the cycle will repeat itself, the first hour falling to Saturn (Saturday). Such is the origin of our week (Bouché-Leclercq 478 ff.; Colson).

3. 560-617: Length of life

The stars also foretell how long we have to live. The first part of the chapter reveals the years awarded us by the signs of the zodiac (the effective sign presumably being that of the Horoscope). Figure 21 sets forth the amounts, which are based on a gradation from 10 at the vernal equinox to 20 at

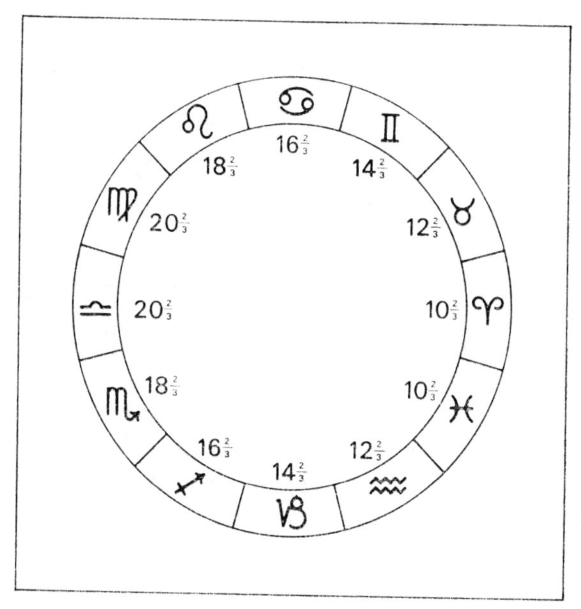

FIG. 21. THE YEARS OF THE ZODIAC, 3. 560-580

the autumnal, and then in each case augmented by two-thirds of a year. That this prediction of age (endorsed by no other astrologer) is not unconditionally guaranteed is shown by the fact that not everybody reaches the age of ten as well as by a second paragraph, which records (again without confirmation from other authorities) the grant made by that temple of the dodecatropos which the Moon occupied

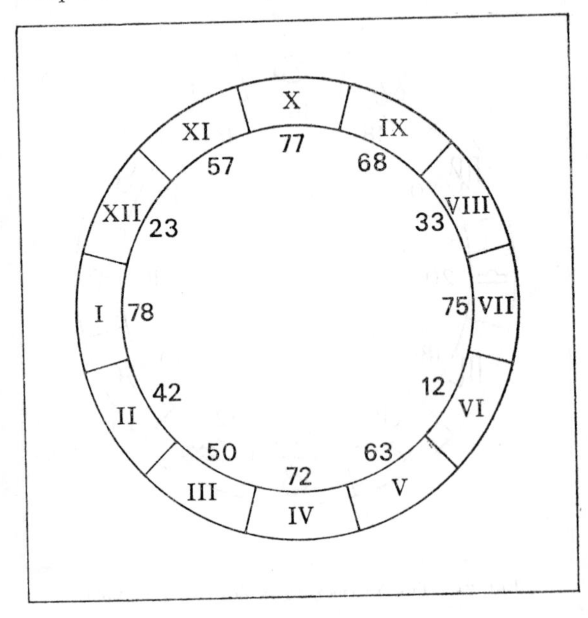

Fig. 22. The Years of the Dodecatropos, 3. 581-617

at the nativity. How these forecasts (evidently the most favourable) are affected by an untoward location of the planets Manilius does not disclose.

But the amounts bestowed by the twelve temples are interesting in that they form a mathematical progression which diminishes by successive triangular numbers : *a*, *a*—1, *a*—3, *a*—6, *a*—10, and so on until the 12th term (which is 12, i.e. *a* is 78). The order of precedence is : the Horoscope and the other three signs of its square ; the other two signs of its trigon ; the two nearest signs of its hexagon ; the two nearest signs of its dodecagon together with their respective opposites. But the fact of the matter is that what Manilius is putting into hexameters is figure 22.

3. 618-682 : Tropic signs

Housman aptly describes as "terminal ornament" this closing chapter, which contains a poetical description of the changes occurring at the mid-points of the four seasons but has no connection with any theme of Book 3, which it brings to a graceful close.

For all its irrelevance, however, the chapter raises a question of some moment. Whereas Manilius mostly locates the actual tropic degree at the beginning of a tropic sign (1. 622 ; 2. 178 ; 3. 278-293 ; 395-436 ; and 443-482), here he places it within the sign (3. 625-628 ; 637-640) and specifically in the eighth (as at 3. 257) or tenth degree. Indeed, his last line, referring to the authority who assigned the tropic to the first degree, even suggests eccentricity on that person's part. Manilius's inconsistent position reflects his use of different sources and need not specially exercise us. But Achilles too (*Isag.* 23, p. 54, 17 ff.) knows of several alternatives : 1° and 8° and 12° and 15°. No Greek or Roman seems to share Manilius's knowledge of the 10th as the tropic degree, but there is some reason to suppose that this identification enjoyed Babylonian support. The variety of opinion over such a factual matter as the nodes of ecliptic and equator clamours for explanation, and of course this lies ready to hand: the precession of the equinoxes.

Hitherto it has been implied that the positions of both the celestial equator and the ecliptic, those imaginary lines determined by the Earth's axis and the plane of its orbit respectively, are fixed. In fact this is not so: the plane of the Earth's orbit (and hence the ecliptic) does not change, but the Earth's axis does. Maintaining its angle of inclination to the ecliptic (23½°), it very slightly wobbles, like a fast spinning top, completing one wobble in 26,000 years. The consequence is that, whereas against the background of the fixed stars the ecliptic remains fixed, the great circle of the equator—maintaining its angle of inclination—is gradually sliding backwards round it. The vernal equinox, that is the point of the spring intersection of ecliptic and equator, as well of course as every other point on the equator, retreats along the line of the ecliptic at a rate of 50″.29 a year.

It is the great glory of Hipparchus to have discovered the precessional shift (apparently by comparing his observations of Spica with those of Timocharis, *cf.* Ptolemy, *Synt.* 7. 3 pp. 28-32), and it detracts little from his feat that he somewhat underestimated the amount (though the error will have led to greater inaccuracy in Ptolemy's star-catalogue, if, as Dreyer [202] assumes, the latter was simply Hipparchus's updated to take account of precession). We are now in a position, reversing the precession, to calculate when the equinoctial point coincided exactly with the beginning or the middle of the zodiacal sign, in the hope that this might give us the date of establishment of the Greek zodiac. Unhappily we are frustrated. It is certain that constellations along the ecliptic had been evolved before the conception of the duodecimal zodiac: when, then, this momentous stage was achieved, it was too late to introduce a new schematization based on tropic points. Eudoxus, we are told (Hipparchus 2. 1. 18), placed the vernal point at Aries 15°, Aratus at the beginning of the sign. Now since their chronological difference corresponds to a precessional shift of not much more than a degree, it is obvious that both are struggling to preserve conventions that do not fit the phenomena. Occasional references to the 8th degree may be related, as Neugebauer (188) suggests, to the vernal point of System B of the Babylonian lunar theory; and the 8th degree or thereabouts may well have marked the equinox when the zodiac as we know it was devised (the Romans—Caesar, Vitruvius, Columella, Pliny—generally adopted the 8th degree tropic).

Thus, if the retrospective search for the location of the vernal point at ♈0° focuses on the time of Hipparchus, one can hardly resist the suspicion that Hipparchus himself has taken a hand in modifying the constellation outlines to bring this about. For this there is some evidence. Inspection of the zodiacal figurations within their thirty-degree compartments (most conveniently depicted in Bouché-Leclercq 131 ff.) reveals that some displacement of the zodiacal figures has taken place. For example: Cancer's territory once obviously extended into Gemini, just as on the other side Leo's head and front paws protrude ten degrees or more into Cancer: the sign's boundary has been adjusted to secure ♋0° as the solstitial point. Exactly the same has happened in respect of Capricorn: no wonder Manilius calls it a cramped constellation, for whilst it has plenty of room to stretch in Sagittarius, it is outrageously elbowed aside by Aquarius: but an explanation lies to hand if its boundaries have been redrawn to secure ♑0° as the autumnal point.

Awkward as precession is for the astronomer, for the astrologer it is fatal. Or so one would have supposed. If in the time of Hipparchus the vernal equinox occurred at the first point of Aries (♈0°), then in the time of Ptolemy it must have occurred at about ♓26°; and today it must occur at about ♓1°. Today in fact the effect of precession has been to move every zodiacal sign twenty-nine degrees away from where, according to astrological doctrine, it ought to be. Oddly enough it is Ptolemy who has saved the day for the astrologer: in *Tetrabiblos* 1. 22 the astronomer, for he was that first and prophet second, virtually says that for astronomical purposes he will define the first point of Aries as the vernal equinox: if that moves, then the whole zodiac will just have to move with it; for astrological purposes men had better look to this movable, artificial zodiac. And so it has come to pass. When today's readers of almanacs are informed that the Sun travels through Aries from March 21st to April 20th, the name Aries denotes not the group of stars so identified and marked in our star-atlases, but thirty degrees of the ecliptic measured off from the vernal equinox, a length of line constantly moving and today almost entirely contained in the astronomical constellation of Pisces. And after another half century, when the nodes have precessed still further, we shall, during the Sun's course in Pisces, start getting Taurus babies.

BOOK FOUR

Consideration of this book may suggest that in it Manilius is following an Egyptian source (much as Firmicus, *Math.* 8. 5-17 follows Manilius, Book 5): the system of decans derives from Egypt, and with Egypt the *partes damnandae* may have a connection (Bouché-Leclercq 235 and note 2); then the doctrine of zodiacal geography which forms the main section (and in our poet's source probably led on to ecliptic signs) has been argued by Bartalucci as reflecting a Ptolemaic source.

4. 1-118: Prooemium

This, the most successful of the exordiums, asserts the futility of hopes and fears in the face of immutable destiny, the existence of which is inferred from the constant occurrence of the unexpected and improbable—we are regaled with a surfeit of evidence—and also from the constant fulfilment of predictions—this we are evidently required to take on trust. Although the future is determined, insists the poet, neither the wickedness of sin nor the excellence of virtue is on that account diminished.

4. 122-293: Zodiacal influences on the native

Opening the astrological portion of the book is a simple and non-technical account of the qualities and skills imparted to those born under each sign of the zodiac. One might think it reasonable to assume (Manilius has omitted to clarify the matter) that these effects are produced by the horoscoping sign, since, when in the fifth book we come upon the identical treatment of extra-zodiacal constellations, we are repeatedly informed that the influences are precipitated at the moment of rising. Nevertheless, it is odd that he has nowhere let slip a confirmation of this view, especially as in a chapter shortly to come (4. 502-585), when dealing with the influences of special degrees, he specifies their position in the ascendant eleven times out of twelve. We should therefore accept Housman's considered opinion (add. [1930]) that, if the poet has chosen his words carefully, he must mean us to understand that the influences ascribed in this chapter to signs, in the next to decans, and in that following to degrees all depend on the Moon's presence therein at the natal hour.

4. 294-407 : Decans

Each sign is equally divided into three units or decans ; and these 36 decans are allotted to the signs in regular order, the first (*i.e.* of Aries : we always begin with Aries) to Aries himself, the second to Taurus, and so on until we reach the thirteenth (= the first of Leo), when the sequence is repeated, as it is again at the twenty-fifth (= the first of Sagittarius). Human frailty, however, has asserted itself at 4. 359 f., where the poet has mistakenly given the first and second decans of Pisces to Aries and Taurus instead of Capricorn and Aquarius, a slip corrected in figure 23.

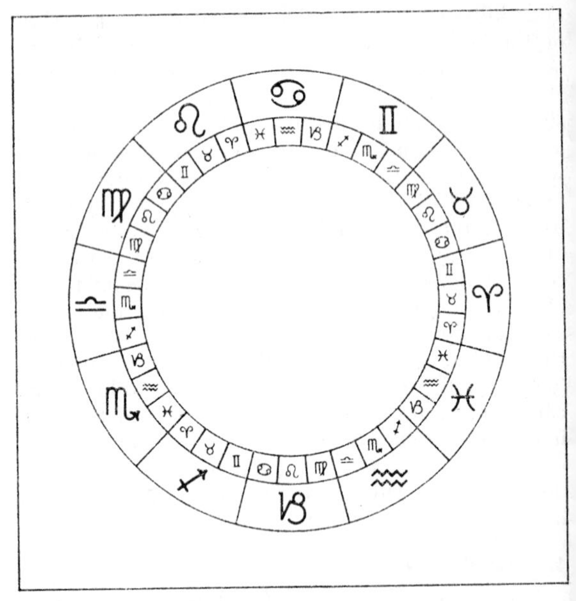

Fig. 23. Decanica, 4. 294-407

In origin the decans are 36 segments of the celestial circle, which from early times provided the Egyptians with convenient 10-day sequences of star-risings. Later, when the zodiac competed with decanal divisions as a method of calibrating the Sun's annual orbit, it was natural that the two systems, arithmetically compatible, should be amalgamated for astrological ends. And it is possible that this amalgamation was first accomplished in the popular handbook (*c.* 150 B.C.) which claimed the authorship of the legendary Nechepso and Petosiris (*cf.* Firmicus, *Math.* 4. 22. 2). Only Manilius allots the decans to the signs of the zodiac; elsewhere they are given to the planets in descending order (*i.e.* Saturn, Jupiter, Mars, Sun, Venus, Mercury, Moon), starting however with Mars, who by getting the first decan of Aries thereby receives one decan more than his fellow-planets when the last of Pisces' falls to his lot.

4. 408-501 : Partes damnandae

The 360 degrees of the zodiac are now considered individually : some are too cold, others too hot, some are too wet, others too dry. Actually Manilius never gives chapter and verse about such matters, but launches into a *tour de force* specifying by number each of the degrees deemed to be insalubrious. Scaliger, Bentley, and Housman all applaud the poet's virtuosity in contriving to maintain an elegant variation in this versification of numbers, the list of which is set forth in table 4.

That some pattern lies behind this tally is suggested by the gradually increasing frequency which marks their recurrence the nearer one approaches the end of each sign. Possibly there is some connection with an Egyptian calendar; possibly variation in the number of days each month accounts for the displacement, for instance, of the 24th degree on four occasions from the otherwise standard malignity of the 25th. Again, one would expect the total of injurious degrees to amount to a round number like 100 rather than the 102 specified by the text. But even allowing some margin for error, whether on the poet's part or in the course of the tradition' I have been no more successful than Housman in attempting to detect an underlying principle.

Like his doctrine of decans, what Manilius has to say about the *partes damnandae* finds no precise parallel elsewhere.

4. 502-584 : Influences of certain zodiacal degrees

This short chapter, dealing, like the preceding, with individual degrees of the ecliptic, forms a colourful contrast to it and seems to foreshadow a

♈	♉	♊	♋	♌	♍	♎	♏	♐	♑	♒	♓
		1	1	1	1		1			1	
		3	3				3				3
4				4				4			
						5					5
6			6		6		6				
7		7				7			7		7
			8					8			
	9								9		
10				10			10				
			11		11					11	11
12								12			
	13								13	13	
14					14	14					
		15	15	15			15			15	
								16			
	17		17			17			17		17
18					18						
		19							19	19	
			20					20			
21		21			21					21	
	22			22			22				
	24				24	24		24			
25		25	25	25			25		25	25	25
	26							26			
27		27	27			27			27		27
	28			28			28	28			
		29	29			29	29			29	
	30			30	30	30		30			

Table 4. Partes Damnandae, 4. 408-501

second time the descriptive treatment the poet will lavish on his final book. Astrologically considered, the information is unsatisfying, for nothing like a systematic schedule is attempted. In fact we only hear of the initial degrees of Aries, Taurus, Leo,

Virgo, Libra, Aquarius, and Pisces ; the middle degrees of Gemini, Cancer, and Sagittarius ; and the last degrees of Scorpio and Capricorn.

4. 585-710 : A description of the world

Before expounding the partition of the world among the signs of the zodiac Manilius sets himself the task of constructing a mappemonde in verse, thus reminding us that it is not only arithmetical tables and geometrical figures that he loves transmuting into hexameters but diagrams as well.

In the frontispiece I have ventured to reverse the procedure, believing that such a schematic illustration will prove the most useful commentary upon the text. Considerations of clarity have prompted me to take some liberties with scale and latitude (the frontispiece of the Loeb Strabo, volume I, may in some respects better represent the poet's proportions), and after some hesitation I have rejected the idea of a circular frame in spite of Manilius's compass, which seems to require it : instead I have made the poet consistent with himself at 1. 246 and restricted his mappemonde to the northern hemisphere (indeed, just conceivably it should be restricted to half of that : see above p. xx).

Whether, as many have thought, Posidonius's περὶ ὠκεανοῦ served as Manilius's source cannot be determined with certainty. It is enough that the closeness of [Aristotle], *De Mundo* 3 and Strabo 2. 121 ff. attests the orthodoxy of his geography. The chapter begins with Manilius's compass, the details of which serve to warn us against leaning too heavily upon a poet's verbal precision. He adopts the twelve-point compass of Timosthenes, naming the cardinal but not the intermediate points (thereby avoiding metrical problems). Let it be said on Manilius's behalf that both *Boreas* (= N) and *Eurus* (= E)—which enjoy Homeric warrant—are permissible generic names for North and East (*cf.* [Aristotle], *De Mundo* 4). But in the twelve-point compass due North and due East are denoted by *Septentrio* (*Aparctias*) and *Subsolanus* (*Apeliotes*) respectively (*e.g.* Seneca, *Nat. Quaest.* 5. 16), whilst *Boreas* is the Greek counterpart of *Aquilo* as is *Eurus* of *Volturnus*. In the frontispiece Manilius's cardinals are retained in large letters, Pliny's intermediates (*N.H.* 2. 119 f.) being added for the sake of completeness.

The world is described as follows. First, the sea, which encircles the earth but bursts into the continent on the west : (a) the southern shore of the Mediterranean as far as the Nile ; (b) the northern shore as far as the Nile ; (c) the Mediterranean islands ; (d) the other inroads of Ocean. Second, the continent, which is divided into three : (a) Libya ; (b) Asia ; (c) Europe. All this, says Manilius, is dispensed among the signs of the zodiac just as are the parts of the body, the scheme of which, already presented to us at 2. 456 ff., he for sheer love of versifying repeats in an elegantly abbreviated form.

4. 711-743 : National differences

A transitional paragraph impresses upon us that individual nations possess distinctive characteristics, Germans, for example, being fair, Ethiopians dark. Not only do peoples vary physically : they speak different languages, cultivate different crops, and breed different animals. There are climatic reasons for this; but these in turn depend upon the signs of the zodiac, each of which exercises paramount dominion over special areas.

4. 744-817: Zodiacal geography

Teaching on this article of astrological doctrine ranges from an early scheme preserved by Paul of Alexandria, devised when the world was young and small, and allotting one country to one sign, to the complicated systems of Odapsus and Valens, who in a later age subdivided the signs of the zodiac so as to create extra accommodation for the more numerous regions then clamouring for attention. The Greek astrologers contradict one another to a degree one would have thought positively embarrassing. Manilius's arrangement is as follows, agreements with Dorotheus of Sidon (the only rival with whom he has much in common) being indicated by asterisks.

ARIES:	Hellespont (*which it swam*); Propontis; Syria; Persia; Egypt.
TAURUS:	Scythia; Asia (*because of Mount Taurus*); Arabia.
GEMINI:	Euxine; Thrace; India.
CANCER:	Ethiopia.*
LEO:	Phrygia* (*because of Cybele's lions*); Cappadocia; Armenia; Bithynia*; Macedonia.*
VIRGO:	Rhodes*; Ionia*; Greece*; Caria.
LIBRA:	Italy.*
SCORPIUS:	Carthage*; Libya* (*land of reptiles*, 4. 662 ff.); Hammonia*; Cyrene; Sardinia and other islands.

SAGITTARIUS :	Crete* ; Sicily ; Magna Graecia.
CAPRICORN :	Spain ; Gaul ; Germany.
AQUARIUS :	Phoenicia ; Tyre ; Cilicia ; Lycia.
PISCES :	Euphrates (*cf. story told at* 4. 579 ff.) ; Tigris ; Red Ocean* ; Parthia ; Bactria ; Asiatic Ethiopia ; Babylon ; Susa ; Nineveh.

4. 818-865 : Ecliptic signs

In Ptolemy's *Tetrabiblos* 2. 4 mention of eclipses as affecting countries and cities follows naturally upon his zodiacal geography, a sequence doubtless shared by other astrological handbooks, including our author's source. Manilius, however, gives no indication of a logical connection, embarking upon what is to all appearances a totally different subject.

At an eclipse of the Moon the sign which it then occupies suffers an eclipse of its influences, and so does the sign diametrically opposed to it. This crippling of their powers has effect not just for the duration of the eclipse but continues for varying periods, which sometimes exceed a year (we are not apprised of the details) ; and even thereafter the debility is not terminated, but passes on to the preceding signs, *i.e.* from Aries (and Libra) to Pisces (and Virgo), and so on.

4. 866-935 : Peroration

The book closes with an affirmation of Stoic belief : man has the power to discover the secrets of destiny, for God dwells in him and by granting him a share of the divine intellect, which governs all things, elevates him to godhead.

BOOK FIVE

5. 1-29: Prooemium

This is not as studied an introduction-piece as those of the other books. Another poet, says Manilius, would have stopped here, but before returning to Earth via the seven planets he will make a tour of the extra-zodiacal constellations, specifying their influences when rising (*paranatellonta*) and when setting and the degree of the ecliptic which rises simultaneously with them.

5. 32-709 . . . : Paranatellonta

These undertakings are not diligently performed: we miss the setting influences entirely and in no less than eight instances numeration of the degree of the ecliptic. Again, the promised accounts of Perseus, Hydra, and Flumina are not forthcoming, and we look in vain for the influences of Deltoton and Corvus.

The degree of the ecliptic rising with a given extra-zodiacal star varies, of course, with the latitude of the observer; and it makes a difference whether one is retailing this information to Romans, Rhodians, or Alexandrians. Manilius's readers may at once abandon any hope of getting sound instruction on this score. The astronomical detail given by him is for the most part either impossible or significantly inaccurate, and in so few cases are his statements acceptable that to accord him credit even for these is misplaced scruple. The fact is that this book provides not so much a manual for the technical astrologer as for the literati a bouquet of character and vocational sketches such as were suggested by the natures of the extra-zodiacal constellations. It is clear that

	Southern Stars rising on left	Ecliptic	Northern Stars rising on right	Type of Native
32- 56	Argo	ARIES 4°		seamen
57- 66	Orion	10°		bustlers
67-101		15°	Heniochus	charioteers
102-117		20°	Haedi	wantons
118-127	Hyades	27°		swineherds
128-139		30°	Capella	browsers
140-156		TAURUS 6°	Pleiades	effeminates
157-173	Lepus	GEMINI 7°		jugglers
174-196	Iugulae	CANCER —		hunters
197-205	Procyon	27°		dog-fanciers
206-233	Canicula	LEO —		hotheads
234-250	Crater	30°		drinkers
251-269		VIRGO 5°	Corona	gardeners
270-292	Spica	10°		farmers
293-310		LIBRA 8°	Sagitta	archers
311-323		—	Haedus	busybodies
324-338		26°	Lyra	musicians

TABLE 5. THE PARANATELLONTA (1), 5. 32-338

Manilius took not the faintest pains over the finer points of συναvατολαί or ἀντικαταδύσεις, as emerges from the compendious arrangement of the *paranatellonta* given in tables 5 and 6.

In this section appear two constellations unknown to the Greek Sphaera: at 5. 311 there rises *Haedus* ("the Kid") with Libra and at 5. 409 *Fides* ("the

	Southern Stars rising on left	Ecliptic	Northern Stars rising on right	Type of Native
		SCORPIUS		
339-347	Ara	8°		priests
348-356	Centaurus	12°		veterinaries
		SAGITTARIUS		
357-363		5°	Arcturus	custodians
364-388		30°	Cycnus	fowlers
		CAPRICORNUS		
389-393		—	Ophiuchus	snake-handlers
394-408	Piscis Notius	—		pearl-divers
409-415		—	Fides	torturers
416-448		—	Delphinus	swimmers
		AQUARIUS		
449-485		—	Cepheus	tragedians
486-503		12°	Aquila	plunderers
504-537		20°	Cassiepia	goldsmiths
		PISCES		
538-630		12°	Andromeda	executioners
631-644		21°	Equus	horsemen
645-655		last degrees	Engonasin	prowlers
656-692	Cetus	last degrees		fishmongers
693-709		*	Helice	bear-tamers
(......)		*	*Draco*	*snake-charmers*

TABLE 6. THE PARANATELLONTA (2), 5. 339-709

Lute") with Capricorn. Now occasionally we find in the Latin poets the singular *Haedus* used for *Haedi*, ζη Aur, the Kids held in the Charioteer's left hand; and so heinous are Manilius's violations of astronomical accuracy in this book that it is no lack of charity to hold him guilty of a grievous confusion here, notwithstanding his mention of the Kids in

5. 102 (rising there, as they should, with Aries). A similar lapse has occurred with *Fides*, an appellation sometimes applied (*e.g.* by Cicero and Varro) to the constellation of the Lyre : *Lyra* (as the constellation is regularly termed) has, however, been dealt with earlier, at 5. 324-338 ; moreover, our author seems to have been aware of this awkward circumstance and taken pains to conceal his repetition by using the different word *fides* and interpreting it (or rather its synonym *fidicula*) as an instrument of torture, just as at 5. 175 he calls Orion *Iugulae* to avoid exposure of his mistake at 5. 57-66.

Firmicus, it may be said, adopts from the poet (whose existence, let alone his name, he totally suppresses) the rising of Haedus and its concomitant influences upon the native, conjures up a rising degree (the 15th of Libra), and in accordance with his practice paints a horrific picture of its effects at the occident (*Math.* 8. 12. 3 f.). He also adopts the poet's *Fides*, specifying its rising in the 10th degree of Capricorn and its occidental effects ; but, evidently dissatisfied with Manilius's double treatment of the constellation, he calls it *Lyra* (*Math.* 8. 15. 3) and passes over what the poet had expounded at 5. 324 ff. Here we may agree with Boll (*Sphaera* 397 ff.) that Firmicus was not assisted by another source.

After 709 the text is interrupted by a lacuna of substantial dimensions, and Firmicus, who has been following Manilius throughout this book, becomes our only source for the poet's conclusion to his treatment of *paranatellonta*. We cannot, however, have lost much of the influences of Helice (= Ursa Major), to judge from *Math.* 8. 17. 6 :

Hoc itaque signo oriente, quicumque nati fuerint, erunt mansuetarii ferarum, id est qui ursos vel tauros vel leones deposita feritate humanis conversationibus socient.	Those born at the rising of this constellation will be tamers of wild beasts, that is men to teach bears, bulls, and lions to lay aside their fierceness and to share in human ways.

In the next section (8. 17. 7) Firmicus proceeds to deal with Draco, and the colourful language exemplified in *Marsi* (tribesmen noted for their wizardry, *cf.* Aulus Gellius, *N.A.* 16. 11. 2) and the antithesis of *venenis* and *salutaria* suggest that Manilius is his source here, too :

Quicumque hoc sidere oriente nati fuerint, erunt Marsi vel qui venenis ex herbarum pigmentis salutaria soleant remedia comparare.	Those born at the rising of this constellation will be snake-charmers and men to procure beneficial medicines from the juices of poisonous herbs.

Without much doubt Draco marked the end of our poet's chapter on *paranatellonta*, for whilst Firmicus goes on to speak of a constellation *Lychnus*, "the Lamp," this is not a part of the Greek Sphaera nor anything he could have found in Manilius, since its influences are transferred from Cassiepia's (5. 522-536, which he had ignored under that constellation).

Planetary influences

After the section on *paranatellonta* (of which only about ten lines have been lost) a passage of some 140 lines seems to have dealt with planetary influences.

This we may infer from its final two lines, omitted at their proper place and wrongly inserted after line 29. Housman's transposition of them after 3. 155 (*cf.* his addenda [1930]) involves altering *ab* to the improbable *nam* without securing any special point: if the lines are genuine, Scaliger's *has* and removal to some place after 709 seem certain. It is true that the space of a mere hundred or so lines is ridiculously insufficient to treat of planetary influences in astrology. But a perfunctory teacher might make the attempt, and it must be remembered that unlike other astrologers Manilius makes the zodiac, not the planets, paramount in his system. He obviously avoids coming to grips with them: in his Sphaera, Book 1, he almost slightingly dismisses them in four lines (805-808), a similar condescension marks his reference to them at the beginning of Book 5 (5-7), and his admissions that he must sometime deal with them (2. 965 and 3. 156 ff.) read suspiciously like the postponement of an irksome labour.

Although the contents of Manilius's presumed chapter are now lost to us and no clues survive on which to base a reconstruction, a summary of basic planetary lore deserves mention.

[Ptolemy, *Tetr.* 1. 4 f., 17-19; Sextus Empiricus 29-40; Firmicus Maternus 2. 2 f., 20; Bouché-Leclercq, chapter IV; Neugebauer-Van Hoesen 1-13]

The planets fall into three classes: Jupiter ("jovial") and Venus are the *Benefics*, exercising benign influences if favourably situated, while Saturn ("saturnine") and Mars are the *Malefics*, possessing similar power for ill; Mercury ("mercurial") is a turncoat, benefic in company with benefics, malefic with malefics. Ptolemy (1. 5) regards the Moon as a benefic and the Sun as a second neutral like Mercury (both are deemed benefic by later astrologers), but in general they are not classified.

The influences of the planets vary according to their position. In the dodecatropos they exert special potency at the cardinal temples (*i.e.* 1, 4, 7, 10) and, situated in any temple, exercise special influence over its particular sphere of activity: Manilius has in fact spoken of these matters in 2. 856-967 (see also figure 13).

The zodiacal location of a planet involves many considerations. Here it is enough to mention *Houses* and *Exaltations*, wherein the planets intensify their effects. Except the Sun and the Moon, which are lords (or rulers) over the same House both day and night, each planet has a diurnal and a nocturnal House, the scheme being as follows:

	(diurnal)	(nocturnal)
SATURN	Capricorn ;	Aquarius
JUPITER	Sagittarius ;	Pisces
MARS	Scorpio ;	Aries
VENUS	Libra ;	Taurus
MERCURY	Virgo ;	Gemini
MOON	Cancer ;	Cancer
SUN	Leo ;	Leo

The Exaltations are signs or more usually particular degrees of signs (the position diametrically opposite being the planet's *Depression*, wherein its influence was reduced to a minimum), thus:

SATURN	+21° Libra ;	−21° Aries
JUPITER	+15° Cancer ;	−15° Capricorn
MARS	+28° Capricorn ;	−28° Cancer
VENUS	+27° Pisces	−27° Virgo
MERCURY	+15° Virgo ;	−15° Pisces
MOON	+ 3° Taurus ;	− 3° Scorpio
SUN	+19° Aries ;	−19° Libra

Just as the signs of the zodiac form special groupings according to their geometrical relationships (*cf.* Manilius 2. 270-432), so the planets, when similarly positioned, are said to be in trine, quartile, or sextile aspect, or in opposition or in conjunction. In particular, one planet dominates another if it precedes it (*i.e.* rises before it) in quartile aspect.

Such is the general framework of principles (which Manilius could have given in a moderately brief compass). Of course, elaborating these principles for each of the possible combinations of planets leads at once to a very large number of separate prognostications; the *Apotelesmatica* of Manetho is largely composed of them; so too are sizable parts of Firmicus. In *Math.* 3. 5. 1-11, for example, over twenty predictions are made for the Sun at the Horoscope (*i.e.* ascendant or first temple), according to whether it is aspecting this planet or that or some combination of them. The rest of the chapter similarly treats of the Sun in other temples. Other chapters deal with the other planets, and yet others with special combinations in the several temples successively (*e.g. Math.* 3.10 Mercury and Jupiter). Manifestly Manilius never embarked or intended to embark upon this amount of detail.

The discovery in recent centuries of three planets invisible to the unaided eye (Uranus, Neptune, and Pluto) has not noticeably discommoded modern astrologers, who have incorporated them into their systems with varying degrees of enthusiasm.

. . . 5. 710-745: Stellar magnitudes

To introduce his peroration, the theme of which is the celestial sphere considered as a heavenly commonwealth, Manilius described how the stars are arranged in six orders of magnitude. But only the end of this account now remains: 710-715 deal with constellations comprising stars of the 3rd magnitude, and 716-717 merely state the existence of the 4th, 5th, and 6th magnitudes without particular identifications.

The passage proves that stellar classification in six magnitudes is over a century older than Ptolemy, whose splendid star-catalogue (*Synt.* 7. 5—8. 1, vol. 2, pp. 38-163) is certain to be in this respect, too, dependent on Hipparchus. It is evident that the lost portion (perhaps of some 30 lines) dealt with constellations possessing stars of the first two magnitudes. We shall see that Manilius executed his plan in a cavalier and careless manner. It seems that, having dealt with a constellation under the 1st or 2nd magnitude, he did not mention it again: this will explain why, for instance,

Gemini's five 3rd-magnitude stars are not alluded to in 710-715. Even so, we miss notice of several signs: Aries, the first sign of the zodiac, which has no 1st or 2nd magnitude stars, has two, if not three, of the 3rd magnitude and therefore required a listing.

In the translation I have given a conjectural summary of the information Manilius may have decked out in hexameters, and I have marked on the star-charts (with a few simplifications and additions) those stars of the first three magnitudes which may be reasonably held to have appeared on the globe or planisphere or other source he used. They may be tabulated as follows (in the order of a: zodiacal, b: northern, and c: southern signs).

First magnitude constellations

(a) TAURUS — 1st: α (*Aldebaran*), brightest of the Hyades;
3rd: γδεθ = Hyades ([1. 371] 5. 119, 127); βζ.
See also PLEIADES below (Third magnitude).

LEO — 1st: α (*Regulus*), β (*Denebola*);
2nd: γδ;
3rd: εζηιμ.

VIRGO — 1st: α = Spica (5. 271);
3rd: βγδε (*Vindemiatrix*) ζη.

(b) BOOTES — 1st: α = Arcturus (1. 318, 5. 358);
3rd: γεζη;
4th: βδ.

LYRA — 1st: α (*Vega*);
3rd: βγ.

HENIOCHUS — (= Auriga) 1st: α = Capella (1. 366, 5. 130);
2nd: β;
3rd: ι (γ = β Tau);
4th: ζη = Haedi (1. 365, 5. 103).

(c) ORION — 1st: α (*Betelgeuse*) = right shoulder (1. 390), β (*Rigel*);
2nd: γ (*Bellatrix*) = left shoulder (1. 390), δ (*Mintaka*), ε (*Alnilam*), ζ (*Alnitak*) = Iugulae (5. 175);
3rd: $\theta^1\theta^2\iota$ = sword (1. 391), $\eta\kappa\pi^3\pi^4\pi^5\pi^6$;
nebulous: $\lambda\phi^1\phi^2$ = head (1. 393).

CANICULA (= Canis Major) 1st: α (*Sirius*) = face (1. 408);
3rd: βδεζη.

PROCYON (= Canis Minor) 1st: α (*Procyon*).

ARGO (now subdivided into four constellations: Carina, Puppis, Pyxis, Vela) 1st: α Car = Canopus (1. 216);
2nd: ζ Pup; γλψ Vel;
3rd: αβ Pyx (Ptolemy, though probably not Manilius, knew of 18 stars of the first three magnitudes).

PISCIS NOTIUS (= Piscis Austrinus) 1st: α (*Fomalhaut*).

Second magnitude constellations

(a) GEMINI 2nd: α (*Castor*), β (*Pollux*);
3rd: γδεζλ;
4th: η.

LIBRA 2nd: αβ.

SCORPIUS 2nd: α (*Antares*);
3rd: βδεηθικλμπρστ;
4th: ζ.

SAGITTARIUS 2nd: αβ (these perhaps not known to Manilius);
3rd: γδελσ (and four others according to Ptolemy).

(b) HELICE (= Ursa Major) 2nd: α (*Dubhe*), β (*Merak*), γ (*Phecda*), ε (*Alioth*), ζ (*Mizar*), η (*Alkaid*);
3rd: δ (*Megrez*). These are the "seven stars" (1. 297, 620). Ptolemy gives also θικλμνξ.

CORONA (= Corona Borealis) 2nd: α (*Alphecca*: 1. 320 ff.).

CYCNUS (= Cygnus) 2nd: α (*Deneb*);
3rd: βγδεζ.

EQUUS (= Pegasus) 2nd: αβ (*Scheat*: 1. 349) γ;
3rd: εζηθ.

ANDROMEDA 2nd: α;
3rd: βγδο.

PERSEUS 2nd: α (*Mirfak*), β (*Algol* "The Ghoul");
3rd: γδεζο.

Third magnitude constellations specified

(a) PLEIADES (in Taurus): actually no member of this cluster attains the 3rd magnitude: η Tau (*Alcyone*), the brightest, is classified by Ptolemy as 5th. Though they are popularly supposed to be seven in number, only six can be clearly distinguished (*cf.* Aratus, *Phaen.* 258 f.). See figure 24.

(b) CYNOSURA (= Ursa Minor) 3rd: α (*Polaris*), βγ (Ptolemy classifies these two as 2nd, and I have so shown them in star-chart 1).

DELPHINUS 3rd: αβγδ. So Aratus, *Phaen.* 317 f. (though ε, in the tail, is also 3rd).

DELTOTON (= Triangulum) 3rd: αβγ.

AQUILA 3rd: βγζο (Manilius thus seems to have omitted α, *Altair*, which is a very bright 2nd).

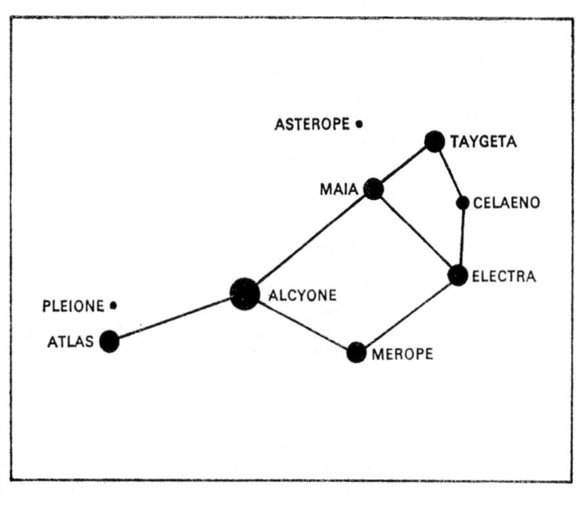

FIG. 24. THE PLEIADES, 5. 710

DRACO 3rd : αβγζηικλ ;
4th : ε.
OPHIUCHUS (now subdivided into Ophiuchus and Serpens) 3rd : αδεζη Oph and αβγδε Ser.
(c) HYDRA 3rd : βμν (if Manilius, like Servius, *Georg.* 1. 205, did include this constellation in 5. 715, he took no account of α, *Alphard*, which is 2nd magnitude).

Third magnitude constellations unspecified

Some of these constellations (*e.g.* Pisces, Cepheus) might more aptly be classified as 4th magnitude, but perhaps the poet had tired of cataloguing and simply cut his list short.

(a) ARIES 3rd : αβγ (4th : 5).
CAPRICORNUS 3rd : αβγδ (4th : 9).
AQUARIUS 3rd : αβγδζηπ and two others (4th : 18). But Ptolemy lists α PsA (*Fomalhaut*) also under Aquarius.
PISCES 3rd : αη (4th : 22).
(b) ENGONASIN (= Hercules) 3rd : αβγδζ (4th : ηπ and 15 others ; 5th : ε).
CEPHEUS 3rd : α (4th : 7).
CASSIEPIA (= Cassiopeia) 3rd : αβγδ (4th : ε and 5 others).
(c) LEPUS 3rd : αβ (4th : 6).
CORVUS 3rd : αβγδε (4th : 1).
CENTAURUS 3rd : γζηθι (4th : at least 10). Ptolemy, but hardly Manilius, knew the Antarctic brilliants (all 1st) α Cen (*Rigil Kent*), β Cen (*Agena*), α Cru (*Acrux*, forming with βγδ Cru the Southern Cross).
CETUS 3rd : αβγδζηθιπτ (4th : 8).
FLUMINA (= Eridanus and the stars forming the water of Aquarius) 3rd : γδεζρ Eri (4th : β and at least 20 others). Ptolemy, but hardly Manilius, knew the Antarctic brilliant θ Eri (*Achernar* : 1st).

Constellations of lower than third magnitude

These constellations are subsumed under the general statement made in 5. 716 f.

(a)	CANCER	4th: 7;
		nebulous: ϵ = Praesaepe (4. 530).
(b)	SAGITTA	4th: γ;
		5th: 3.
(c)	CRATER	4th: α and 6 others.
	ARA	4th: $\alpha\gamma$ and 3 others.

INTRODUCTION

THE MANUSCRIPTS

[Housman, ed. 1[2] vii ff., 83 ff., 5[2] v ff.; Garrod, ed. 2 xv ff.; Goold[1]; Goold[2] 96 ff.; Gain[4]]

There are extant about twenty manuscripts of Manilius, but of these only three preserve an independent tradition. **G, codex Gemblacensis** (Bruxellensis 10012), and **L, codex Lipsiensis** 1465, are both of the 11th century and descend from a common parent, **α**. Though not written till the year 1417, and badly written at that, **M, codex Matritensis** 3678 (formerly M31), is perhaps the most valuable MS. of Manilius: it was a direct copy of the archetype, and a more faithful copy than *α*, which seems to have been another.

The archetype (like that of Lucretius) carried not only the verses of the poet but also chapter-headings inserted at intervals through the text, not of course stemming from the poet but seemingly attesting some ancient edition. Guided by certain significant errors, notably the loss or displacement of sheets (again there is a curious resemblance to the Lucretian tradition), we are able to detect traces of the pagination of an ancestral MS. (let us call it **A**) containing 22 lines to the page. The most striking example affords an explanation of the MS. order of 1. 355-611: two double-leaves in the second quire were mistakenly placed in the reverse order, as may be seen from table 7 (which also shows the basis for the computation of the lacuna at 5. 709). For the sake of clarity chapter-headings (*e.g.* one line after 1. 482) and subsequent interpolations (*e.g.* 5. 686-688) involved in the reconstruction are not specified in the table. These, together with much else, are

folio	*recto*	*verso*	
9	1. 310-331	332-354	Quire 2
11	1. 399-420	421-442	
10	1. 355-376	377-398	
12	1. 443-464	465-485	
13	1. 486-507	508-529	
15	1. 568-589	590-611	
14	1. 530-550	551-567	
16	1. 612-633	634-655	
97	5. 619-640	641-662	Quire 13
98	5. 663-684	685-709	
99	*44 lines*		
100	*44 lines*		
101	*44 lines*		
102	*44 lines*		
103	5. 710-732	733-745	
104	*blank*		

TABLE 7. CONJECTURAL RECONSTRUCTION OF THE MS. A

given in my article in *Rheinisches Museum.* I have, however, changed my mind about some points of detail (*e.g.* I now make 5. 439 the last line of fol. 92v and include 5. 515 on fol. 94v); and I should also notice another scholar's article (Gain[4]) proposing an alternative pagination for Book 2.

This is the place to mention a more important issue affecting our conception of Manilius's poem, one which deserves a bibliographical docket of its own.

[Ellis 217-233; Housman, ed. 1[2] lxix, lxxii; Garrod, ed. 2 lix ff.; Thielscher; Ullman; Gain[3]]

The savant Gerbert, who became Pope Sylvester II, twice refers in his letters to the text of our author. Writing in 983 (*Letter* 8 : Migne, *PL* 139 col. 203) from Bobbio to the Archbishop of Rheims he mentions that he has recently lighted on *viii volumina Boetii de astrologia.* Five years later he writes (*Letter* 130 : Migne, *PL* 139 col. 233) from Rome to a confidant at Bobbio requesting the copying of a list of books : *M. Manlius de astrologia* ; *Victorius de rethorica* ; *Demosthenis optalmicus.* Now from the order of titles in the list the first can be identified with entry 387 in a 10th-century Bobbio catalogue, which with its preceding entry forms a single item as follows : (384-386) *libros Boetii iii de aritmetica et* (387) *alterum de astronomia*—for there follow (388) *librum Marii Victoris de rethorica* and shortly after (399) *librum i Demosthenis.*

In view of the fact that the chief MS. of Manilius bears as title of the poem (before Book 3) *M. Manlii Boeiii* (*sic* : manifestly for *Boetii*) *astronomic*[*a*], there can hardly be any doubt that Gerbert refers in *Letter* 130 to our Manilius. And there should be no doubt that, in speaking of *viii volumina Boetii*, he is, exactly like the Bobbio catalogue, compendiously referring to a single manuscript containing three books of Boethius's *Arithmetica* and five books of Marcus Manilius's *Astronomica.*

Unfortunately, if understandably, the reference has misled others into speculating that the poem once consisted of eight books. Thielscher's belief that three books of the *Astronomica* have been lost after Book 5 is simply refuted by the first line of that book, sufficient warning that we have reached the end of the undertaking. For Gain the loss of three books has occurred before the beginning of Book 5, where the scribe of M has written, without mention of either author or work, *explicit liber ii incipit liber iii.* But sooner than amend these numerals to *vii* and *viii* we should accept Thielscher's explanation, namely that Poggio's scribe (*ignorantissimus omnium viventium*, the humanist calls him) was "at the end of Book 4 so tired and weary that he no longer remembered the names [of author and work : to judge from LM they did not recur in the tradition after the first book] and even forgot what book he had ended and what book he was about to begin."

The possibility that better MSS. somewhere exist and await discovery must be accounted remote in the extreme ; our present evidence for the tradition is not likely to be augmented. Its corruptions, says Housman, are multitudinous but not profound. There is justice in the remark, for a great many obviously true emendations restore through the slightest of changes perfect sense from utter nonsense ; and few passages can be cited where the general purport is in doubt. Moreover, dislocations are few, and interpolations, although more numerous, are also easier to detect and discount.

Whilst it must be admitted that the *ipsissima verba* of the poet have not everywhere been recovered with certainty, we can today recognize throughout the work his distinctive and unmistakable voice, and if here and there it momentarily fades or is drowned by an alien sound, coherence is seldom threatened and continuity is broken only at the great lacuna.

EDITORIAL PRINCIPLES

After his death Housman's photographs of the MSS. were presented to the Cambridge University Library: I have collated his reports against them, later examining M in Madrid. His accuracy was truly phenomenal. The Latin text I have thence constructed is presented according to the following principles.

All significant departures from the MS. tradition are noted beneath the text. Thus, even where a possibly incorrect conjecture has been adopted, both that and the transmitted reading are clearly specified. Conjecturers' names, however, are not given: these may be discovered from Housman's large edition or, if not to be found there, from the Appendix p. 365. For example, at 1. 12 the Latin text gives *census*, a conjecture of Scaliger's for *sensus*, the reading of the MSS.: this is signified by the note "12 sensus."

The reader should particularly bear in mind that where the text has MS. authority, variant readings in the tradition are almost always ignored: thus no note is given at 1. 756 *densa* (the reading of G), for which LM have *densat*. Since the three independent MSS. are intimately related and the reading of the archetype is generally recoverable with certainty, it is seldom necessary to record more than one reading; moreover, it has been judged preferable to indicate the extent of conjecture adopted rather than the range of corruption in the MSS.: so at 1. 843 f. *artosque capellas / mentitur parvas ignis*, where the MSS. readings are: *partasque capellas m. parvos signis* (M), *partosque capellos m. parvis signis* (L), and *partosque capillos m. parvis signis* (G), the words lacking MS. authority are indicated by the notes "843 partasque" and "844 parvis signis," whether or no these words (rather than *partosque* and *parvos signis*) stood in the archetype.

Two types of departure from the MSS., however, are not noticed: (1) minute and obviously correct conjectures, made not later than the 16th century and universally accepted, are ignored (*e.g.* 1. 105 *sonitum*—so the cod. Flor.—where the archetype had *solitum*); (2) orthography is standardized, regardless of what the MSS. give, so that the form *sequuntur* (*cf.* 2. 208) is consistently printed at 3. 278, 3. 516, and 4. 269 (where Housman gave *sequntur*, *secuntur*, and *sequontur* respectively). Assimilated forms and Greek spellings (such as the nominative *horoscopos*) have generally been preferred. Even so, standardization has not been indiscriminately applied: the reader will encounter 1. 122 *ex nullis* but 5. 717 *e numero*; 1. 509 *quotiens* but 3. 487 *decies*. The MSS. seem to point to a principle of spelling the accusative plural of *i*-stems in *-es* for nouns and *-is* for adjectives and participles: Breiter's standardization of this practice has been followed, though it may not go back to the poet.

Apart from a few special places (like 1. 38 f., where the contrary is explicitly indicated) line-numbers mark the order of the verses as they appear in the MSS.: where, as for example in the case of 1. 30-33, the numerical sequence is disturbed, this is to be imputed to conjecture. Verses italicized in the text (as they also are in the translation) have no MS. authority; they are editorial supplements which attempt to restore the sense of a conjectured lacuna.

Square brackets in the text enclose those verses which, though exhibited by the MSS., are judged to be interpolated : they are not translated in the body of the text but in footnotes. Square brackets in the critical notes, on the other hand, draw attention to words not attested by the MSS. : thus, at 1. 156 *aequoraque effudit*, the note "[que] perfudit" signifies that the MSS. read *aequora perfudit.*

For the most part Housman's punctuation is adopted (the main difference being that, as at 3. 5, commas are employed to signal postponed conjunctions); his paragraphing and line numeration are accepted as definitive.

SELECT BIBLIOGRAPHY

The abbreviations employed are those in general use among classical scholars. Apart from two asterisked items included for their historical significance the list is restricted to works used by the editor and in the case of secondary sources to works cited by him.

[Pauly-Wissowa *RE* XIV (1928) 1115-1133 *s.v.* Manilius 6 (†J. van Wageningen); Schanz-Hosius, *RLG* 4 2 (1935) 441-447; for 1925–1942, *Lustrum* 1 (1956) 129-158 (R. Helm); thereafter see *L'Année Philologique*]

Editions of Manilius

* Regiomontanus (about 1472): the editio princeps.

* Bonincontrius (1484): with the first commentary.

Scaliger (1579[1], 1600[2], 1655[3]): a vastly improved text with a commentary which was until this century "the only avenue to a study of the poem" (Housman).

Fayus (1679): the Delphin edition, with copious annotation, paraphrase (Latin), *index vocabulorum*, and Huet's *Animadversiones*.

Bentley (1739): the text marks a great advance even on Scaliger; exegesis (wonderfully lucid) is given only where a change of text is proposed; includes a projection of the Farnese globe.

Pingré (1786): with an elegant (French prose) translation; "in no edition of Manilius is there so little that calls for censure" (Housman).

Jacob (1846): the standard text until superseded by Housman's.

Bechert (1900): contributed to Postgate's *Corpus*

Poetarum Latinorum (*Prolegomena* in *CR* 14 [1900] 296-304).

BREITER (1907): text with apparatus; (1908) (German) commentary.

GARROD (1911): Book 2—introduction, text, apparatus, (English prose) translation, commentary.

VAN WAGENINGEN (1915): the Teubner edition; (1914) (Dutch prose) translation; (1921) copious (Latin) commentary.

The work of A. E. HOUSMAN on Manilius eclipsed everything that had gone before and deserves special mention:

"Emendations of Book 1," *JP* 26 (1899) 60-63;
"Emendations of Book 5," *JP* 27 (1901) 162-165 (Housman's other Manilian articles are subsumed in his Editio Maior, and are therefore not recorded here);
Book 1 (1903): general introduction, text, (Latin) commentary, emendations of Books 2, 3, and 4, index;
Book 2 (1911): astrological introduction, text, (Latin) commentary, index;
Book 3 (1916): as for Book 2;
Book 4 (1920): as for Book 2;
Book 5 (1930): survey, introduction, text, (Latin) commentary, addenda to Books 1-4, capitula, orthographical appendix, index to all five volumes;
Editio Minor (1932): text and apparatus, index nominum.
In 1937 (after Housman's death) a second edition of the individual books was produced under

the direction of A. S. F. Gow: in this only minor additions and corrections were made, the chief change being the incorporation of the 1930 addenda in the volumes to which they refer.

English translations

SHERBURNE, EDWARD (1674): Book 1 only, in verse.
CREECH, THOMAS (1697): in verse.

Ancient astrological and astronomical sources
(all unspecified dates are A.D.)

Greek

ACHILLES (3rd cent.): *Isagoge* to the *Phaenomena* of Aratus; fragments collected in *Commentariorum in Aratum Reliquiae* pp. 26-85 Maass.

ALBUMASAR (Abu Ma'shar of Baghdad = Greek Apomasar: 9th cent.): *De Revolutionibus Nativitatum* (= a late 10th-cent. Greek translation of the Arabic original, this dependent on Greek sources and perhaps ultimately on Dorotheus, *Pentateuch*, Book 4), ed. D. Pingree (Teubner 1968).

AMMON: see MAXIMUS.

ANTIOCHUS ATHENIENSIS (*c.* 200; was excerpted by Rhetorius and other late writers, and very extensively in Arabic texts, *cf.* Pingree, *Gnomon* 40 [1968] 278): fragments of his *Treasury*, ed. Boll, *CCAG* I.

ANUBION (1st cent.: author of an astrological poem in elegiacs, a source of Manetho, Book 1): see Koechly's Manetho, and *CCAG* VIII. 1 and VIII. 2.

ARATUS (3rd cent. B.C.) : *Phaenomena*, ed. E. Maass (1893) ; ed. with trans. G. R. Mair (Loeb 1921) ; ed. with commentary and (French) translation, Jean Martin (1956) ; *Commentariorum in Aratum Reliquiae* (contains Scholia), ed. E. Maass (1898).

CCAG (= *Catalogus Codicum Astrologorum Graecorum*) : 12 volumes (Brussels 1898-1953), begun under direction of Franz Cumont ; catalogues by libraries all Greek MSS. containing astrological material, with new or important texts edited in appendices, which consequently form the bulk of the volumes.

I	(MSS. in Florence) ed. Olivieri, 1898
II	(Venice) Kroll, Olivieri, 1900
III	(Milan) Martini, Bassi, 1901
IV	(other Italian) Bassi, etc., 1903
V.1	(Rome) Cumont, Boll, 1904
V.2	(Vatican) Kroll, 1906
V.3	(Vatican) Heeg, 1910
V.4	(Rome) Weinstock, 1940
VI	(Vienna) Kroll, 1903
VII	(Germany) Boll, 1908
VIII.1	(Paris) Cumont, 1929
VIII.2	(Paris) Ruelle, 1911
VIII.3	(Paris) Boudreaux, 1912
VIII.4	(Paris) Boudreaux, 1921
IX.1	(Oxford) Weinstock, 1951
IX.2	(other British) Weinstock, 1953
X	(Athens) Delatte, 1924
XI.1	(Escorial) Zuretti, 1932
XI.2	(other Spanish) Zuretti, 1934
XII	(Russia) Sangrin, 1936

CLEOMEDES (2nd cent.) : *De Motu Circulari Corporum Caelestium*, ed. H. Ziegler (Teubner 1891) : with Latin translation.

DOROTHEUS SIDONIUS (*c.* 50) : *Fragments*, ed. V. Stegemann (Heft 1—all published—1943 : for a conspectus of the contents of the *Pentateuch*, see Hephaestio I p. XXV); *Carmen Astrologicum*, ed. D. Pingree (Teubner 1976): contains Arabic version (with English translation) of a 3rd-century Pahlavi version of the whole *Pentateuch*, followed by Greek and Latin fragments.

Doxographi Graeci: see DIELS.

ERATOSTHENES (3rd cent. B.C.): *Catasterismorum Reliquiae* (what we have is a late epitome), ed. C. Robert (1878; repr. 1963): with parallel passages from the Aratus scholia, the Germanicus scholia (BP and G versions), and the so-called Hyginus; ed. A. Olivieri (Teubner 1897).

GEMINUS (1st cent. B.C.): *Elementa Astronomiae* (= *Isagoge*), ed. C. Manitius (Teubner 1898; repr. 1974); with German translation.

HELIODORUS (*c.* 500): *In Paulum Alexandrinum Commentarius*, ed. Ae. Boer (Teubner 1962).

HEPHAESTIO THEBANUS (*c.* 415): ed. D. Pingree (Teubner): I *Apotelesmatica* (1973); II *Apotelesmaticorum Epitomae Quattuor* (Byzantine) (1974).

HIPPARCHUS (2nd cent. B.C.): *In Arati et Eudoxi Phaenomena Commentarii*, ed. C. Manitius (Teubner 1894): with German translation.

MANETHO (the pseudonym of a compiler, 4th century, of heterogeneous verse material, some of it early imperial): *Apotelesmatica*, ed. A. Koechly (in the Didot *Bucolici Graeci*, 1862): with Latin translation.

MAXIMUS (4th cent.): *De Actionum Auspiciis*, ed. Ludwich (Teubner 1877): with Ammon; text and Latin translation included with Koechly's Manetho.

PAULUS ALEXANDRINUS (second half, 4th cent.):

Elementa Apotelesmatica (= *Isagoge*), ed. Ae. Boer (Teubner 1958).

PORPHYRY (3rd cent.): *Introduction to Ptolemy's Tetrabiblos* (= *Isagoge*), edd. S. Weinstock, E. (= Ae.) Boer, *CCAG* V. 4. Ed. with other commentaries, H. Wolf (1559): with Latin translation.

PTOLEMY (first half, 2nd cent.): *Syntaxis*, ed. J. L. Heiberg (2 vols, Teubner: 1898, 1903); German translation by Karl (= C.) Manitius (2 vols, Teubner: 1912, 1913; repr. 1963); *Tetrabiblos*, ed. and trans. F. E. Robbins (Loeb 1940); edd. F. Boll, Ae. Boer (Teubner 1940); see also Porphyry.

RHETORIUS (*c.* 500): large fragments of his *Treasury* (quoting earlier writers like Teucer and Antiochus) have been published in *CCAG* I, V. 4, VII, VIII. 1 and VIII. 4.

SEXTUS EMPIRICUS (*c.* 200): *Adversus Astrologos* (= *Adv. Math.* 5), ed. and trans. R. G. Bury (Loeb, vol. IV, 1949).

TEUCER BABYLONIUS (1st cent.): see Boll's *Sphaera* and *CCAG* IX. 2.

VETTIUS VALENS (2nd cent.): *Anthologiae*, ed. W. Kroll (1908); see also *CCAG* VIII. 1 and Neugebauer-Van Hoesen.

Latin

AVIENIUS (4th cent.)—so we must now call him: Cameron, *CQ* 17 (1967) 392 ff.: *Aratea*, ed. A. Breysig (Teubner 1882).

CICERO (1st cent. B.C.): *Aratea*, ed. A. Traglia (1952).

FIRMICUS MATERNUS (*c.* 335): *Mathesis*, edd. W. Kroll, F. Skutsch, K. Ziegler (2 vols, Teubner: 1897,

1913; repr. 1968); translated as *Ancient Astrology* by Jean Rhys Bram (1975).
GERMANICUS (early 1st cent.): *Aratus*, ed. with trans. and commentary, D. B. Gain (1976); scholia, ed. A. Breysig (1867; repr. 1967).
HYGINUS (pseudo- ?, 2nd cent. ?): *Poetica Astronomica*: see Eratosthenes: also included in the Van Staveren variorum edition of *Auctores Mythographi Latini* (1742: with notes of Muncker and others); *Fabulae*, ed.[3] H. J (=I). Rose (1967).

Secondary sources

ALLEN, R. H.: *Star-Names and Their Meanings* (1899; repr. 1963).
ALTON, E. H.: review Manilius 5, ed. Housman, *Hermathena* 46 (1931) 258-265.
BAILEY, D. R. SHACKLETON: "Maniliana," *CQ* 6 (1956) 81-86.
BARTALUCCI, ALDO: "Una fonte egizia di età tolemaica nella geografia zodiacale di Manilio," *SIFC* 33 (1961) 91-100.
BOLL, FRANZ: *Sphaera* (1903; repr. 1967): new Greek astrological texts (especially Teucer, Antiochus, and Valens) and Egyptian and oriental influences on the Greek Sphaera.
BOUCHÉ-LECLERCQ, A.: *L'Astrologie Grecque* (1899; repr. 1963): not yet superseded as the standard work on the subject.
BÜHLER, WINFRIED: "Maniliana," *Hermes* 87 (1959) 475-494.
COLSON, F. H.: *The Week* (1926).
CRAMER, FREDERICK H.: *Astrology in Roman Law and*

Politics (1954): contains an introduction to Greek astrology.

Dicks, D. R.: *Early Greek Astronomy to Aristotle* (1970).

Diels, H.: *Doxographi Graeci* (1879).

Dreyer, J. L. E.: *History of the Planetary Systems from Thales to Kepler* (1905; repr. 1953).

Ellis, Robinson: *Noctes Manilianae* (1891).

Fletcher, G. B. A.: "Manilius," *Durham University Journal* 65 (1973) 129-150.

Flores, E.: *Contributi di Filologia Maniliana* (1966) 59.

Gain[1] (D. B.): "Notes and Conjectures on the *Astronomica* of Manilius," *Antichthon* 2 (1968) 63-67.

Gain[2]: "Lucubrationes Manilianae," *AC* 38 (1969) 162 f.

Gain[3]: "Gerbert and Manilius," *Latomus* 29 (1970) 128-132.

Gain[4]: "*De Fonte Codicum Manilianorum* reviewed," *RhM* 114 (1971) 261-264.

Garrod[1] (H. W.): "Two Editions of Manilius," *CQ* 2 (1908) 130.

Garrod[2]: *Oxford Book of Latin Verse* (1912) 225.

Goethe: see Grumach.

Goold[1] (G. P.): "De Fonte Codicum Manilianorum," *RhM* 97 (1954) 359-372.

Goold[2]: "Adversaria Maniliana," *Phoenix* 13 (1959) 93-112.

Goold[3]: "A Greek Professorial Circle at Rome," *TAPA* 91 (1961) 168-192.

Grumach, Ernst: *Goethe und die Antike* (1949) I 390 f.

Gundel, W. and Gundel, H. G.: *Astrologumena* (1966).

Heath[1] (T. L.): *Aristarchus of Samos* (1913): a history of Greek astronomy to Aristarchus.

Heath[2]: *Greek Astronomy* (1932): a collection of sources in translation.

HERRMANN, L. : "Hypothèse sur L. et M. Manilius," *AC* 31 (1962) 82-90.
HOUSEHOLDER, F. W. : see NAIDEN.
JEBB, R. C. : *Bentley* (1882) 143.
LINDSAY, JACK : *Origins of Astrology* (1971).
MALCHIN, F. : *De auctoribus quibusdam qui Posidonii libros meteorologicon adhibuerunt* (1893) 19.
MOELLER, J. : *Studia Maniliana* (1901).
MUELLER, E. : *De Posidonio Manilii Auctore* (1901).
NAIDEN, J. R. and HOUSEHOLDER, F. W. : "A Note on Manilius i. 431-42," *CP* 37 (1944) 187-191.
NEUGEBAUER, O. : *The Exact Sciences in Antiquity*[2] (1957).
NEUGEBAUER, O. and VAN HOESEN, H. B. : *Greek Horoscopes* (1959) : contains a wealth of astrological exposition in addition to translations and commentary.
NORTON, ARTHUR P. : *Norton's Star Atlas*[16] (1973) : for Epoch 1950.
SALMASIUS, CLAUDIUS : *De Annis Climactericis et Antiqua Astrologia* (1648) : "le livre classique . . . le premier—après le commentaire de Scaliger sur Manilius—et le dernier effort de l'érudition indépendante, s'exerçant à comprendre l'astrologie" (Bouché-Leclercq).
SCHWARZ, WOLFGANG : "Praecordia mundi," *Hermes* 100 (1972) 601-614.
SIKES, E. E. : *Roman Poetry* (1923) 172-181.
THIELE, G. : *Antike Himmelsbilder* (1898).
THIELSCHER, PAUL : "Ist *M. Manilii Astronomicon Libri V* richtig ? " *Hermes* 84 (1956) 353-372.
THOMAS, P. : "Notes et conjectures sur Manilius," *Mém. acad. roy. Belg.* 46 (1892) 25.
ULLMAN, B. L. : "Geometry in the Mediaeval

Quadrivium," *Studi di bibliografia e di storia in onore di Tammaro de Marinis*, volume IV (Verona 1964) 278.

Van Hoesen: see Neugebauer.

Webb, E. J.: *The Names of the Stars* (1952).

THE ASTRONOMICA OF MARCUS MANILIUS

BOOK ONE

The *Astronomica opens with a proud assertion of its author's originality and continues with the early history of the study of the heavens; this leads to a sketch of primitive man and a cosmology which alike embrace the beliefs of Stoicism. The central and most substantial part of the first book is the poet's Sphaera, a versified atlas of the sky. He gives pride of place to the signs of the zodiac, and proceeds to catalogue first the northern and then the southern constellations, following this with a fine declamation on the eternal and immutable nature of the stars. After a brief mention of the planets he passes on to an account of the various celestial circles. The last of these, the Milky Way, has inspired the poet to embark upon a long and ambitious digression culminating in a catalogue of heroes who have received apotheosis. Comets and meteors occupy the final pages of the book, though the astrological relevance of this section is slight, its real function being to pave the way for the theme of the peroration, heavenly portents of disaster. Here Manilius handles with some felicity the plague at Athens and Rome's civil wars brought to an end by the battle of Actium and the triumph of Augustus.*

M. MANILII
ASTRONOMICON
LIBER PRIMUS

Carmine divinas artes et conscia fati
sidera diversos hominum variantia casus,
caelestis rationis opus, deducere mundo
aggredior primusque novis Helicona movere
cantibus et viridi nutantis vertice silvas
hospita sacra ferens nulli memorata priorum.
hunc mihi tu, Caesar, patriae princepsque paterque,
qui regis augustis parentem legibus orbem
concessumque patri mundum deus ipse mereris,
das animum viresque facis ad tanta canenda.
iam propiusque favet mundus scrutantibus ipsum
et cupit aetherios per carmina pandere census.
hoc sub pace vacat tantum. iuvat ire per ipsum
aera et immenso spatiantem vivere caelo
signaque et adversos stellarum noscere cursus.

[12] sensus

THE ASTRONOMICA OF MARCUS MANILIUS

BOOK 1

By the magic of song to draw down from heaven god-given skills and fate's confidants, the stars, which by the operation of divine reason diversify the chequered fortunes of mankind ; and to be the first to stir with these new strains [a] the nodding leaf-capped woods of Helicon, as I bring strange lore untold by any before me : this is my aim. You, Caesar,[b] First Citizen and Father of your Country, who rule a world obedient to your august laws and merit the heaven granted to your sire, yourself a god, are the one who inspires my design and gives me strength for such lofty themes. Now is heaven the readier to favour those who search out its secrets, eager to display through a poet's song the riches of the sky. Only in time of peace is there leisure for this task. It is my delight to traverse the very air and spend my life touring the boundless skies, learning of the constel-

[a] Astrological poetry.

[b] Augustus. As Octavian he had assumed the style *princeps* ; the title *pater patriae* was formally conferred on him by the Senate in 2 B.C., but *pater* had long since been anticipated (*e.g.* Horace, *Carm.* 1. 2. 50).

quod solum novisse parum est. impensius ipsa
scire iuvat magni penitus praecordia mundi,
quaque regat generetque suis animalia signis
cernere et in numerum Phoebo modulante referre.
bina mihi positis lucent altaria flammis,
ad duo templa precor duplici circumdatus aestu
carminis et rerum : certa cum lege canentem
mundus et immenso vatem circumstrepit orbe
vixque soluta suis immittit verba figuris.

Quem primum interius licuit cognoscere terris
munere caelestum. quis enim condentibus illis
clepsisset furto mundum, quo cuncta reguntur ?
quis foret humano conatus pectore tantum,
invitis ut dis cuperet deus ipse videri,
sublimis aperire vias imumque sub orbem,
et per inane suis parentia finibus astra ?
tu princeps auctorque sacri, Cyllenie, tanti ;
per te iam caelum interius, iam sidera nota
nominaque et cursus signorum, pondera, vires,
maior uti facies mundi foret, et veneranda
non species tantum sed et ipsa potentia rerum,
sentirentque deum gentes qua maximus esset.

[18] quaeque
[37] quam

38 f.: *not in the MSS.*

---

[a] The fixed stars of the sky revolve from east to west, and so do the planets (seven, including the Sun and Moon) ; but the speed of the latter is somewhat slower, so that their movement appears to be contrary to that of the stars.

[b] Mercury.

[c] Here Bonincontrius inserted two verses (38 f.) of his own composition (*qui sua disposuit per tempora, cognita ut*

lations and the contrary motions of the planets.[a] But this knowledge alone is not enough. A more fervent delight is it to know thoroughly the very heart of the mighty sky, to mark how it controls the birth of all living beings through its signs, and to tell thereof in verse with Apollo tuning my song. Two altars with flame kindled upon them shine before me ; at two shrines I make my prayer, beset with a twofold passion, for my song and for its theme. The poet must sing to a fixed measure, and the vast celestial sphere rings in his ears besides, scarce allowing even words of prose to be fitted to their proper phrasing.

[25]Deeper knowledge of heaven was first granted to earth by the gift of the gods. For who, if the gods wished to conceal it, would have guilefully stolen the secret of the skies, by which all things are ruled ? Who of but human understanding would have essayed so great a task as to wish against heaven's wish to appear a god himself ; to reveal paths on high and paths beneath the bottom of the earth and stars obedient to appointed orbits through the void ? You, God of Cyllene,[b] are the first founder of this great and holy science ; through you has man gained a deeper knowledge of the sky—the constellations, the names and courses of the signs, their importance and influences—that the aspect of the firmament might be enhanced, that awe might be roused not only by the appearance but by the power of things, and that mankind might learn wherein lay God's greatest power.[c] Moreover, nature proffered her

essent / omnibus et mundi facies caelumque supernum : "who arranged heaven's face and the sky above in special cycles of time that these be known to all").

et natura dedit vires seque ipsa reclusit
regalis animos primum dignata movere
proxima tangentis rerum fastigia caelo,
qui domuere feras gentes oriente sub ipso,
[quas secat Euphrates, in quas et Nilus abundat]
qua mundus redit et nigras super evolat urbes.
tum qui templa sacris coluerunt omne per aevum
delectique sacerdotes in publica vota
officio vinxere deum ; quibus ipsa potentis
numinis accendit castam praesentia mentem,
inque deum deus ipse tulit patuitque ministris.
hi tantum movere decus primique per artem
sideribus videre vagis pendentia fata.
singula nam proprio signarunt tempora casu,
longa per assiduas complexi saecula curas :
nascendi quae cuique dies, quae vita fuisset,
in quas fortunae leges quaeque hora valeret,
quantaque quam parvi facerent discrimina motus.
postquam omnis caeli species, redeuntibus astris,
percepta, in propias sedes, et reddita certis
fatorum ordinibus sua cuique potentia formae,
per varios usus artem experientia fecit
exemplo monstrante viam, speculataque longe
deprendit tacitis dominantia legibus astra
et totum aeterna mundum ratione moveri
fatorumque vices certis discernere signis.
  Nam rudis ante illos nullo discrimine vita
in speciem conversa operum ratione carebat

64 alterna

---

[a] Zoroaster and Belus may be alluded to.

[b] [44] ". . . whose lands are severed by the Euphrates or flooded by the Nile, . . ."

aid and of her own accord opened up herself, deigning first to inspire those kings [a] whose minds reached out to heights bordering on heaven, kings who civilized savage peoples beneath the eastern sky,[b] where the stars return to view and soar above the cities of dusky nations. Then priests [c] who all their lives offered sacrifice in temples and were chosen to voice the people's prayer secured by their devotion the sympathy of God ; their pure minds were kindled by the very presence of the powerful deity, and the God of heaven brought his servants to a knowledge of heaven and disclosed its secrets to them. These were the men who founded our noble science and were the first by their art to discern the destinies dependent on the wandering stars. Embracing long ages in unremitting toil, they assigned to each period of time its particular events, noting an individual's nativity and the subsequent pattern of his life, the influence of each hour on the laws of fate, and the great differences effected by small moments. After every aspect of the sky had been observed, as the stars returned to their customary positions, and the unvarying sequences of fate had assigned to each figuration of the planets its peculiar influence, by repeated practice and with examples pointing the way experience built up the science ; and from wide observation discovered that by hidden laws the stars wield sovereign power and that all heaven moves to the eternal spirit of reason and by sure tokens distinguishes the vicissitudes of fate.

[66]Before their times man lived in ignorance : he looked without comprehension at the outward appearance and saw not the design of nature's

[c] Nechepso and Petosiris may be alluded to.

et stupefacta novo pendebat lumine mundi,
tum velut amisso maerens, tum laeta ren*ato,*
*surgentem neque enim totiens Titana fug*atis
sideribus, variosque dies incertaque noctis
tempora nec similis umbras, iam sole regresso
iam propiore, suis poterat discernere causis.
necdum etiam doctas sollertia fecerat artes,
terraque sub rudibus cessabat vasta colonis ;
tumque in desertis habitabat montibus aurum,
immotusque novos pontus subduxerat orbes,
nec vitam pelago nec ventis credere vota
audebant ; se quisque satis novisse putabant.
sed cum longa dies acuit mortalia corda
et labor ingenium miseris dedit et sua quemque
advigilare sibi iussit fortuna premendo,
seducta in varias certarunt pectora curas
et, quodcumque sagax temptando repperit usus,
in commune bonum commentum laeta dederunt.
tunc et lingua suas accepit barbara leges,
et fera diversis exercita frugibus arva,
et vagus in caecum penetravit navita pontum,
fecit et ignotis iter in commercia terris.
tum belli pacisque artes commenta vetustas ;
semper enim ex aliis alias proseminat usus.
ne vulgata canam, linguas didicere volucrum,
consultare fibras et rumpere vocibus angues,
sollicitare umbras imumque Acheronta movere,
in noctemque dies, in lucem vertere noctes.

[69] amissis [72] poterant

works ; he gazed in bewilderment at the strange new light of heaven, now sorrowing at its loss, now joyful at its rebirth ; *nor why the Sun so often rose, putting to flight* the stars, why days varied in duration and the period of darkness fluctuated, and why the lengths of shadows differed according as the Sun retreated or drew nearer, could he understand from the true causes. Nor as yet had ingenuity taught man arts and crafts : the earth lay waste and idled under ignorant husbandmen ; gold then had its abode in unvisited mountains, and the sea disturbed by none concealed the existence of unsuspected worlds, for man dared not to entrust his life to the deep or his hopes to the winds, and deemed his little knowledge adequate. But long ages sharpened human wits, the struggle for survival endowed those wretched folk with ingenuity, and the burden of each man's lot forced him to look to himself to better it : then they took thought to divide their tasks and vied with each other in performing them, and whatever shrewd experience had found by trial, they joyfully communicated and surrendered to the common weal. Their barbarous speech, too, then subjected itself to laws of its own ; rough soils were tilled for a variety of crops ; and the roving sailor made his way into the uncharted sea and established trade-routes between lands unknown to each other. Then the passage of time led to the arts of war and peace, for practice ever breeds one art from another. Not to tell the commonplace, men learnt to understand the utterance of birds ; to divine from the entrails of animals ; to burst snakes asunder by incantations ; to summon up the dead and rouse the depths of Acheron ; to turn day into night and night into the

omnia conando docilis sollertia vicit.
nec prius imposuit rebus finemque modumque
quam caelum ascendit ratio cepitque profundam
naturam rerum causis viditque quod usquam est.
nubila cur tanto quaterentur pulsa fragore,
hiberna aestiva nix grandine mollior esset,
arderent terrae solidusque tremesceret orbis ;
cur imbres ruerent, ventos quae causa moveret
pervidit, solvitque animis miracula rerum
eripuitque Iovi fulmen viresque tonandi
et sonitum ventis concessit, nubibus ignem.
quae postquam in proprias deduxit singula causas,
vicinam ex alto mundi cognoscere molem
intendit totumque animo comprendere caelum,
attribuitque suas formas, sua nomina signis,
quasque vices agerent certa sub sorte notavit
omniaque ad numen mundi faciemque moveri,
sideribus vario mutantibus ordine fata.

Hoc mihi surgit opus non ullis ante sacratum
carminibus. faveat magno fortuna labori,
annosa et molli contingat vita senecta,
ut possim rerum tantas emergere moles
magnaque cum parvis simili percurrere cura.

Et quoniam caelo descendit carmen ab alto
et venit in terras fatorum conditus ordo,
ipsa mihi primum naturae forma canenda est
ponendusque sua totus sub imagine mundus.

[96] manumque [104] tonantis

---

[a] It would seem that the poet had already attained middle-age.

brightness of day. Thus, ever ready to learn, ingenuity by its endeavours overcame all obstacles. Nor did man's reason set bound or limit to its activities until it scaled the skies, grasped the innermost secrets of the world by its understanding of their causes, and beheld all that anywhere exists. It perceived why clouds were shaken and shattered by so loud a crash; why winter's snowflakes were softer than summer's hail; why volcanoes blazed with fire and the solid earth quaked; why rain poured down and what cause set the winds in motion. It freed men's minds from wondering at portents by wresting from Jupiter his bolts and power of thunder and ascribing to the winds the noise and to the clouds the flame. After reason had referred these several happenings to their true causes, it ventured beyond the atmosphere to seek knowledge of the neighbouring vastness of heaven and comprehend the sky as a whole; it determined the shapes and names of the signs, and discovered what cycles they experienced according to fixed law, and that all things moved to the will and disposition of heaven, as the constellations by their varied array assign different destinies.

[113]This is the theme that rises before me, a theme hitherto unhallowed in verse. May fortune favour my grand emprise, and may a life of many years crowned with a serene old age be mine,[a] enabling me to surmount the great vastness of the subject and pursue my course with equal care through mighty things and small.

[118]And since from the heights of heaven my song descends and thence comes down the established rule of fate, first must I sing of nature's true appearance and describe the whole universe after its own

quem sive ex nullis repetentem semina rebus
natali quoque egere placet, semperque fuisse
et fore, principio pariter fatoque carentem ;
seu permixta chaos rerum primordia quondam
discrevit partu, mundumque enixa nitentem
fugit in infernas caligo pulsa tenebras ;
sive individuis, in idem reditura soluta,
principiis natura manet post saecula mille,
et paene ex nihilo summa est nihilumque futurum,
caecaque materies caelum perfecit et orbem ;
sive ignis fabricavit opus flammaeque micantes,
quae mundi fecere oculos habitantque per omne
corpus et in caelo vibrantia fulmina fingunt ;
seu liquor hoc peperit, sine quo riget arida rerum
materies ipsumque vorat, quo solvitur, ignem ;
aut neque terra patrem novit nec flamma nec aer
aut umor, faciuntque deum per quattuor artus
et mundi struxere globum prohibentque requiri
ultra se quicquam, cum per se cuncta crearint,
frigida nec calidis desint aut umida siccis,
spiritus aut solidis, sitque haec discordia concors
quae nexus habilis et opus generabile fingit
atque omnis partus elementa capacia reddit :
semper erit pugna ingeniis, dubiumque manebit
quod latet et tantum supra est hominemque deumque
sed facies quacumque tamen sub origine rerum

130 summum
140 creantur
144 rapacia
145 genus in pugna

[a] So Xenophanes. [b] So Hesiod.

[c] The doctrine of Leucippus, elaborated by his more famous pupil Democritus and by Epicurus.

[d] So Heraclitus. [e] So Thales.

likeness. Some hold that the universe does not derive its elements from any source but is devoid of origin, that it ever was and ever shall be, without beginning as it is without end [a]; it may be that ages ago chaos in travail separated the mingled elements of matter and that, having given birth to the shining universe, the darkness fled, banished to infernal gloom [b]; perhaps nature after a thousand ages remains an aggregate of indivisible atoms, though doomed to dissolve and return to the same form, and the total sum is made up of practically nothing and will become nothing, and it is lifeless matter which has produced heaven and earth [c]; possibly the universe was constructed out of fire and flickering flames,[d] which have formed the eyes of heaven and dwell throughout the whole system and shape the lightning which flashes in the skies; the skies, perchance, were born of water,[e] without which matter is parched and hardened and which quenches the very fire by which it is destroyed; else it may be that neither earth nor fire nor air nor water acknowledges a begetter, but themselves constitute a godhead of four elements,[f] which have formed the sphere of the universe and ban all search for a source beyond them, having created all things from themselves, so that cold combines with hot, wet with dry, and airy with solid, and the discord is one of harmony, allowing apt unions and generative activity and enabling the elements to produce all things. These questions will always cause dispute among men of genius, and uncertainty is bound to attend that which is hidden from us and is so far above the ken of man and god. But, however obscure its origin, all are

[f] So Empedocles.

convenit, et certo digestum est ordine corpus.
ignis in aetherias volucer se sustulit oras
summaque complexus stellantis culmina caeli
flammarum vallo naturae moenia fecit.
proximus in tenuis descendit spiritus auras
aeraque extendit medium per inania mundi.
tertia sors undas stravit fluctusque natantis,
aequoraque effudit toto nascentia ponto,
ut liquor exhalet tenuis atque evomat auras
aeraque ex ipso ducentem semina pascat,
ignem flatus alat vicinis subditus astris.
ultima subsedit glomerato pondere tellus,
convenitque vagis permixtus limus harenis
paulatim ad summum tenui fugiente liquore ;
quoque magis puras umor secessit in undas
et saccata magis struxerunt aequora terram
adiacuitque cavis fluidum convallibus aequor,
emersere fretis montes, orbisque per undas
exsiliit, vasto clausus tamen undique ponto.
idcircoque manet stabilis, quia totus ab illo
tantundem refugit mundus fecitque cadendo
undique, ne caderet medium totius et imum.
[ictaque contractis consistunt corpora plagis
et concurrendo prohibentur longius ire]

Quod nisi librato penderet pondere tellus,
non ageret currus, mundi subeuntibus astris,
Phoebus ab occasu et numquam remearet ad ortus,

149 auras
156 [que] perfudit
154 alit
163 siccata
167: *see after* 214
170 imum est
172 prohibent in
174 cursus
175 ad occasum

---

[a] Here follows the Stoic view of the creation of the universe from the four elements.

agreed about the outward appearance of the universe, and the orderly arrangement of its structure is fixed. Winged fire [a] soared aloft to ethereal reaches and, compassing the rooftops of the starry sky, fashioned the walls of the world with ramparts of flame. Air next sank down to become the tenuous breezes and spread out the atmosphere midway through the empty spaces of the sky. The third place was allotted to the expanse of the waters and floating billows, as the level sea at its birth was poured abroad to form the whole of ocean, to the end that water might breathe out and expel the subtle vapours and so feed air which draws the seeds of its being from water, whilst, set beneath the neighbouring stars, the breath of air might nourish the fire. Lastly to the bottom sank earth, moulded into a ball by its weight, and mud, mixed with shifting sand, collected as the light liquid gradually made its way to the top; more and more the moisture withdrew to form clear waters, the filtering of the liquid built up land, and the fluid plains came to rest in hollow valleys; so by degrees mountains emerged from the deep, and the round world sprang forth from the waves, but closed in on every side by the vast ocean. And it remains stable for the reason that every part of the firmament is equally distant from it, and, by falling from every direction, has made it impossible for its central and lowest part to fall.[b]

[173]And did not the Earth's weight hang poised, the Sun would not drive his car past his setting, as the stars of heaven appeared, and would never return to

[b] [171 f.] "For bodies struck by inward blows all round remain stationary, and equal pressure to the centre prevents them from moving."

lunave summersos regeret per inania cursus,
nec matutinis fulgeret Lucifer horis,
Hesperos emenso dederat qui lumen Olympo.
nunc, quia non imo tellus deiecta profundo
sed medio suspensa manet, sunt pervia cuncta, 18
qua cadat et subeat caelum rursusque resurgat.
nam neque fortuitos ortus surgentibus astris
nec totiens possum nascentem credere mundum
solisve assiduos partus et fata diurna,
cum facies eadem signis per saecula constet, 18
idem Phoebus eat caeli de partibus isdem
lunaque per totidem luces mutetur et orbes
et natura vias servet, quas fecerat ipsa,
nec tirocinio peccet, circumque feratur
aeterna cum luce dies, qui tempora monstrat 1
nunc his nunc illis eadem regionibus orbis,
semper et ulterior vadentibus ortus ad ortum
occasumve obitus, caelum et cum sole perennet.

Nec vero admiranda tibi natura videri
pendentis terrae debet. cum pendeat ipse 1
mundus et in nullo ponat vestigia fundo,
quod patet ex ipso motu cursuque volantis,
cum suspensus eat Phoebus currusque reflectat
huc illuc agilis, et servet in aethere metas,
cum luna et stellae volitent per inania mundi, 2
terra quoque aerias leges imitata pependit.

[178] immenso [194] tibi natura admiranda
[192] ad ortus [198] cursum

[a] Venus is not seen as both morning-star and evening-star within the same day, though this fanciful notion was popular with the ancient poets (see Ellis on Catullus 62. 34).

his rising, or the Moon pursue through space a path under the horizon, nor at the hour of dawn would shine the Morning Star, which earlier as the Star of Eve had sent forth its light after traversing the sky.[a] As it is, however, since the Earth has not been cast down to the lowest depth but remains poised in mid-air, all the space about it affords passage, so that the heavens may set, pass under the Earth, and duly rise again. For I cannot believe that when stars appear to view their risings are the product of chance or that the firmament is so often created anew and the Sun dies daily and is continually reborn : for centuries the constellations have displayed the same features, the selfsame Sun has emerged from the selfsame quarter of the sky, and the Moon has waxed and waned over a constant number of days and phases. Nature adheres to the paths which she herself has made and commits not the errors of inexperience : the daytime revolves about the globe with its never-failing light, heralding the selfsame hours now to these regions, now to those, and ever does dawn move farther away if one travels towards the dawn, or sunset if towards the sunset ; and as with the Sun's, so too with heaven's rising and setting.

[194]But the principle of the Earth's suspension should cause you no surprise. The firmament itself hangs thus and does not rest on any base, as is clear from the actual movement of its swift career ; the Sun moves unsupported, as it wheels its chariot nimbly now this way and now that, keeping to its turning points in heaven ; and the Moon and the stars wing their way through empty regions of the sky : therefore the Earth, too, in obedience to celestial laws, has hung suspended. Thus it is that

est igitur tellus mediam sortita cavernam
aeris, e toto pariter sublata profundo,
nec patulas distenta plagas, sed condita in orbem
undique surgentem pariter pariterque cadentem. 20
haec est naturae facies : sic mundus et ipse
in convexa volans teretis facit esse figuras
stellarum ; solisque orbem lunaeque rotundum
aspicimus tumido quaerentis corpore lumen,
quod globus obliquos totus non accipit ignes. 21
haec aeterna manet divisque simillima forma,
cui neque principium est usquam nec finis in ipsa,
sed similis toto ore sibi perque omnia par est.
sic tellus glomerata manet mundumque figurat
imaque de cunctis mediam tenet undique sedem. 16
Idcirco terris non omnibus omnia signa 21
conspicimus. nusquam invenies fulgere Canopon
donec ad Heliacas per pontum veneris oras ;
sed quaerunt Helicen, quibus ille supervenit ignis,
quod laterum tractus habitant, medioque tumore
eripiunt terrae caelum visusque coercent. 22
te testem dat, luna, sui glomeraminis orbis,
quae, cum mersa nigris per noctem deficis umbris,
non omnis pariter confundis sidere gentes,
sed prius eoae quaerunt tua lumina terrae,

[203] et
[212] ipso
[213] ore sibi] remanet
[214] stellis
manent . . . figurant
[217] adeiacas *M* : niliacas *GL*
[218] quaerent
[221] glomerabilis
[224] terrae] gentes

[a] The conjecture *Heliacas*, a word unattested elsewhere in

Earth has been allotted a hollow space in mid-air, equidistant from every quarter of heaven's depths, not spread into flat plains but fashioned into a sphere which rises and falls equally at every point. This is the shape of nature: so even the universe, itself in circular movement, gives round shapes to the stars; round, we see, is the orb of the Sun, and round is the orb of the Moon, which looks in vain for light for its rotund body inasmuch as the sun's oblique rays do not fall on the whole of its globe. This is the shape that continues for ever and most resembles that of the gods: nowhere in it is there beginning or end, but it is like unto itself over all its surface, identical at every point. So too the Earth is rounded and reflects the shape of the heavens, and being the lowest of all heavenly bodies occupies a completely central location.

[215]That is why it is impossible to behold all the constellations in every part of the world. You will never find Canopus shining until you have crossed the seas and come to the shores of Rhodes [a]; but those who dwell beneath that brilliant star will look in vain for the Bear, because they inhabit regions on the Earth's lower flanks and because lands which bulge in between rob them of this part of heaven and block their view of it. The Earth makes you, O Moon, a witness to its roundness: when at night your star is plunged into utter darkness and suffers eclipse, it does not bewilder all nations at the same time; but first the lands of the orient miss your light,

Latin, reflects a Greek counterpart which normally means "of the Sun" but could legitimately signify "Rhodian," since the *Heliadae* (οἱ Ἡλιάδαι) were an old Rhodian family: see Introduction p. xix and figure 1.

post medio subiecta polo quaecumque coluntur,
[ultima ad hesperios infectis volveris alis]
seraque in hesperiis quatiuntur gentibus aera.
quod si plana foret tellus, semel orta per omnem
deficeres pariter toti miserabilis orbi.
sed quia per teretem deducta est terra tumorem,
his modo, post illis apparet Delia terris
exoriens simul atque cadens, quia fertur in orbem
ventris et acclivis pariter declivia iungit
atque alios superat gyros aliosque relinquit.
[ex quo colligitur terrarum forma rotunda]
Hanc circum variae gentes hominum atque ferarum
aeriaeque colunt volucres. pars eius ad arctos
eminet, austrinis pars est habitabilis oris
sub pedibusque iacet nostris supraque videtur
ipsa sibi fallente solo declivia longa
et pariter surgente via pariterque cadente.
hanc ubi ad occasus nostros sol aspicit actus,
illic orta dies sopitas excitat urbes
et cum luce refert operum vadimonia terris ;
nos in nocte sumus somnosque in membra vocamus.
pontus utrosque suis distinguit et alligat undis.
Hoc opus immensi constructum corpore mundi
membraque naturae diversa condita forma

227 extremis
229 deficeret
242 ortus
245 locamus

a [226] "... last to western nations you wheel with blackened wings . . ."

b Some people believed that lunar eclipses were caused by magicians whose incantations charmed down the moon from heaven ; so they raised a tremendous din by clashing together

then the places situated beneath the middle sky,[a] and late is the brass clashed among the peoples of the west.[b] If the Earth were flat, you would rise for the whole world only once and the failure of your light would be lamented by every land at the same time. But since the shape of the Earth follows a smooth curve, the Moon appears to these lands first, and then to those, rising and setting simultaneously, because it is carried along a bellying orbit and equally combines a downward with an upward slope, as it rises above some horizons, leaving others behind.[c]

[236]All over the Earth dwell the countless tribes of man and beast and fowl of the air. One region which they inhabit reaches towards the northern bears, another is situated in southern climes : the latter lies beneath our feet, but believes itself on top, because the ground conceals its gradual curvature and one's path may equally well be said to rise as to fall. When the Sun reaches our occident and looks upon this lower world,[d] there the dawn of day is rousing cities from slumber and is bringing to those lands with daylight the appointed round of work ; we here are wrapped in night and summon sleep to our limbs. Both worlds are at once separated and connected by ocean's waves.[e]

[247]This fabric which forms the body of the boundless universe, together with its members composed of

brass objects to prevent the Moon hearing the incantations and thus nullify the magic.

[c] [235] "From this is inferred the round shape of the Earth."

[d] With Virgil, *Georg.* 1. 249 ff., and perhaps following him, the poet has confused the western with the southern hemisphere.

[e] See Introduction p. xx.

aeris atque ignis, terrae pelagique iacentis,
vis animae divina regit, sacroque meatu
conspirat deus et tacita ratione gubernat
mutuaque in cunctas dispensat foedera partes,
altera ut alterius vires faciatque feratque
summaque per varias maneat cognata figuras.

Nunc tibi signorum lucentis undique flammas
ordinibus certis referam. primumque canentur
quae media obliquo praecingunt ordine mundum
solemque alternis vicibus per tempora portant
atque alia adverso luctantia sidera mundo,
omnia quae possis caelo numerare sereno,
e quibus et ratio fatorum ducitur omnis,
ut sit idem mundi primum quod continet arcem.

Aurato princeps Aries in vellere fulgens
respicit admirans aversum surgere Taurum
summisso vultu Geminos et fronte vocantem,
quos sequitur Cancer, Cancrum Leo, Virgo Leonem.
aequato tum Libra die cum tempore noctis
attrahit ardenti fulgentem Scorpion astro,
in cuius caudam contento derigit arcu
mixtus equo volucrem missurus iamque sagittam.
tum venit angusto Capricornus sidere flexus.
post hunc inflexa defundit Aquarius urna

252 multa quod **253** alter

[a] *Cf.* 2. 60-83, 3. 48-55, 4. 888-890.

[b] The signs of the zodiac, which of course an astrologer needs to be sure of accurately observing, though they are not all well-defined or conspicuous groups. For the interpretation of this paragraph, see Schwarz.

nature's divers elements, air and fire, earth and level sea, is ruled by the force of a divine spirit ; by sacred dispensation the deity brings harmony and governs with hidden purpose, arranging mutual bonds between all parts, so that each may furnish and receive another's strength and that the whole may stand fast in kinship despite its variety of forms.[a]

[255]Now shall I tell you in their fixed ranks of the fiery signs which gleam in every part of heaven. And first my song will be of those [b] that with their slanting array girdle the heavens in the midst thereof and bear in succession through the seasons the Sun and the other planets which struggle against the movement of the celestial sphere, signs all of which you will be able to count in a cloudless sky and from which the whole scheme of destiny is derived : thus shall that part of heaven be first which holds the vaults of heaven together.

[263]Resplendent in his golden fleece the Ram leads the way and looks back with wonder at the backward rising of the Bull, who with lowered face and brow summons the Twins ; these the Crab follows, the Lion the Crab, and the Virgin the Lion. Then the Balance, having matched daylight with the length of night, draws on the Scorpion, ablaze with his glittering constellation,[c] at whose tail the man with body of a horse aims with taut bow a winged shaft, ever in act to shoot. Next comes Capricorn, curled up within his cramped [d] asterism, and after him from urn upturned the Waterman pours forth

[c] *Astro* refers to the whole constellation, a particularly conspicuous one, and not just to the brilliant α Sco (= Antares).

[d] Capricorn does not completely fill up thirty degrees.

Piscibus assuetas avide subeuntibus undas,
quos Aries tangit claudentis ultima signa.
At qua fulgentis caelum consurgit ad Arctos,
omnia quae summo despectant sidera mundo
nec norunt obitus unoque in vertice mutant
in diversa situm caelumque et sidera torquent,
aera per gelidum tenuis deducitur axis
libratumque regit diverso cardine mundum;
sidereus circa medium quem volvitur orbis
aetheriosque rotat cursus, immotus at ille
in binas Arctos magni per inania mundi
perque ipsum terrae derectus constitit orbem.
nec vero solidus stat robore corporis axis
nec grave pondus habet, quod onus ferat aetheris alti,
sed cum aer omnis semper volvatur in orbem
quoque semel coepit totus volet undique in ipsum,
quodcumque in medio est, circa quod cuncta moventur,
usque adeo tenue ut verti non possit in ipsum
nec iam inclinari nec se convertere in orbem,
hoc dixere axem, quia motum non habet ullum
ipse, videt circa volitantia cuncta moveri.
Summa tenent eius miseris notissima nautis
signa per immensum cupidos ducentia pontum.
maioremque Helice maior decircinat arcum
(septem illam stellae certantes lumine signant),

[277] tantum
[278] situ
[284] conspicit
[285] e solido . . . eius

[a] Malchin suggests that the poet is rendering some contrived Greek etymology such as Achilles gives, *Isag.* 28 p. 61, 23 ff.: ὠνόμασται δ' ἄξων διὰ τὸ περὶ αὐτὸν ἄγεσθαι καὶ

the wonted stream for the Fishes which swim eagerly into it ; and these as they bring up the rear of the signs are joined by the Ram.

[275]Now where heaven reaches its culmination in the shining Bears, which from the zenith of the sky look down on all the stars and know no setting and, shifting their opposed stations about the same high point, set sky and stars in rotation, from there an insubstantial axis runs down through the wintry air and controls the universe, keeping it pivoted at opposite poles : it forms the middle about which the starry sphere revolves and wheels its heavenly flight, but is itself without motion and, drawn straight through the empty spaces of the great sky to the two Bears and through the very globe of the Earth, stands fixed. Yet the axis is not solid with the hardness of matter, nor does it possess massive weight such as to bear the burden of the lofty firmament ; but since the entire atmosphere ever revolves in a circle, and every part of the whole rotates to the place from which it once began, that which is in the middle, about which all moves, so insubstantial that it cannot turn round upon itself or even submit to motion or spin in circular fashion, this men have called the axis, since, motionless itself, it yet sees everything spinning about it.[a]

[294]The top of the axis is occupied by constellations well known to hapless mariners, guiding them over the measureless deep in their search for gain. Helice,[b] the greater, describes the greater arc ; it is marked by seven stars which vie with each other

περιδινεῖσθαι τὸν οὐρανόν "and it is called *axis* since the sky *is carried* and spins about it."

[b] Ursa Major.

qua duce per fluctus Graiae dant vela carinae.
angusto Cynosura brevis torquetur in orbe,
quam spatio tam luce minor ; sed iudice vincit
maiorem Tyrio. Poenis haec certior auctor
non apparentem pelago quaerentibus orbem.
nec paribus positae sunt frontibus : utraque caudam
vergit in alterius rostro sequiturque sequentem.
has inter fusus circumque amplexus utramque
dividit et cingit stellis ardentibus Anguis,
ne coeant abeantve suis a sedibus umquam.

Hunc inter mediumque orbem, quo sidera septem
per bis sena volant contra nitentia signa,
mixta ex diversis consurgunt viribus astra,
hinc vicina gelu, caelique hinc proxima flammis ;
quae quia dissimilis, qua pugnat, temperat aer,
frugiferum sub se reddunt mortalibus orbem.
proxima frigentis Arctos boreanque rigentem
nixa venit species genibus, sibi conscia causae.
a tergo nitet Arctophylax idemque Bootes,
cui verum nomen vulgo posuere, minanti
quod similis iunctis instat de more iuvencis ;
Arcturumque rapit medio sub pectore secum.
at parte ex alia claro volat orbe Corona
luce micans varia ; nam stella vincitur una
circulus, in media radiat quae maxima fronte
candidaque ardenti distinguit lumina flamma.
Cnosia desertae fulgent monumenta puellae,

[311] gelu] poli

---

[a] Ursa Minor.
[b] *Cf.* 1. 15, note.
[c] Engonasin (commonly known as Hercules).
[d] α CrB = Alphecca.

in radiance : under its guidance the ships of Greece set sail to cross the seas. Cynosura [a] is small and wheels round in a narrow circle, less in brightness as it is in size, but in the judgement of the Tyrians it excels the larger bear. Carthaginians count it the surer guide when at sea they make for unseen shores. They are not set face to face : each with its muzzle points at the other's tail and follows one that follows it. Sprawling between them and embracing each the Dragon separates and surrounds them with its glowing stars lest they ever meet or leave their stations.

[308]Between this northern and the middle zone, wherein the seven planets fly through the twelve signs that resist their passage,[b] there rise constellations which have part in opposite qualities, here close to heaven's cold and here to its flames. Since these constellations are tempered by an atmosphere which is unlike its neighbours in so far as it is at variance with them, they render fertile for mortals the lands situate beneath them. Next to the chill Bears and the frozen north comes a figure on bended knee,[c] the reason for whose posture is known to none but him. In his rear shines the Bearward, called also Bootes ; *true is the name men have widely given him, threatening*-like since he presses forward as one does over a team of bullocks ; and he pulls along with him the star Arcturus beneath the middle of his breast. But on the other side of Bootes floats the Crown's lustrous ring, which twinkles with varying luminosity ; for the circle is dominated by a single star,[d] which with passing splendour sparkles in the mid forehead and enhances with its blazing flame the bright lights of the constellation. They shine as the

et Lyra diductis per caelum cornibus inter
sidera conspicitur, qua quondam ceperat Orpheus 3:
omne quod attigerat cantu, manesque per ipsos
fecit iter domuitque infernas carmine leges.
hinc caelestis honos similisque potentia causae:
tunc silvas et saxa trahens nunc sidera ducit
et rapit immensum mundi revolubilis orbem. 3
serpentem magnis Ophiuchus nomine gyris
dividit et torto cingentem corpore corpus,
explicet ut nodos sinuataque terga per orbes.
respicit ille tamen molli cervice reflexus
et redit effusis per laxa volumina palmis. 3
semper erit, paribus bellum quia viribus aequant.
proxima sors Cycni, quem caelo Iuppiter ipse
imposuit, formae pretium, qua cepit amantem,
cum deus in niveum descendit versus olorem
tergaque fidenti subiecit plumea Ledae. 3
nunc quoque diductas volitat stellatus in alas.
hinc imitata nitent cursumque habitumque sagittae
sidera. tum magni Iovis ales fertur in altum,
assueta evolitans gestet ceu fulmina mundi,
digna Iove et caelo, quod sacris instruit armis. 3
tum quoque de ponto surgit Delphinus ad astra,
oceani caelique decus, per utrumque sacratus.
quem rapido conatus Equus comprendere cursu
festinat pectus fulgenti sidere clarus

331 gyris] signis 344 assueto volitans
332 toto

---

[a] The eagle (= Aquila).
[b] See Introduction p. xxvii.

memorial of deserted Ariadne ; and one may see among the stars the Lyre, its arms spread apart in heaven, with which in time gone by Orpheus charmed all that his music reached, making his way even to the ghosts of the dead and causing the decrees of hell to yield to his song. Wherefore it has honour in heaven and power to match its origin : then it drew in its train forests and rocks ; now it leads the stars after it and makes off with the vast orb of the revolving sky. One called Ophiuchus holds apart the serpent which with its mighty spirals and twisted body encircles his own, that so he may untie its knots and back that winds in loops. But, bending its supple neck, the serpent looks back and returns ; and the other's hands slide over the loosened coils. The struggle will last for ever, since they wage it on level terms with equal powers. Hard by is the place allotted to the Swan : Jupiter himself placed it in the sky as a reward for the shape with which he snared the admiring Leda, when, a god changed into a snow-white swan, he came down and offered his feathered form to the unsuspecting woman. Now too with outspread wings it flies among the stars. Next shines a constellation which resembles the appearance and flight of an arrow. Then soars to the heights the bird of mighty Jupiter [a] as though in its flight it carried the thunderbolts of heaven, its wonted weapons ; it is a bird worthy of Jupiter and the sky, which it furnishes with awful armaments. Then the Dolphin too rises starward from the deep, the pride of sea and sky, in each revered.[b] Him the Horse in swift career strives to overtake and speeds along, his front distinguished by a resplendent star[c] : this

[c] β Peg = Scheat.

et finitur in Andromeda. [quam Perseus armis 350
eripit et sociat sibi. cui] succedit iniquo
divisis spatio, quod terna lampade crispans
conspicitur, paribus Deltoton nomine sidus
ex simili dictum, Cepheusque et Cassiepia
in poenas resupina suas iuxtaque relictam 355
Andromedan, vastos metuentem Pristis hiatus,
[expositam ponto deflet scopulisque revinctam]
ni veterem Perseus caelo quoque servet amorem
auxilioque iuvet fugiendaque Gorgonis ora
sustineat spoliumque sibi pestemque videnti. 360
tum vicina ferens nixo vestigia Tauro
Heniochus, studio mundumque et nomen adeptus,
quem primum curru volitantem Iuppiter alto
quadriiugis conspexit equis caeloque sacravit.
hunc subeunt Haedi claudentes sidere pontum, 365
nobilis et mundi nutrito rege Capella,
cuius ab uberibus magnum ille ascendit Olympum
lacte fero crescens ad fulmina vimque tonandi.
hanc ergo aeternis merito sacravit in astris
Iuppiter et caeli caelum mercede rependit. 370
[Pleiadesque Hyadesque, feri pars utraque Tauri,
in borean scandunt. haec sunt aquilonia signa]

[352] divisus
tertia lampada
dispas
355-398 *after* 399-442
[355] resupina] signata
[356] piscis
[360] testemque

[a] [350b, 351a] interpolated to secure mention of Perseus: "Whom by force of arms Perseus rescues and unites to himself. After whom . . ."

[b] *I.e.* called after its deltoid shape.

[c] For the spelling of the queen's name, see Introduction p. xxviii.

constellation is bounded by Andromeda.[a] There follows, with two equal sides parted by one unequal, a sign seen flashing with three stars and named Deltoton, called after its likeness.[b] Next come Cepheus and Cassiepia,[c] her face upturned to witness the sacrifice she caused and Andromeda abandoned hard by, who would shrink before the enormous gaping jaws of the Sea-monster,[d] did not Perseus even in heaven maintain his former love and come with timely aid, holding up the dread face of the Gorgon, a triumph for him but death for the beholder. Next, bearing his footsteps near the crouching Bull comes the Charioteer whose calling won him heaven and a name : first of men to speed in a high chariot behind a team of four, he was seen by Jupiter and hallowed in the skies. Him follow the Kids[e] that with their constellation close the seas,[f] and the goat famed for having suckled the king of heaven ; nourished from her udders he seized the throne of Olympus, drawing strength from that wild milk to wield the lightning and the power of thunder. Therefore Jupiter hallowed it, as was its due, with a place among the eternal stars, repaying heaven gained with the gift of heaven.[g]

[d] [357] "... weeps for her left to the mercy of the sea and shackled to the rocks, ..."

[e] Haedi (the Kids) and Capella (the Goat) are not properly constellations but simply individual stars in Auriga. See Introduction p. xxviii.

[f] Referring to the matutinal setting of the Kids in mid-November, from when until early March navigation was for the most part suspended.

[g] [371 f.], possibly interpolated to ensure inclusion of the two star-groups referred to in 5. 118 ff. and 128 ff.: "And the Pleiades and the Hyades, each a part of the wild Bull, mount towards the north. These are the northern signs."

Aspice nunc infra solis surgentia cursus
quae super exustas labuntur sidera terras ;
quaeque inter gelidum Capricorni sidus et axe
imo subnixum vertuntur lumina mundum,
altera pars orbis sub quis iacet invia nobis
ignotaeque hominum gentes nec transita regna
commune ex uno lumen ducentia sole
diversasque umbras laevaque cadentia signa
et dextros ortus caelo spectantia verso.
nec minor est illis mundus nec lumine peior,
nec numerosa minus nascuntur sidera in orbem.
cetera non cedunt : uno vincuntur in astro,
Augusto, sidus nostro qui contigit orbi,
legum nunc terris post caelo maximus auctor.
cernere vicinum Geminis licet Oriona
in magnam caeli tendentem bracchia partem
nec minus extento surgentem ad sidera passu,
singula fulgentis umeros cui lumina signant
et tribus obliquis demissus ducitur ensis,
at caput Orion excelso immersus Olympo
per tria subducto signatur lumina vultu.
[non quod clara minus sed quod magis alta recedant]
hoc duce per totum decurrunt sidera mundum.
subsequitur rapido contenta Canicula cursu,
qua nullum terris violentius advenit astrum
nec gravius cedit. nunc horrida frigore surgit,
nunc vacuum soli fulgentem deserit orbem :

375 intra . . . axem
383 orbe
385 quod
386 legum] caesar
398 nunc] nec
399 ne . . . solis

[a] [394] ". . . not since they are less brilliant, but since they recede to greater heights." The opinion, mentioned by Geminus 1. 23, that some fixed stars are more distant than others, is foreign to Manilius.

[373]Look now at the constellations which rise to the south of the Sun's path and glide above torrid lands; look, too, at the lights whose orbits wheel between the chill sign of Capricorn and the sky which rests on the bottom pole. Beneath these stars lies the other part of Earth we may not tread and unknown nations of mankind. Inaccessible to us, their realms draw from the same Sun light that we share and shadows which fall contrary to ours; in an inverted sky they behold signs which set on the left and rise on the right. Their heaven is neither smaller nor inferior in brightness, and no fewer are the stars which rise upon their world. All else yield they not: yet they admit defeat over a single luminary, Augustus, who like a star has fallen to the fortune of our world: greatest lawgiver is he now on earth, in heaven will be hereafter. Near neighbour to the Twins, Orion may be seen stretching his arms over a vast expanse of sky and rising to the stars with no less huge a stride. A single light marks each of his shining shoulders, and three aslant trace the downward line of his sword; but three mark Orion's head, which is imbedded in high heaven with his countenance remote.[a] It is Orion who leads the constellations as they speed over the full circuit of heaven. At his heels follows the Dog outstretched in full career: no star comes on mankind more violently or causes more trouble when it departs. Now it rises shivering with cold, now it leaves a radiant world open to the heat of the Sun [b]: thus it moves the world to either

[b] In ancient times the Dogstar's evening rising occurred in early January, its evening setting in early May (for an explanation of these terms see the Loeb Aratus, Introduction E, or Dicks 12 f.).

sic in utrumque movet mundum et contraria reddit. 40
hanc qui surgentem, primo cum redditur ortu,
montis ab excelso speculantur vertice Tauri,
eventus frugum varios et tempora discunt,
quaeque valetudo veniat, concordia quanta.
bella facit pacemque refert, varieque revertens 40
sic movet, ut vidit, mundum vultuque gubernat.
magna fides hoc posse color cursusque micantis
ignis ad os. vix sole minor, nisi quod procul haerens
frigida caeruleo contorquet lumina vultu.
cetera vincuntur specie, nec clarius astrum 41
tingitur oceano caelumque revisit ab undis.
tum Procyon veloxque Lepus ; tum nobilis Argo
in caelum subducta mari, quod prima cucurrit,
emeritum magnis mundum tenet ante periclis,
servando dea facta deos. cui proximus Anguis 41
squamea dispositis imitatur tergora flammis ;
et Phoebo sacer ales et una gratus Iaccho
Crater et duplici Centaurus imagine fulget,
pars hominis, tergo pectus commissus equino.
ipsius hinc mundo templum est, victrixque solutis 42
Ara nitet sacris, vastos cum Terra Gigantas
in caelum furibunda tulit. tum di quoque magnos
quaesivere deos ; eguit Iove Iuppiter ipse,

403 dicunt *LM*: ducunt *G*
408 ignis ad os] in radios
414 acta
416 tergora] lumina
422 tumidi
423 esurcione *M*: dubitavit *GL*

extreme and brings opposite effects. Those who from Mount Taurus' lofty peak [a] observe it ascending when it returns at its first rising learn of the various outcomes of harvests and seasons, what state of health lies in store, and what measure of harmony. It stirs up war and restores peace, and returning in different guise affects the world with the glance it gives it and governs with its mien. Sure proof that the star has this power are its colour and the quivering of the fire that sparkles in its face. Hardly is it inferior to the Sun, save that its abode is far away and the beams it launches from its sea-blue face are cold. In splendour it surpasses all other constellations, and no brighter star is bathed in ocean or returns to heaven from the waves. Then come Procyon and the swift Hare. Then famed Argo, raised to the skies from the sea which it was the first to cross, occupies the heaven it earned through grievous perils in a bygone age, made a god for having given safety to gods. Next to it is the Water-snake, whose stars are so arranged as to represent its scaly back. Then shines the bird that is sacred to Phoebus [b] and with it the Bowl beloved of Bacchus and the Centaur of twofold form ; he is half man, joined at the waist to the body of a horse. Next has heaven a temple of its own, where, its rites now paid, the Altar gleams after victory gained when Earth in rage bore forth the monstrous Giants against the skies. Then even the gods sought aid of mighty gods, and Jupiter himself felt the need of another Jupiter, fearing lest

[a] Scaliger observes that the mention of Mount Taurus is a compliment to Aratus, who was a Cilician.

[b] The raven (= Corvus) : Phoebus adopted its shape to elude Typhon (Apollodorus, *Bibl.* 1. 6. 3 ; Ovid, *Met.* 5. 329).

quod poterat non posse timens, cum surgere terram
cerneret, ut verti naturam crederet omnem,
montibus atque altis aggestos crescere montes,
et iam vicinos fugientia sidera colles
arma importantis et rupta matre creatos,
discordis vultum permixtaque corpora partus.
nec di mortiferum sibi quemquam aut numina norant
siqua forent maiora suis. tunc Iuppiter Arae
sidera constituit, quae nunc quoque maxima fulget.
quam propter Cetos convolvens squamea terga
orbibus insurgit tortis et fluctuat alvo,
[intentans similem morsum iam iamque tenenti]
qualis ad expositae fatum Cepheidos undis
expulit adveniens ultra sua litora pontum.
tum Notius Piscis venti de nomine dictus
exsurgit de parte Noti. cui iuncta feruntur
flexa per ingentis stellarum Flumina gyros:
alterius capiti coniungit Aquarius undas,
alter ab exserto pede profluit Orionis
amnis; et in medium coeunt et sidera miscent.

His inter solisque vias Arctosque latentis,
axem quae mundi stridentem pondere torquent,
orbe peregrino caelum depingitur astris,
quae notia antiqui dixerunt sidera vates.
ultima, quae mundo semper volvuntur in imo,
quis innixa manent caeli fulgentia templa,

425 et
426 aliis
427 tam
429 vultu
430 necdum hostiferum [aut]

[a] An error of the poet's, for Cetus lies on the other side of the sky: see Introduction pp. xxx f.

[b] [435] an interpolation based on Virgil, *Aen.* 12. 754 f.: "threatening to bite like one in the very act of seizing its prey."

his power prove powerless. He saw the earth rising up, so that he deemed all nature was being overthrown; mountains piled on lofty mountains he saw growing; he saw the stars retreating from heights which were now neighbours, heights which brought up the armed Giants, brood of a mother they tore apart, deformed creatures of unnatural face and shape. Nor did the gods know whether anyone could inflict death upon them or whether forces existed greater than their own. Then it was that Jupiter set up the constellation of the Altar, which of all altars shines brightest even now. Next [a] to it Cetus undulates its scaly body; it rises aloft upon a spiral of coils and splashes with such a belly [b] as drove the sea beyond its proper shores when it appeared from the waves to destroy the daughter of Cepheus exposed upon the cliffs. Then rises the Southern Fish in the quarter of the wind after which it is named. To it are joined the Rivers [c] which make their winding way along great curves of stars: Aquarius connects his waters with the upper reaches of the one stream, *whilst the other flows from Orion's out-thrust foot*; they meet each other and blend their stars together.

443 Such are the stars with which the sky is stippled in the part of the firmament foreign to us between the pathways of the Sun and those unseen Bears which turn the axis that groans beneath heaven's weight: these stars the ancient poets have called southern. The most distant stars, which everlastingly circle at the bottom of heaven and for ever support the shining temples of the firmament, and which never, with a

[c] Manilius thinks of the stream from Aquarius's urn and the River Eridanus as forming a composite constellation.

nusquam in conspectum redeuntia cardine verso,
sublimis speciem mundi similisque figuras
astrorum referunt. aversas frontibus Arctos
uno distingui medias claudique Dracone
credimus exemplo, quia mens fugientia visus
hunc orbem caeli vertentis sidera cursu
tam signo simili fultum quam vertice fingit.

Haec igitur magno divisas aethere sedes
signa tenent mundi totum deducta per orbem.
tu modo corporeis similis ne quaere figuras,
omnia ut aequali fulgentia membra colore
deficiat nihil aut vacuum qua lumine cesset.
non poterit mundus sufferre incendia tanta,
omnia si plenis ardebunt sidera membris.
quidquid subduxit flammis, natura pepercit
succubitura oneri, formas distinguere tantum
contenta et stellis ostendere sidera certis.
linea designat species, atque ignibus ignes
respondent ; media extremis atque ultima summis
creduntur : satis est si se non omnia celant.
praecipue, medio cum luna implebitur orbe,
certa nitent mundo tum lumina : conditur omne
stellarum vulgus ; fugiunt sine nomine turba.
pura licet vacuo tum cernere sidera caelo,
nec fallunt numero, parvis nec mixta feruntur.

Et, quo clara magis possis cognoscere signa,
non varios obitus norunt variosque recursus,

451 et versas
453 quamvis fulgentia
455 cardine tam [signo]
460 et
464 disiungere
470 cum
471 turba] signa
472 plura

change of pole, come back to our sight, reflect the appearance of the upper sky and like shapes of stars. We assume by analogy that Bears with faces averted from each other are separated and encircled by a single Dragon, since the mind imagines that this southern circle of heaven, turning in its rotation those constellations which shun our gaze, is supported by similar signs just as it is by a similar pole.

[456]These then are the stars which are spread over the whole sphere of heaven and occupy in the vast ether their several abodes. Only look not for shapes like bodily shapes or think to see all the members shining with equal brilliance with nothing missing or anywhere left devoid of light. The heavens will be unable to endure the heat of so intense a conflagration, if the full form of every constellation is figured in flame. Whatever nature has removed from such fires she has subtracted from a burden to which she would have proved unequal. She is satisfied with merely indicating the forms of the constellations and depicting them by certain stars. An outline describes the appearance and along it beacon answers to beacon ; the centre is to be inferred from the edges and the rear from the surface : it is enough if not all is hidden. When in mid-course the Moon is full, then most of all do the princely luminaries shine conspicuous in the heavens ; the whole stellar populace fades from sight, and they flee, an innominate throng. Then may one see the constellations immaculate in the open spaces of the sky ; they neither bewilder us with their numbers nor move encumbered by a multitude of petty stars.

[474]And that you may better recognize the bright constellations, mark this : they know no variation

certa sed in proprias oriuntur singula luces
natalesque suos occasumque ordine servant.
nec quicquam in tanta magis est mirabile mole
quam ratio et certis quod legibus omnia parent.
nusquam turba nocet, nihil ullis partibus errans
laxius aut brevius mutatove ordine fertur.
quid tam confusum specie, quid tam vice certum est?

Ac mihi tam praesens ratio non ulla videtur,
qua pateat mundum divino numine verti
atque ipsum esse deum, nec forte coisse magistra,
ut voluit credi, qui primus moenia mundi
seminibus struxit minimis inque illa resolvit;
e quibus et maria et terras et sidera caeli
aetheraque immensis fabricantem finibus orbes
solventemque alios constare, et cuncta reverti
in sua principia et rerum mutare figuras.
quis credat tantas operum sine numine moles
ex minimis caecoque creatum foedere mundum?
si fors ista dedit nobis, fors ipsa gubernet.
at cur dispositis vicibus consurgere signa
et velut imperio praescriptos reddere cursus
cernimus ac nullis properantibus ulla relinqui?
cur eadem aestivas exornant sidera noctes
semper et hibernas eadem, certamque figuram
quisque dies reddit mundo certamque relinquit?
iam tum, cum Graiae verterunt Pergama gentes,

476 sidera
480 errant
481 et levius
489 immensos

[a] Epicurus.

in their settings and returnings, but each without fail rises to display its proper stars, regularly keeping to the same times of coming up and going down. Nothing in this mighty edifice is more wonderful than its design and the obedience of all to immutable law. Nowhere does confusion do harm ; nothing in any of its parts moves randomly to enlarge or shorten its course or change the order of its movement. What is there so complex in its appearance, and yet so sure in all its ways ?

[483]For my part I find no argument so compelling as this to show that the universe moves in obedience to a divine power and is indeed the manifestation of God, and did not come together at the dictation of chance. Yet this is what he [a] would have us believe who first built the walls of the heavens from minute atoms and into these resolved them again ; he held that from these atoms are formed the seas, the lands, and the stars in the sky, and the air by which in its vast space worlds are created and dissolved ; and that all matter returns to its first origins and changes the shapes of things. Who could believe that such massive structures have been created from tiny atoms without the operation of a divine will, and that the universe is the creature of a blind compact ? If chance gave such a world to us, chance itself would govern it. Then why do we see the stars arise in regular succession and duly perform as at the word of command their appointed courses, none hurrying ahead, none left behind ? Why are the summer nights and the nights of winter ever made beautiful with the selfsame stars ? Why does each day of the year bring back to the sky a fixed pattern and a fixed pattern leave at its departure ? Already when the

Arctos et Orion adversis frontibus ibant,
haec contenta suos in vertice flectere gyros,
ille ex diverso vertentem surgere contra
obvius et toto semper decurrere mundo.
temporaque obscurae noctis deprendere signis
iam poterant, caelumque suas distinxerat horas.
quot post excidium Troiae sunt eruta regna!
quot capti populi! quotiens fortuna per orbem
servitium imperiumque tulit varieque revertit!
Troianos cineres in quantum oblita refovit
imperium! fatis Asiae iam Graecia pressa est.
saecula dinumerare piget, quotiensque recurrens
lustrarit mundum vario sol igneus orbe.
omnia mortali mutantur lege creata,
nec se cognoscunt terrae vertentibus annis
exutas variam faciem per saecula ferre.
at manet incolumis mundus suaque omnia servat,
quem neque longa dies auget minuitque senectus
nec motus puncto curvat cursusque fatigat;
idem semper erit quoniam semper fuit idem.
non alium videre patres aliumve nepotes
aspicient. deus est, qui non mutatur in aevo.
numquam transversas solem decurrere ad Arctos
nec mutare vias et in ortum vertere cursus
auroramque novis nascentem ostendere terris,
nec lunam certos excedere luminis orbes
sed servare modum, quo crescat quove recedat,

[517] ferre] gentes [519] quae

[a] Rome, founded by the line of Trojan Aeneas.
[b] *I.e.* has been displaced from primacy in the world.

Greeks overthrew Troy the Bear and Orion moved with fronts opposed, she content to describe her gyrations at the pole, he rising to face her in opposition as she wheels on the other side and to traverse for ever the whole sky. Even then could men measure by the constellations the passage of the dark night, even then had the sky recorded the succession of its hours. How many the realms that have collapsed since the sack of Troy! How many the peoples led into captivity! How oft has Fortune brought slavery and sovereignty in turn and returned in a different guise! To what high sovereignty has it rekindled the ashes of Troy, forgetful of the past! [a] And Greece in its turn has been overtaken by the fate of Asia.[b] It were a toilsome task to chronicle the centuries and tell how oft the fiery Sun has returned to traverse the heavens in its varied course. Everything born to a mortal existence is subject to change, nor does the earth notice that, despoiled by the passing years, it bears an appearance which varies through the ages. The firmament, however, conserving all its parts, remains intact, neither increased with length of time nor diminished by old age; it is neither the least bit warped by its motion nor wearied by its speed: it will remain the same for ever, since the same has it always been. No different heaven did our fathers see, no different heaven will our posterity behold. It is God, and changes not in time. That the Sun never deviates to the crosswise-lying Bears and never changes direction, setting course for the orient and bringing forth a dawn born of unwonted lands; that the Moon does not exceed her appointed orbs of light, but preserves the regularity of her waxing and waning;

nec cadere in terram pendentia sidera caelo
sed dimensa suis consumere tempora gyris,
non casus opus est, magni sed numinis ordo.

Haec igitur texunt aequali sidera tractu
ignibus in varias caelum laqueantia formas.
altius his nihil est ; haec sunt fastigia mundi ;
publica naturae domus his contenta tenetur
finibus, amplectens pontum terrasque iacentis.
omnia concordi tractu veniuntque caduntque,
qua semel incubuit caelum versumque resurgit.
sunt alia adverso pugnantia sidera mundo,
quae terram caelumque inter volitantia pendent,
Saturni, Iovis et Martis Solisque, sub illis
Mercurius Venerem inter agit Lunamque volatus.

Ipse autem quantum convexo mundus Olympo
obtineat spatium, quantis bis sena ferantur
finibus astra, docet ratio, cui nulla resistunt
claustra nec immensae moles caecive recessus ;
omnia succumbunt, ipsum est penetrabile caelum.
nam quantum terris atque aequore signa recedunt,
tantum bina patent. quacumque inciditur orbis
per medium, pars efficitur tum tertia gyri
exiguo dirimens solidam discrimine summam.
summum igitur caelum bis bina refugit ab imo
astra, bis e senis ut sit pars tertia signis.

530-567 *after* 568-611
[530] gyris] signis
[808] locatus
[542] caeduntque
[549] [e]

---

[a] *Cf.* 1. 15, note.

[b] *I.e.* the circumference of a circle is equal to three times the diameter plus a small increment (or c = πd where π = approximately $3\frac{1}{7}$). The poet speaks of units of "two signs" in this passage because such a unit is the length both of the

that the stars poised in heaven fall not upon Earth but take fixed periods of time to accomplish their orbits : all this is not the result of chance, but the plan of a God most high.

[532]These then are the constellations which decorate the sky with even spread, their fires panelling the ceiling of heaven with various designs. Higher than these is there nothing, for they are the roof of the universe ; they are the limits within which the common abode of nature is content to be held, embracing the sea and lands that lie beneath. They all move on a consistent course, coming into view and setting where heaven ever sinks and, turning, reappears. There exist other stars, which strive against the contrary movement of the sky [a] and in their swift orbits are poised between heaven and earth : Saturn, Jupiter, Mars, and the Sun, and beneath them Mercury performing its flight between Venus and the Moon.

[539]How great is the space occupied by the vault of the heavens and how great the territory within which the twelve signs of the zodiac move, we learn from reason, reason that no barriers or huge masses or dark recesses withstand ; all things yield to reason, and it can penetrate the sky itself. Now the distance of the signs from land and sea is equal to the extent of two signs : for, where a circle is cut through the middle, the line formed amounts to a third of the circumference, a line so dividing the whole as to leave a small difference.[b] Therefore the highest point of heaven is twice two signs distant from the lowest, a distance, that is, of one third of the twelve

radius of a circle and of each side of a hexagon inscribed therein (Euclid, *Elem.* 4. 15).

sed quia per medium est tellus suspensa profundum,
binis a summo signis discedit et imo.
hinc igitur quodcumque supra te suspicis ipse,
qua per inane meant oculi quaque ire recusant,
binis aequandum est signis ; sex tanta rotundae
efficiunt orbem zonae, qua signa feruntur
bis sex aequali spatio texentia caelum,
ne mirere vagos partus eadem esse per astra
et mixtum ingenti generis discrimine fatum,
singula cum tantum teneant tantoque ferantur
tempore, sex tota surgentia sidera luce
nec spatio noctis linquentia plura profundum.

Restat ut aetherios fines tibi reddere coner
filaque dispositis vicibus comitantia caelum,
per quae derigitur signorum flammeus ordo.

* * * * *

primus et aetheria succedens proximus arce
circulus ad borean fulgentem sustinet Arcton
sexque fugit solidas a caeli vertice partes.
alter ad extremi decurrens sidera Cancri,

557 nec   564 f.: *see after* 611

[a] *I.e.* below the horizon.

[b] In the following passage (to verse 602) Manilius adopts Eudoxus's system of degrees, which apportioned 60 degrees to one complete revolution of a circle. Thereafter, at 2. 307 for example, he employs the ordinary system of 360 degrees.

signs. But since the Earth is suspended in the middle of space, it is two signs separated from the zenith and two from the nadir. Consequently, whatever glance you cast heavenward from the Earth, both where the eyes pass through space and where they cannot do so,[a] must be equal to a length of two signs ; six of these lengths make up the circle of the round belt, where move the twelve signs which weave in equal lengths the pattern of heaven : so wonder not that men born under the same stars experience different fortunes and that destinies vary with enormous differences of kind, since the space that each sign occupies is so large, and so great the period of time in which it revolves, six signs rising in the course of the day *and as many emerging from ocean in the hours of night.*

561I have still to try to expound to you the regions of the sky and the lines marking heaven at fixed intervals through which are directed the fiery ranks of the constellations.

563A*First must I describe the five parallel circles, so called because each so traverses the heavens as ever to remain at a constant distance from the poles and from each other. That you may plot them accurately, remember that the sky's curve from zenith to nadir stretches through thirty full degrees.*[b] *And coming nearest the pinnacle of heaven, the first* circle [c] upholds the Bear gleaming at the north and is distant from the celestial pole six full degrees. Below it a second,[d] touching

[c] The Arctic Circle (all these terms apply to the sphere of heaven, *i.e.* are celestial), 6° Eud. (= 36°) south of the North Pole (54° from the Equator).

[d] The Summer Tropic or Tropic of Cancer, 5° Eud. (= 30°) south of the Arctic Circle (24° from the Equator).

in quo consummat Phoebus lucemque moramque
tardaque per longos circumfert lumina flexus, 57
aestivum medio nomen sibi sumit ab aestu,
temporis et titulo potitur, metamque volantis
solis et extremos designat fervidus actus,
et quinque in partes aquilonis distat ab orbe.
tertius in media mundi regione locatus 57
ingenti spira totum praecingit Olympum
parte ab utraque videns axem, qua lumine Phoebus
componit paribus numeris noctemque diemque
veris et autumni currens per tempora mixta,
cum medium aequali distinguit limite caelum ; 58
quattuor et gradibus sua fila reducit ab aestu.
proximus hunc ultra brumalis nomine limes
ultima designat fugientis limina solis,
invida cum obliqua radiorum munera flamma
dat per iter minimum nobis, sed finibus illis, 58
quos super incubuit, longa stant tempora luce
vixque dies transit candentem extenta per aestum ;
bisque iacet binis summotus partibus orbis.
unus ab his superest extremo proximus axi
circulus, austrinas qui stringit et obsidet Arctos. 59
hic quoque brumalem per partes quinque relinquit,
et, quantum a nostro sublimis cardine gyrus,
distat ab adverso tantundem proximus illi.

577 quo
582 timens
583 fulgentis lumina
584 inviaque [cum]

the Crab's extreme sign, wherein Phoebus completes the lingering of his light and slowly swings his rays through long curves, adopts from the heat of midsummer the name of summer tropic, acquiring the title of the season; with its warmth it marks out the turning-point and northerly limit of the Sun's winged career, and is five degrees distant from the northern circle. A third circle,[a] set in the midmost region of the sky, girds with a huge ring the full circuit of heaven and sees a pole on either side; here with his light the Sun matches night and day in hours of equal number as he passes through the tempered seasons of spring and autumn and divides the middle of heaven with impartial boundary; this circle withdraws its track four degrees away from that of the summer solstice. Next after this a boundary named the winter tropic[b] marks the furthest station of the retreating Sun, when, as he describes his briefest arc, he gives us scant bounty with the slanting fires of his rays, although in the lands over which he hangs time halts, light tarries long, and, prolonged by the glowing heat, the day scarce comes to an end; this boundary is removed from the last by twice two degrees of the circumference. After these one circle remains[c]: it is the nearest to the distant pole and confines and beleaguers the southern Bears. Likewise it leaves the winter circle five degrees behind, and is as distant from the opposite pole, which it adjoins, as is the

[a] The Equator, 4° Eud. (= 24°) south of the Summer Tropic (90° from the North Pole).

[b] The Winter Tropic or Tropic of Capricorn, 4° Eud. (= 24°) south of the Equator.

[c] The Antarctic Circle, 5° Eud. (= 30°) south of the Winter Tropic.

[sic per tricenas vertex a vertice partes
divisus duplici summa circumdat Olympum 59
et per quinque notat signantis tempora fines]
his eadem est via quae mundo, pariterque rotantur
inclines, sociosque ortus occasibus aequant,
quandoquidem flexi, quo totus volvitur orbis,
fila trahunt alti cursum comitantia caeli, 60
intervalla pari servantes limite semper
divisosque semel fines sortemque dicatam.

Sunt duo, quos recipit ductos a vertice vertex,
inter se adversi, qui cunctos ante relatos
seque secant gemino coeuntes cardine mundi 60
transversoque polo rectum ducuntur in axem,
tempora signantes anni caelumque per astra
quattuor in partes divisum mensibus aequis.
alter ab excelso decurrens limes Olympo
Serpentis caudam siccas et dividit Arctos 61
et iuga Chelarum medio volitantia gyro,
[circulus a summo nascentem vertice mundum 56
permeat Arctophylaca petens per terga Draconis,
tangit et Erigonen, Chelarum summa recidit] 56
extremamque secans Hydram mediumque sub austris 61
Centaurum adverso concurrit rursus in axe,
et redit in caelum, squamosaque tergora Ceti
Lanigerique notat fines clarumque Trigonum 61

594 sic tibi per binas
599 sexti
612 astris

---

[a] Here someone interpolated [594-596] to secure that mention of the Eudoxean division of the circle into sixty degrees which has been lost in the gap after 563: "Thus each pole is parted thirty degrees from the other and with twice this total encompasses the sky and marks it with five boundaries which signal the seasons." Of course, only three circles (the two tropics and the equator) signal the seasons.

northern circle from our pole.[a] The path of these circles is the same as the sky's, and they wheel along the same orbit ; they match their risings and settings uniformly, since, curved along the spin of the whole sphere, they trace lines which follow the rotation of high heaven, ever preserving equidistant intervals from each other, the boundaries for all time assigned them, and their appointed stations.

603 There are two circles,[b] placed crosswise to each other, which are drawn from one pole and received by the other : they cut all the circles mentioned above and cut each other, converging at the two poles of heaven ; thence they traverse the sky and are drawn straight to the pole. They mark the seasons of the year and the division of heaven along the zodiac into four portions of equal months. One line,[c] descending from the summit of the sky, passes through the Dragon's tail and the Bears that shun ocean, and the yoke of the Balance[d] which revolves in the midmost circle.[e] It cleaves the extremity of the Water-snake and the middle of the southern Centaur, and again converges upon the opposing circle at the pole ; returning thence heavenward, it marks the scaly back of the Whale, the boundary of the Ram, the bright Triangle, the lowest folds of

[b] The Colures.

[c] The Equinoctial Colure.

[d] Literally "the yoke of the Claws": see Introduction p. xxv.

[e] [564-565A], interpolated to restore the substance of 609-611, when these verses had been displaced by the great transposition (see table 7, p. cvii): "The circle traverses that part of the sky which begins at the northern pole ; seeking the Bearward, it passes through the back of the Dragon, touches the Virgin, and cuts off the tip of the Balance."

Andromedaeque sinus imos, vestigia matris,
principiumque suum repetito cardine claudit.
alter in hunc medium summumque incumbit in axem
perque pedes primos cervicem transit et Ursae,
quam septem stellae primam iam sole remoto 6
producunt nigrae praebentem lumina nocti,
et Geminis Cancrum dirimit stringitque flagrantem
ore Canem clavumque Ratis, quae vicerat aequor,
inde axem occultum per gyri signa prioris
transversa atque illo rursus de limite tangit 6
te, Capricorne, tuisque Aquilam designat ab astris,
perque Lyram inversam currens spirasque Draconis
posteriora pedum Cynosurae praeterit astra
transversamque secat vicino cardine caudam :
hic iterum coit ipse sibi, memor unde profectus. 6

Atque hos aeterna fixerunt tempora sede,
immotis per signa modis, statione perenni :
hos volucris fecere duos. namque alter ab ipsa
consurgens Helice medium praecidit Olympum
discernitque diem sextamque examinat horam 6
et paribus spatiis occasus cernit et ortus.
hic mutat per signa vices ; et seu quis eoos
seu petit hesperios, supra se circinat orbem
verticibus super astantem mediumque secantem
caelum et diviso signantem culmine mundum,
cumque loco terrae caelumque et tempora mutat,
quando aliis aliud medium est. volat hora per orbem,
atque, ubi se primis extollit Phoebus ab undis,

631 hoc 637 [et] seu si

[a] The Solstitial Colure. [b] The Meridian.

[c] In this passage an hour denotes one twelfth of the period between sunrise and sunset: by the sixth hour we are to understand of course the end of the sixth hour.

Andromeda's robe, and her mother's feet, and, the pole regained, ends with its beginning. The other circle [a] rests upon the middle of this one at the upper pole, whence it passes through the forefeet and neck of that Bear which with the setting of the Sun seven stars bring first to view as it offers its lights to the blackness of night ; it parts the Crab from the Twins and grazes the Dog with the blazing face and the rudder of the Ship which conquered ocean ; thence it touches the hidden pole, cutting at right angles the path of the former circle. On its way back from this line it touches Capricorn and, leaving Capricorn's stars, marks the Eagle ; it runs through the inverted Lyre and the coils of the Dragon, and passes by the stars of Cynosure's hind feet, whose tail it cuts at right angles close to the pole. Here it meets itself again, mindful whence it came.

631These circles the seasons have fixed in a permanent abode ; their paths through the signs do not change, and their position remains the same for ever : but there are two circles to which they have given wings. One,[b] starting from the northern pole, cuts the sky in the middle and divides the day into two, determines the sixth hour [c] and beholds at equal distances sunrise and sunset. It changes position among the constellations ; and, whether one moves towards east or west, he finds above him a circle passing directly overhead so as to part the sky in the middle and mark a bisection of heaven's vault ; with his position on the Earth a man changes both his view of the sky and the time of day, because different peoples have different skies and times along their meridians. The mid-hour revolves about the world : when Phoebus rises from the surface of the ocean, it

illis sexta manet, quos tum premit aureus orbis,
rursus ad hesperios sexta est, ubi cedit in umbras : 6
nos primam ac summam sextam numeramus utramque
et gelidum extremo lumen sentimus ab igni.
alterius fines si vis cognoscere gyri,
circumfer facilis oculos vultumque per orbem.
quidquid erit caelique imum terraeque supremum, 6
qua coit ipse sibi nullo discrimine mundus
redditque aut recipit fulgentia sidera ponto,
praecingit tenui transversum limite mundum.
haec quoque per totum volitabit linea caelum,
nunc tractum ad medium vergens mundique tepentem 6
orbem, nunc septem ad stellas nec mota sub astra ;
seu quocumque vagae tulerint vestigia plantae
has modo terrarum nunc has gradientis in oras,
semper erit novus et terris mutabitur arcus.
quippe aliud caelum ostendens aliudque relinquens 6
dimidium teget et referet, varioque notabit
fine et cum visu pariter sua fila movente.
[hic terrestris erit, quia terram amplectitur, orbis ;
et mundum plano praecingit limite gyrus
atque a fine trahens titulum memoratur horizon] 6

His adice obliquos adversaque fila trahentis
inter se gyros, quorum fulgentia signa
alter habet, per quae Phoebus moderatur habenas
subsequiturque suo solem vaga Delia curru

655 non tantum
mediumque repente
656 nec] nunc
657 sed
661 tegit et refert

---

[a] The Horizon.
[b] [663-665] "This circle must be regarded as belonging to

is the sixth hour for those who dwell immediately beneath the golden orb ; and again, it is the sixth hour for the people of the west, when the Sun retires to the shadows : these two sixth hours we reckon as first and last of day respectively, and at such times we receive but chill rays from the Sun's distant fires. Should you wish to learn the bounds of the other circle,[a] let your eyes and gaze freely rotate in a circle about yourself : where heaven's lowest edge and Earth's uppermost rim meet, where the universe comes together without an intervening space and restores to the sea or receives therefrom the shining stars, there you have the fine boundary-line which runs like a girdle across the universe. This line, too, can fly all over the sky, now moving to heaven's central region and torrid zone, now to the Bear's seven twinklers and the stars that never disappear from view ; and, wherever a man's wandering feet have trodden, as he travels now to these lands and now to those, he will always be finding a new circle, which will change with his position on the Earth. In fact, this circle will show you one part of heaven and leave hidden the other ; it will mark heaven with a variable boundary, on which moves its position precisely as moves the observer's eye.[b]

[666]To these you must add two circles which lie athwart and trace lines that cross each other. One [c] contains the shining signs through which the Sun plies his reins, followed by the wandering Moon in

the Earth, since it embraces the Earth ; the circle girdles the sky with a level boundary-line and, taking its name from the fact that it limits our sight, is called horizon" (the word *ὁρίζων* means "limiting").

[c] The Zodiac.

et quinque adverso luctantia sidera mundo
exercent varias naturae lege choreas.
hunc tenet a summo Cancer, Capricornus ab imo,
bis recipit, lucem qui circulus aequat et umbras,
Lanigeri et Librae signo sua fila secantem.
sic per tris gyros inflexus ducitur orbis
rectaque devexo fallit vestigia clivo.
nec visus aciemque fugit tantumque notari
mente potest, sicut cernuntur mente priores,
sed nitet ingenti stellatus balteus orbe
insignemque facit latò caelamine mundum.
[et ter vicenas partes patet atque trecentas
in longum, bis sex latescit fascia partes
quae cohibet vario labentia sidera cursu]

Alter in adversum positus succedit ad Arctos
et paulum a boreae gyro sua fila reducit
transitque inversae per sidera Cassiepiae,
inde per obliquum descendens tangit Olorem
aestivosque secat fines Aquilamque supinam
temporaque aequantem gyrum zonamque ferentem
solis equos inter caudam, qua Scorpios ardet,
extremamque Sagittari laevam atque sagittam,
inde suos sinuat flexus per crura pedesque
Centauri alterius rursusque ascendere caelum

680 caelato lumine

---

[a] *Cf.* 1. 15, note.

[b] The Equator.

[c] The Equator and the two Tropics.

[d] [681-683] "And in length it covers three hundred and three score degrees, whilst in width it forms a band of twice six degrees which contains within it the planets that glide in

her chariot, and wherein the five planets which struggle against the opposite movement of the sky [a] perform the dances of their orbits that nature's law diversifies. This circle is held by the Crab at the top, at the bottom by Capricorn, and is twice crossed by the circle which balances day and night,[b] whose line it cuts in the signs of the Ram and the Balance. Thus the curve of the circle is drawn through three circles [c] and conceals by its downward slope the straightness of its path. Nor does it elude the sight of the eye, as if it were a circle to be comprehended by the mind alone, even as the previous circles are perceived by the mind: nay, throughout its mighty circuit it shines like a baldric studded with stars and gives brilliance to heaven with its broad outline standing out in sharp relief.[d]

[684]The other circle [e] is placed crosswise to it. It approaches the Bears but bends back its outline a little way from the circle of the north; passing through the constellation of the inverted [f] Cassiepia it thence descends by a slanting path to reach the Swan; it cuts the summer boundary, the supine Eagle, the circle of equal day and night, and the zone which carries the horses of the Sun, passing between the blazing tail of the Scorpion and the tip of the Archer's left hand and arrow; from there it winds its tortuous trail through the legs and hoofs of the southern Centaur, and begins once more to climb

their diverse orbits." Numeration by the 360-degree circle, though given us at 2. 307 ff., is out of place here: the interpolator was doubtless led by 679 f. to specify the breadth of the zodiac.

[e] The Milky Way.

[f] Cassiepia hangs upside down in heaven as further punishment for her impiety (*cf.* Aratus, *Phaen.* 654 ff.).

incipit Argivumque ratem per aplustria summa
et medium mundi gyrum Geminosque per ima
signa secat, subit Heniochum, teque, unde profectus,
Cassiepia, petens super ipsum Persea transit
orbemque ex illa coeptum concludit in ipsa;
trisque secat medios gyros et signa ferentem
partibus e binis, quotiens praeciditur ipse.
nec quaerendus erit: visus incurrit in ipsos
sponte sua seque ipse docet cogitque notari.
namque in caeruleo candens nitet orbita mundo
ceu missura diem subito caelumque recludens,
ac veluti viridis discernit semita campos
quam terit assiduo renovans iter orbita tractu.
[inter divisas aequabilis est via partes]
ut freta canescunt sulcum ducente carina,
accipiuntque viam fluctus spumantibus undis
quam tortus verso movit de gurgite vertex,
candidus in nigro lucet sic limes Olympo
caeruleum findens ingenti lumine mundum.
utque suos arcus per nubila circinat Iris,
sic super incumbit signato culmine limes
candidus et resupina facit mortalibus ora,
dum nova per caecam mirantur lumina noctem
inquiruntque sacras humano pectore causas:
num se diductis conetur solvere moles
segminibus, raraque labent compagine rimae
admittantque novum laxato tegmine lumen;

719 seminibus

---

[a] Argo.

[b] [707], interpolated by one who missed an apodosis to the *veluti* clause: ". . . so is the road between the sundered parts uniform."

the sky ; it cuts the ship of the Greeks [a] through the top of the stern-post, heaven's middle circle, and the Twins through the bottom of their sign ; then it enters the Charioteer and, making for Cassiepia, whence it set out, passes over the figure of Perseus ; and it completes in Cassiepia the circuit which it began with her. At two points it cuts the three middle circles and the circle which carries the signs and is as often cleft itself. One need not search to find it : of its own accord it strikes the eyes ; it tells of itself unasked, and compels attention. It shines like a glowing path in the dark-blue of the heavens, as though it would suddenly open up the sky and let forth the light of day, or like a track running through green fields, a track worn by the constant passage of cart-wheels repeating the same journey.[b] As the sea whitens where a vessel draws the furrow of its wake and, whilst the waters foam, the surge forms a road which churned eddies have roused from the upturned depths, so the track shines bright in the blackness of heaven, cleaving with a huge band of light the dark-blue sky. And just as the rainbow describes its arc through the clouds, even thus the white track marking the vault of heaven lies overhead ; and it draws the gaze of mortals upwards, as they marvel at the strange glow through night's darkness and search, with mind of man, the cause of the divine. Perchance,[c] they wonder, the firmament is seeking to split into separate segments ; with the slackening of the framework cracks are opening and admit new light through a split in the ceiling : what

[c] The poet here presents a list of different theories about the Milky Way. See Introduction p. xxxv.

quid sibi non timeant, magni cum vulnera caeli
conspiciant feriatque oculos iniuria mundi ?
an coeat mundus, duplicisque extrema cavernae
conveniant caelique oras et segmina iungant,
perque ipsos fiat nexus manifesta cicatrix
suturam faciens mundi, stipatus et orbis
aeriam in nebulam densa compagine versus
in cuneos alti cogat fundamina caeli.
an melius manet illa fides, per saecula prisca
illac solis equos diversis cursibus isse
atque aliam trivisse viam, longumque per aevum
exustas sedes incoctaque sidera flammis
caeruleam verso speciem mutasse colore,
infusumque loco cinerem mundumque sepultum ?
fama etiam antiquis ad nos descendit ab annis
Phaethontem patrio curru per signa volantem,
dum nova miratur propius spectacula mundi
et puer in caelo ludit curruque superbus
luxuriat nitido, cupit et maiora parente,
deflexum solito cursu, curvisque quadrigis
monstratas liquisse vias orbemque recentem
imposuisse polo, nec signa insueta tulisse
errantis meta flammas currumque solutum.
quid querimur flammas totum saevisse per orbem
terrarumque rogum cunctas arsisse per urbes ?
cum vaga dispersi fluitarunt fragmina currus,
et caelum exustum est : luit ipse incendia mundus,
et vicina novis flagrarunt sidera flammis

721 sibi] quasi
724 sidera
726 fusuram
727 densa] clara
739 nitido] mundo
740 regentem
742 meta] natu
746 fragmina] lumina
748 nova vicinis

would men not fear might befall them, when they behold the great firmament damaged, and hurt done to heaven strikes their eyes? Possibly the skies are coming together and the bases of two vaults meet and fasten the rims of celestial segments; out of the connection is formed a conspicuous scar marking a suture of the skies, and transformed by its dense structure into ethereal mist the compressed seam causes the foundations of high heaven to harden into a solid joint. Perhaps that belief is more justly held which asserts that here in ages past the horses of the Sun pursued a different course and wore another path; over long centuries the track was burnt up and the constellations scorched by the flames, losing their sky-dark appearance beneath an altered hue: ash spread over the region and buried heaven beneath it. A story also comes down to us from ancient times that Phaethon sped through the stars in his father's chariot; whilst looking more intently at the unfamiliar spectacle of heaven and boyishly sporting in the sky as he proudly revelled in the glittering chariot, longing to surpass his father, he veered from the accustomed course and with a swerving team left the appointed track and branded a fresh circuit upon the heavens: the inexperienced signs could not withstand the fires which wandered from their guide-post and a chariot out of control. Why complain that flames raged the whole world over and that the earth became a funeral pyre which burned in every city? When the débris of the shattered chariot was scattered far and wide, even the sky caught fire: heaven itself paid dear for that conflagration, and the nearby stars blazed with

nunc quoque praeteriti faciem referentia casus.
nec mihi celanda est vulgata fama vetusta
mollior, e niveo lactis fluxisse liquorem
pectore reginae divum caelumque colore
infecisse suo ; quapropter lacteus orbis
dicitur, et nomen causa descendit ab ipsa.
an maior densa stellarum turba corona
contexit flammas et crasso lumine candet,
et fulgore nitet collato clarior orbis ?
an fortes animae dignataque nomina caelo
corporibus resoluta suis terraeque remissa
huc migrant ex orbe suumque habitantia caelum
aetherios vivunt annos mundoque fruuntur ;
atque hic Aeacidas, hic et veneramur Atridas,
Tydidenque ferum, terraeque marisque triumphis
naturae victorem Ithacum, Pyliumque senecta
insignem triplici, Danaumque ad Pergama reges,
Hectoraque Iliacae gentis columenque decusque,
Auroraeque nigrum partum, stirpemque Tonantis
rectorem Lyciae ? nec te, Mavortia virgo,
praeteream, regesque alios, quos Thracia misit
atque Asiae gentes et Magno maxima Pella ;
quique animi vires et strictae pondera mentis
prudentes habuere viri, quibus omnis in ipsis
census erat, iustusque Solon fortisque Lycurgus,

[750] famae vulgata vetustas
[766] = 2.3
[769] grecia
[771] strictas pondere

---

[a] Whose sons were Peleus and Telamon, fathers respectively of Achilles and Ajax.
[b] Agamemnon and Menelaus.
[c] Ulysses.
[d] Nestor.
[e] Memnon.
[f] Sarpedon.
[g] Penthesilea.

unwonted flames and even now display the marks of past calamity. Nor must I conceal an ancient legend less tragic than the well-known one, that from the snow-white breast of heaven's queen there flowed a stream of milk which left its colour upon the skies; wherefore is it called the Milky Way, and the name derives from its actual origin. Or is it that a greater host of stars has woven its fires in a dense circlet and glows with concentrated light, and that the ring shines the more radiantly for the massing of its brightness? Perhaps the souls of heroes, outstanding men deemed worthy of heaven, freed from the body and released from the globe of Earth, pass hither and, dwelling in a heaven that is their own, live the infinite years of paradise and enjoy celestial bliss. If so, here we honour the line of Aeacus,[a] here the sons of Atreus,[b] and warlike Diomede; the man of Ithaca,[c] too, who by his triumphs on land and sea was nature's conqueror; the Pylian,[d] renowned for a long life of triple span, and the other Greek chiefs who fought at Troy; *Hector, bulwark and glory of the Ilian race*, together with Aurora's dusky son[e] and him[f] of the Thunderer's stock who ruled Lycia; nor let me pass you by, Amazonian maid,[g] and the other kings sent by Thrace, the nations of Asia, and Pella,[h] whose greatest glory lies in him[i] called Great. Nor must I fail to mention the sages, strong of mind, men of exact and weighty judgement, who in their own selves possessed every endowment: the upright Solon and resolute Lycurgus; the inspired

[h] Housman considers that by kings sent to Troy from Pella are meant the leaders of the Paeones: Pyraechmes (Homer, *Il.* 2. 848) and Asteropaeus (*ib.* 21. 152 ff.).
[i] Alexander.

aetheriusque Platon, et qui fabricaverat illum
damnatusque suas melius damnavit Athenas,
Persidos et victor, strarat quae classibus aequor ;
Romanique viri, quorum iam maxima turba est,
Tarquinioque minus reges et Horatia proles,
tota acies partus, nec non et Scaevola trunco
nobilior, maiorque viris et Cloelia virgo,
et Romana ferens, quae texit, moenia Cocles,
et commilitio volucris Corvinus adeptus
et spolia et nomen, qui gestat in alite Phoebum,
et Iove qui meruit caelum Romamque Camillus
servando posuit, Brutusque a rege receptae
conditor, et furti per bella Papirius ultor,
Fabricius Curiusque pares, et tertia palma
Marcellus Cossusque prior de rege necato,
certantes Decii votis similesque triumphis,

786 furti] phirri 789 -que decii

---

[a] Socrates.

[b] Themistocles.

[c] The Horatii (like the Curiatii) were triplets (*cf.* Livy 1. 24 f.).

[d] Gaius Mucius Scaevola (*cf.* Livy 2. 12).

[e] *Cf.* Livy 2. 13.

[f] Upon his shield, suggests Housman (for Horatius Cocles, *cf.* Livy 2. 10).

[g] Marcus Valerius Corvinus : a raven (*cf.* 1. 417, note) perched on his helmet before his combat with a gigantic Gaul, and after his victory he assumed his cognomen from it (Aulus Gellius, *N.A.* 9. 11 : Livy 7. 26 gives the name as Corvus).

[h] From the Gauls (*cf.* Livy 5. 47 to the end).

Plato and the man [a] who had fashioned him, whose condemnation served rather to condemn his native Athens ; and the conqueror [b] of Persia, the land whose fleet covered the sea. There follow the warriors of Rome, whose host is now the largest ; here are the kings save only Tarquin, here the Horatian brothers, a whole battle-line produced at a single birth,[c] here also Scaevola, ennobled by his mutilated arm [d] ; Cloelia, a maid of more than manly courage,[e] and Cocles, bearing as blazon [f] the walls of Rome which he defended ; Corvinus, winner of spoils and a name, aided in combat by a bird which hides beneath a bird's exterior the godhead of Phoebus [g] ; Camillus, who by saving Jove's Capitol [h] won a place in heaven and in saving founded Rome ; Brutus, the establisher of a city rescued from a king ; Papirius, who avenged a treacherous defeat in open battle [i] ; Fabricius and Curius, a well-matched pair [j] ; Marcellus, third to win the palm of honour,[k] and Cossus, who before him had stripped the spoil from a king he killed ; and the Decii, rivals in their self-sacrifice and equals in their glory.[l] Here is

[i] Lucius Papirius Cursor, who avenged the disaster of the Caudine Forks (*cf.* Livy 9. 15).

[j] In their incorruptibility (*cf.* Cicero, *Par. Stoic.* 12, which contains a remarkably similar catalogue of Roman heroes).

[k] Referring to *spolia opima*, spoil personally taken in battle by the victorious from the vanquished general : first won by Romulus from Acron (*cf.* Livy 1. 10 ; Propertius 4. 10), secondly by Cossus from Tolumnius (*cf.* Livy 4. 19), and thirdly by Marcus Claudius Marcellus from Virdomarus (*cf.* Livy, *Perioch.* 20).

[l] Father, son, and grandson all named Publius Decius Mus and said to have devoted (*i.e.* ritually sacrificed) themselves in battle for the Roman cause (*cf.* Cicero, *Tusc. Disp.* 1. 89 ; only father and son are mentioned in *Par. Stoic.* 12).

invictusque mora Fabius, victorque nefandi
Livius Hasdrubalis socio per bella Nerone,
Scipiadaeque duces, fatum Carthaginis unum,
Pompeiusque orbis domitor per trisque triumphos
ante diem princeps, et censu Tullius oris
emeritus fasces, et Claudi magna propago,
Aemiliaeque domus proceres, clarique Metelli,
et Cato fortunae victor, fictorque sub armis
miles Agrippa suae, Venerisque ab origine proles
Iulia. descendit caelo caelumque replebit,
quod reget, Augustus, socio per signa Tonante,
cernet et in coetu divum magnumque Quirinum
quemque novum superis numen pius addidit ipse,
altius aetherii quam candet circulus orbis.
illa deis sedes: haec illis, proxima divum
qui virtute sua similes vestigia tangunt.

Nunc prius incipiam stellis quam reddere vires
signorumque canam fatalia carmine iura,
implenda est mundi facies, corpusque per omne

790 necati
794 deum
795 fasces] caelum
797 fictorque] matrisque
799 replevit
800 regit
801 cernit
803 deum
805-808: *see after* 538
809 nunc] c *LM*: ac *G*

---

[a] Publius Cornelius Scipio Africanus, who defeated Hannibal at Zama in 202 B.C., and his adopted son Aemilianus, who destroyed Carthage in 146 B.C.: an imitation of Virgil, *Aen.* 6. 842 f.

[b] In 79 B.C. over the Marians in Sicily and Africa, in 71 over Sertorius in Spain, and in 61 over the pirates and Mithridates. The poet's chronology is not quite accurate, since Pompey became consul before the legal age in 70 (*i.e.* before his third triumph).

Fabius, whom delay made invincible, and here Livius, with his comrade-in-arms Nero, the vanquisher of ruthless Hasdrubal; the Scipionic generals, a single agent of Carthage's doom [a]; world-conquering Pompey, whom three triumphs [b] made first citizen before the appointed time; and Cicero, who won the consulship by the wealth of his oratory. Here is the great line of Claudius [c]; the leading members of the Aemilian house; and the famed Metelli. Here are Cato and Agrippa, who proved in arms the one the master, the other the maker of his destiny; and the Julian [d] who boasted descent from Venus. Augustus has come down from heaven and heaven one day will occupy, guiding its passage through the zodiac with the Thunderer [e] at his side; in the assembly of the gods he will behold mighty Quirinus *and him whom he himself has dutifully added as a new deity to the powers above*, on a higher plane than shines the belt of the Milky Way. There is the gods' abode, and here is theirs, who, peers of the gods in excellence, attain to the nearest heights.

[809]Now before I proceed to ascribe to the stars [f] their powers and tell in my song of the fateful laws of the signs, I must complete the picture of heaven and note whatever in the whole system shines with

[c] Referring to that Appius Claudius who is said by Livy (2. 16), being a Sabine of Regillum, to have migrated in 504 B.C. to Rome and there founded the *gens Claudia*.

[d] Julius Caesar.

[e] Jupiter.

[f] "Planets" says Housman, addendum to 1. 805-808, but this is too specific. Since the poet does not proceed to deal with the planets, it seems best to interpret the word generally: he means, of course, no more than "before I proceed to the astrological part of my work."

quidquid ubique nitens vigeat quandoque notandum est.
sunt etenim raris orti natalibus ignes,
protinus et rapti. subitas candescere flammas
aera per liquidum natosque perire cometas
rara per ingentis viderunt saecula motus.
sive, quod ingenitum terra spirante vaporem
umidior sicca superatur spiritus aura,
nubila cum longo cessant depulsa sereno
et solis radiis arescit torridus aer,
apta alimenta sibi demissus corripit ignis
materiamque sui deprendit flamma capacem,
et, quia non solidum est corpus, sed rara vagantur
principia aurarum volucrique simillima fumo,
in breve vivit opus coeptusque incendia fine
subsistunt pariterque cadunt fulgentque cometae.
quod nisi vicinos agerent occasibus ortus
et tam parva forent accensis tempora flammis,
alter nocte dies esset, Phoebusque rediret,
immersum et somno totum deprenderet orbem.
tum, quia non una specie dispergitur omnis
aridior terrae vapor et comprenditur igni,
diversas quoque per facies accensa feruntur
lumina, quae ruptis exsistunt nata tenebris.
nam modo, ceu longi fluitent de vertice crines,
flamma comas imitata volat, tenuisque capillos
diffusos radiis ardentibus explicat ignis ;
nunc prior haec facies dispersis crinibus exit,
et glomus ardentis sequitur sub imagine barbae ;
interdum aequali laterum compagine ductus

812 nitet
813 etiam rari sorti natalis euntes
815 tractosque
825 coeptaque
829 caelumque
834 subitis
839 globus

especial brightness at any place or time. For there are fires born at infrequent intervals and forthwith swept away.[a] In times of great upheaval rare ages have seen the sudden glow of flame through the clear air and comets blaze into life and perish. Maybe the earth breathes forth an inborn vapour, and this damper breath is overpowered by an arid air; when clouds are banished for long periods of clear weather so that air grows hot and dry under the rays of the sun, fire then descends and seizes its apt sustenance and a flame takes hold of the matter that suits its nature; and since there is no solid body, but only the wandering elements of the breezes, tenuous and most like to drifting smoke, the action is short-lived and the fires last no longer than the moment of their beginning: the comets perish as they blaze. Were their risings not near neighbour to their sinkings and the duration of the kindled flames not so short, a second daytime would exist by night, and the Sun would return to find the whole world fast asleep. Moreover, since all the drier exhalation from the earth is spread abroad and catches fire in no uniform manner, it is also in different shapes that we see those kindled lights which spring to life and shatter the darkness. For sometimes, as though long tresses were flowing down from a person's head, the flame flies in the guise of hair, and the slender fire lets loose its streaming locks in brilliant rays; and now the tresses are scattered and this former appearance ends, succeeded by a ball in the likeness of a fiery beard. Occasionally the outline of a comet's sides fashions with

[a] In the ensuing passage the poet confuses comets, which are not of momentary duration, and shooting stars, which are.

quadratamve trabem fingit teretemve columnam.
quin etiam tumidis exaequat dolia flammis
procere distenta uteros, artosque capellas
mentitur parvas ignis glomeratus in orbes
hirta figurantis tremulo sub lumine menta,
lampadas et fissas ramosos fundit in ignes.
et tenuem longis iaculantur tractibus ignem
praecipites stellae passimque volare videntur,
cum vaga per liquidum scintillant lumina mundum
exsiliuntque procul volucris imitata sagittas,
ardua cum gracili tenuatur semita filo.
sunt autem cunctis permixti partibus ignes,
qui gravidas habitant fabricantes fulmina nubes
et penetrant terras Aetnamque minantur Olympo
et calidas reddunt ipsis in fontibus undas
ac silice in dura viridique in cortice sedem
inveniunt, cum silva sibi collisa crematur ;
ignibus usque adeo natura est omnis abundans :
ne mirere faces subitas erumpere caelo
aeraque accensum flammis lucere coruscis
arida complexum spirantis semina terrae,
quae volucer pascens ignis sequiturque fugitque,
fulgura cum videas tremulum vibrantia lumen
imbribus e mediis et caelum fulmine ruptum.
sive igitur ratio praebentis semina terrae
in volucris ignes potuit generare cometas ;
sive illas natura faces obscura creavit
sidera per tenuis caelo lucentia flammas,
sed trahit ad semet rapido Titanius aestu
involvitque suo flammantis igne cometas
ac modo dimittit, sicut Cyllenius orbis

843 partasque
844 parvis signis
845 menses
849 crinibus
848 nitidum
850 exuruntque

even dimensions a rectangular beam or a rounded column. Then again, the fire of comets may match with swollen flames casks with greatly distended paunches ; or, massed in compact circles the flickering light of which resembles their shaggy chins, it may feign the appearance of small goats ; or it may produce torches which are split into several branches of flame. There are also shooting stars, which hurl long trails of slender fire and are seen flying everywhere, when wandering lights flash through the clear sky, and dart afar like winged arrows, tracing the slender line of a path on high. Indeed, fire is found mingled with every part of the universe ; it dwells in the laden clouds and forges lightning, it makes its way into the earth and threatens heaven with the flames of Etna, it causes waters to boil at their very sources, and finds a habitation in hard flint and in verdant bark, when trees dashed against trees are set aflame ; to such an extent does all nature abound with fire. So wonder not that torches suddenly burst forth from the skies ; and that the air is kindled and shines with flickering flames after embracing the dry seeds exhaled by the earth, seeds which the swift fire, as it feeds, both pursues and shuns, for you see the lightning hurl its quivering flash from the midst of a rainstorm and the heavens rent with the thunderbolt. Possibly it is the principle of earth supplying seed for fleet fire which has given birth to comets ; or perhaps in those torches nature has created dim stars that shine in heaven with meagre flames, but the Sun with its consuming heat attracts the blazing comets to itself, absorbs them in its own fire, and then releases them : just so do

---

867 ob cuncta

et Venus, accenso cum ducit vespere noctem,
saepe latent falluntque oculos rursusque revisunt;
seu deus instantis fati miseratus in orbem
signa per affectus caelique incendia mittit;
numquam futtilibus excanduit ignibus aether,
squalidaque elusi deplorant arva coloni,
et sterilis inter sulcos defessus arator
ad iuga maerentis cogit frustrata iuvencos.
aut gravibus morbis et lenta corpora tabe
corripit exustis letalis flamma medullis
labentisque rapit populos, totasque per urbes
publica succensis peraguntur iusta sepulcris.
qualis Erectheos pestis populata colonos
extulit antiquas per funera pacis Athenas,
alter in alterius labens cum fata ruebant,
nec locus artis erat medicae nec vota valebant;
cesserat officium morbis, et funera derant
mortibus et lacrimae; lassus defecerat ignis
et coacervatis ardebant corpora membris,
ac tanto quondam populo vix contigit heres.
talia significant lucentes saepe cometae:
funera cum facibus veniunt, terrisque minantur
ardentis sine fine rogos, cum mundus et ipsa
aegrotet natura hominum sortita sepulcrum.
quin et bella canunt ignes subitosque tumultus
et clandestinis surgentia fraudibus arma,

[873] nitent
[883] fata
[895] omnium *M*: novum *GL*

[a] The orbits of Mercury and Venus round the Sun lie within that of the Earth, and consequently their angular distances from the Sun cannot exceed a certain value (roughly 23° for Mercury, 46° for Venus), and as they approach conjunction their light is outshone by the supreme

the orb of Mercury and the planet Venus,[a] when she kindles her evening lamp and brings on night, oft disappear and elude our gaze and oft visit us again. It may well be that by means of these moods and conflagrations of the sky Heaven in pity is sending upon earth tokens of impending doom, for the fires wherewith the heavens blaze have never lacked significance, but farmers, cheated of their hopes, mourn over blighted fields, and amid barren furrows the weary ploughman vainly urges to the yoke his drooping team. Or else, when grievous sickness and a slow decline has gripped men's bodies, a mortal flame burns out the seat of life and sweeps away the stricken peoples; and the obsequies of the community fill whole cities with blazing pyres. Such was the plague [b] which ravaged Erechtheus' folk and bore forth ancient Athens to an unwarlike grave, when, one after another, men collapsed upon the dead and died: no place was there for a doctor's skill, no prayers availed; duty fell a prey to sickness, and none were left to bury, none to weep the dead; the wearied fires sufficed not for their office, and limbs piled on limbs the corpses burnt: hardly was an heir to be found in that nation once so vast. Such are the disasters which the glowing comets oft proclaim. Death comes with those celestial torches, which threaten earth with the blaze of pyres unceasing, since heaven and nature's self are stricken and seem doomed to share men's tomb. Wars, too, the fires portend, and sudden insurrection, and arms

brilliance of the Sun; thus they are mostly visible only for a short time before sunrise or after sunset.

[b] The plague at Athens, no doubt suggested by Lucretius 6. 1138 ff.

externas modo per gentes ut, foedere rupto
cum fera ductorem rapuit Germania Varum
infecitque trium legionum sanguine campos,
arserunt toto passim minitantia mundo
lumina, et ipsa tulit bellum natura per ignes
opposuitque suas vires finemque minata est.
ne mirere gravis rerumque hominumque ruinas,
saepe domi culpa est: nescimus credere caelo.
civilis etiam motus cognataque bella
significant. nec plura alias incendia mundus
sustinuit, quam cum ducibus iurata cruentis
arma Philippeos implerunt agmine campos,
vixque etiam sicca miles Romanus harena
ossa virum lacerosque prius super astitit artus,
imperiumque suis conflixit viribus ipsum,
perque patris pater Augustus vestigia vicit.
necdum finis erat: restabant Actia bella
dotali commissa acie, repetitaque rerum
alea et in ponto quaesitus rector Olympi,
femineum sortita iugum cum Roma pependit
atque ipsa Isiaco certarunt fulmina sistro;
restabant profugo servilia milite bella,

917 pompa rependit

[a] The disaster of the Saltus Teutoburgiensis, inflicted on the Romans by Arminius, occurred in A.D. 9. Housman interprets *modo* (898) . . . *etiam* (906) . . . as "now . . . now . . .," understanding 898-903 and 904 ff. as two examples of the insurgency referred to in 896 f. Bühler (490 f.), however, convincingly identifies the architecture of the passage as 896-903 (treachery of arms) followed by 904 ff. (civil war).

[b] Housman's interpretation ". . . against herself and threatening her own destruction" seems less logical than Bühler's (491), adopted in the translation.

[c] Brutus and Cassius.

uplifted in stealthy treachery; so of late [a] in foreign parts, when, its oaths forsworn, barbarous Germany made away with our commander Varus and stained the fields with three legions' blood, did menacing lights burn in every quarter of the skies; nature herself waged war with fire, marshalling her forces against us and threatening our destruction.[b] Wonder not at the grievous disasters which betide man and man's affairs, for the fault oft lies within us: we have not sense to trust heaven's message. Comets also presage civil discord and strife between kin. At no other time did heaven experience more conflagrations than when the sworn forces of the blood-stained conspirators [c] filled with their ranks the plains of Philippi, and on sand scarcely yet dry of blood [d] the Roman legionary took his stand on the bones of warriors and limbs mangled in former fighting; the empire's armed might engaged in conflict with itself, and Augustus, father of his country, marching in his father's steps, prevailed. Nor was this all: war at Actium was still to come, waged by an army pledged in dowry, when the destiny of the world was again at stake and the ruler of heaven was determined on the sea; the fate of Rome, threatened with a female yoke, hung in the balance, and the very thunderbolt clashed with the sistrum of Isis.[e] Still to come [f] were battles against slaves

[d] That the battles of Pharsalus and Philippi (160 miles apart) were fought on the same spot is a poetic fancy for which Virgil, *Georg.* 1. 489 ff., is responsible.

[e] That is, the Roman gods, whose weapon was the thunderbolt, clashed with the Egyptian deities armed with the sistrum.

[f] Hardly: the resistance of Sextus Pompeius was broken by the year 36 and he himself executed in 35, whereas the battle of Actium did not take place until 31.

cum patrios armis imitatus filius hostes
aequora Pompeius cepit defensa parenti.
sed satis hoc fatis fuerit : iam bella quiescant
atque adamanteis discordia vincta catenis
aeternos habeat frenos in carcere clausa ;
sit pater invictus patriae, sit Roma sub illo,
cumque deum caelo dederit non quaerat in orbe.

---

[a] May Rome not miss the deified Julius, seeing that she has Augustus, still alive, to take care of her.

joined by runaway soldiers, when a son took up arms after the example of his father's foes and a Pompey infested the seas which his parent had swept clean. Let fate content itself with this! May wars now cease and, fettered with bonds of adamant, may discord, prisoned fast, be curbed for evermore! Unconquered be the father of our fatherland; may Rome serve none but him; and for all that she has given a god to heaven, may she miss him not on earth! [a]

BOOK TWO

CONSIDERING *its technical nature the second book is very elegantly written: it deals with the two circles involved in every horoscope—the movable circle of the twelve zodiacal signs and the immovable circle of the twelve temples. First, however, we are regaled with an extended proem celebrating those poets who have inaugurated different genres of hexameter poetry; this naturally leads to Manilius himself, who renews his claim of originality and his confession of the Stoic creed. He thereupon embarks on an exhaustive description of the signs of the zodiac: their groupings, their characteristics, their aspects, their tutelary deities, the parts of the body subject to them, and their tangled relationships. We learn of those portions of the zodiacal signs called dodecatemories, though the poet soon leaves this difficult subject, emphasizing that astrological like any other teaching must be imparted in a progressive and logical order. The last part of the book sets forth the special significance of the four cardinal points and the quadrants between them; and enumerates the twelve segments of this circle together with the aspects of human existence over which they preside.*

LIBER SECUNDUS

Maximus Iliacae gentis certamina vates
et quinquaginta regum regemque patremque
Hectoraque Aeacidae victamque sub Hectore Troiam,
erroremque ducis totidem, quot vicerat, annis
infestum experti dominum maris atque renato
instantem bello geminataque Pergama ponto
ultimaque in patria captisque penatibus arma
ore sacro cecinit ; patriam cui turba petentum,
dum dabat, eripuit, cuiusque ex ore profusos
omnis posteritas latices in carmina duxit
amnemque in tenuis ausa est deducere rivos
unius fecunda bonis. sed proximus illi
Hesiodus memorat divos divumque parentes
et chaos enixum terras orbemque sub illo
infantem et primos titubantia sidera cursus
Titanasque senes, Iovis et cunabula magni

[3] hectoreumque facit
[5] [que] per agmina
[7] patria quae iura petentem
[14] corpus

---

[a] Homer : 1-3 describe the *Iliad*, 4-6 the *Odyssey*.
[b] Priam (son of Laomedon, son of Ilus).
[c] *Cf. Cert. Hom. et Hes.* (Loeb Hesiod p. 566) : "But, as for Homer, you might almost say that every city with its

BOOK 2

The greatest of bards [a] with utterance inspired sang the struggle of Ilus' race, a king and father of fifty kings,[b] and Hector by Achilles and Troy with Hector vanquished; and the roaming of that hero who for as many years as his conquest had taken *encountered Neptune's anger, and in a renewal of* war his assault, and Pergamum resurrected on the sea and the final combat in native land and home held by intruders. The host of claimants [c] for his birthplace, in giving him many, has left him with none. Yet all posterity has for its verse drawn on the rich stream issuing from his lips and, daring to channel his river into their slender rills, has become fertile by the wealth of one. But Hesiod [d] is next to him and tells of the gods and parents of the gods, Chaos in travail with Earth, the childhood of the world beneath its sway, the stars faltering as they first embarked on their courses, the ancient Titans, the cradling of mighty Jove, Jove

inhabitants claims him as her son." Bentley, who with the arresting conjecture *patriam cui Graecia, septem* ("Greece robbed him of his birthplace by giving him seven") first restored the meaning of these lines, quotes the hexameter tag *Smyrna, Rhodos, Colophon, Salamin, Ios, Argos, Athenae.*

[d] 12-23 (= 18) *Theogony*; 19-24 *Works and Days*. The summary given by Manilius, however, is inaccurate and misleading.

et sub fratre viri nomen, sine matre parentis,
atque iterum patrio nascentem corpore Bacchum,
silvarumque deos secretaque numina Nymphas.
quin etiam ruris cultus legesque notavit
militiamque soli, quod colles Bacchus amaret,
quod fecunda Ceres campos, quod Pallas utrumque,
atque arbusta vagis essent quod adultera pomis ;
omniaque immenso volitantia lumina mundo,
pacis opus, magnos naturae condit in usus.
astrorum quidam varias dixere figuras,
signaque diffuso passim labentia caelo
in proprium cuiusque genus causasque tulere ;
Persea et Andromedan poena matremque dolentem
solventemque patrem, raptamque Lycaone natam,
officioque Iovis Cynosuram, lacte Capellam
et furto Cycnum, pietate ad sidera ductam
Erigonen ictuque Nepam spolioque Leonem
et morsu Cancrum, Pisces Cythereide versa,
Lanigerum victo ducentem sidera ponto,
ceteraque ex variis pendentia casibus astra
aethera per summum voluerunt fixa revolvi.
quorum carminibus nihil est nisi fabula caelum
terraque composuit mundum quae pendet ab illo.

[16] matre] fratre
[23] sacrataque . . . nymphis
[21] Pallas] bachus
[28] -ae poenas
[38] caelum

[a] He was both brother and husband of Juno.
[b] Minerva.
[c] Through a ruse of Juno Semele, six months pregnant, was frightened to death by the thunderbolts of her lover, Jupiter ; however he rescued her unborn child, Bacchus, and sewed him in his thigh, whence he was "born again" at full time (*cf.* Apollodorus, *Bibl.* 3. 4. 3).
[d] The vine.

who gained husband's name as brother [a] and father's for one no mother bare,[b] Bacchus born a second time from his father's body,[c] the woodland deities and those retiring spirits, the Nymphs. He further told of tillage of the countryside and its laws and man's warfare with the soil, of the love of Bacchus [d] for the hills, of fertile Ceres [e] for the plains, and of Pallas [f] for both, and of the grafting of errant fruits on trees as though in illicit union; moreover, a task of peace, he establishes the courses of all the luminaries through the vast heavens so as to further the great designs of nature. Some [g] too have told of the varied patterns of the stars and have ascribed to their proper class and origin the signs that glide at large in the far-spread sky; and Perseus, delivering from sacrifice Andromeda and her grieving parents, Lycaon's ravished daughter,[h] and the Cynosure, raised to the stars for her care of Jove, the Goat for her milk, the Swan for her disguise, Erigone for her filial affection, the Scorpion for the stroke it dealt, the Lion for its spoil, the Crab for its bite, the Fishes for the transformation of Venus, and the Ram who leads the signs for his conquest of the sea [i]—these and all the other constellations such poets regard as revolving in fixed orbits through the highest empyrean in consequence of different lots in life. In their songs is heaven naught but fable and earth the fashioner of the skies on which it depends. Then the ways of

[e] Wheat.
[f] The olive.
[g] Including Aratus.
[h] Callisto.
[i] For these and other catasterisms see Introduction pp. xxiv ff.

quin etiam ritus pastorum et Pana sonantem
in calamos Sicula memorat tellure creatus,
nec silvis silvestre canit perque horrida motus
rura serit dulcis Musamque inducit in aulas.
ecce alius pictas volucres ac bella ferarum,
ille venenatos angues aconitaque et herbas
fata refert vitamque sua radice ferentis.
quin etiam tenebris immersum Tartaron atra
in lucem de nocte vocant orbemque revolvunt
interius versum naturae foedere rupto.
omne genus rerum doctae cecinere sorores,
omnis ad accessus Heliconos semita trita est,
et iam confusi manant de fontibus amnes
nec capiunt haustum turbamque ad nota ruentem.
integra quaeramus rorantis prata per herbas
undamque occultis meditantem murmur in antris,
quam neque durato gustarint ore volucres,
ipse nec aetherio Phoebus libaverit igni.
nostra loquar, nulli vatum debebimus orsa,
nec furtum sed opus veniet, soloque volamus
in caelum curru, propria rate pellimus undas.
namque canam tacita naturae mente potentem
infusumque deum caelo terrisque fretoque
ingentem aequali moderantem foedere molem,
totumque alterno consensu vivere mundum
et rationis agi motu, cum spiritus unus

[39] pecorum
[42] auras
[44] hic nata per
[57] ora

[a] Theocritus.

[b] Possibly a Greek source of Macer's *Ornithogonia* and Grattius's *Cynegetica*.

[c] Nicander in his *Theriaca* and *Alexipharmaca* respectively.

shepherds and Pan piping upon his reeds are told by a son [a] of Sicily's isle : to the woods he sings a more than woodland strain, sows tender emotions over the rugged countryside, and brings the Muse to the steading. See, another [b] tells of brightly coloured birds and warfare against beasts, and one [c] of envenomed snakes and again of aconite and herbs which carry life and death in their roots. Next there are those [d] who summon gloom-shrouded Tartarus from the blackness of night into the light of day and, breaking nature's covenant, turn the world inside out and bring forth what lies within. Every kind of theme the learned sisters have hymned, and worn is every path which leads to Helicon ; turbid are the waters which now trickle from those springs, and insufficient are the draughts therefrom for the crowds that flock to the well-known haunts. Fresh meadows let us seek over the dewy grass and streams that practise their murmurings in hidden caves, streams neither tasted by beak of bird nor sipped by celestial fire of Phoebus himself. Mine own theme shall I sing, my words shall I owe to none amongst bards, and there shall emerge no stolen thing, but work of my own contriving ; in a lone car I soar to the heavens, in a ship of my own I sweep the seas. For I shall sing of God, silent-minded monarch of nature,[e] who, permeating sky and land and sea, controls with uniform compact the mighty structure ; how the entire universe is alive in the mutual concord of its elements and is driven by the pulse of reason,

[d] Perhaps referring to the Greek source of Lucan's *Catacthonion*.

[e] *Cf.* 1. 247-254, 3. 48-55, 4. 888-890.

per cunctas habitet partes atque irriget orbem 65
omnia pervolitans corpusque animale figuret.
quod nisi cognatis membris contexta maneret
machina et imposito pareret tota magistro
ac tantum mundi regeret prudentia censum,
non esset statio terris, non ambitus astris, 70
erraretque vagus mundus standove rigeret,
nec sua dispositos servarent sidera cursus
noxque alterna diem fugeret rursusque fugaret,
non imbres alerent terras, non aethera venti
nec pontus gravidas nubes nec flumina pontum 75
nec pelagus fontes, nec staret summa per omnis
par semper partes aequo digesta parente,
ut neque deficerent undae nec sideret orbis
nec caelum iusto maiusve minusve volaret.
motus alit, non mutat opus. sic omnia toto 80
dispensata manent mundo dominumque sequuntur.
hic igitur deus et ratio, quae cuncta gubernat,
ducit ab aetheriis terrena animalia signis,
quae, quamquam longo, cogit, summota recessu,
sentiri tamen, ut vitas ac fata ministrent 85
gentibus ac proprios per singula corpora mores.
nec nimis est quaerenda fides: sic temperat arva
caelum, sic varias fruges redditque rapitque,
sic pontum movet ac terris immittit et aufert,
atque haec seditio pelagus nunc sidere lunae 90

[69] sensum
[71] haereretque standoque
[78] sidera
[87] minus

since a single spirit dwells in all its parts and, speeding through all things, nourishes the world and shapes it like a living creature. Indeed, unless the whole frame stood fast, composed of kindred limbs and obedient to an overlord, unless providence directed the vast resources of the skies, the Earth would not possess its stability, nor stars their orbits, and the heavens would wander aimlessly or stiffen with inertia; the constellations would not keep their appointed courses nor would alternately the night flee day and put in turn the day to flight, nor would the rains feed the earth, the winds the upper air, the sea the laden clouds, rivers the sea and the deep the springs[a]; the sum of things would not remain for ever equal through all its parts, so disposed by the fairness of its creator that neither should the waves of the sea fail nor the land sink beneath them, nor the revolving heavens become larger or smaller than the mean. Motion sustains and does not alter the edifice. In this due order over the whole universe do all things abide, following the guidance of a master. This God and all-controlling reason, then, derives earthly beings from the signs of heaven; though the stars are remote at a far distance, he compels recognition of their influences, in that they give to the peoples of the world their lives and destinies and to each man his own character. Nor is the proof hard to find: thus is it that the sky affects the fields, thus gives and takes away the various crops, puts the sea to movement, casting it on land and fetching it therefrom, and thus this restlessness possesses ocean, now caused by the shining of the

[a] For the replenishment of rivers from the sea (and the whole passage) see Lucretius 5. 261-272, 6. 608-638.

mota tenet, nunc diverso stimulata recessu,
nunc anni spatio Phoebum comitata volantem ;
sic summersa fretis, concharum et carcere clausa,
ad lunae motum variant animalia corpus
et tua damna, tuas imitantur, Delia, vires ; 95
tu quoque fraternis sic reddis curribus ora
atque iterum ex isdem repetis, quantumque reliquit
aut dedit ille, refers et sidus sidere constas ;
denique sic pecudes et muta animalia terris,
cum maneant ignara sui legisque per aevum, 1
natura tamen ad mundum revocante parentem
attollunt animos caelumque et sidera servant
corporaque ad lunae nascentis cornua lustrant
venturasque vident hiemes, reditura serena.
quis dubitet post haec hominem coniungere caelo, 1
cui, cupiens terras ad sidera surgere, munus
eximium natura dedit linguamque capaxque
ingenium volucremque animum, quem denique in unu
descendit deus atque habitat seque ipse requirit ?
mitte alias artes, quarum est permissa facultas
invidiosa adeo, nec nostri munera census : 1
[mitto quod aequali nihil est sub lege tributum,
quo patet auctoris summam, non corporis, esse ;
mitto quod certum est et inevitabile fatum
materiaeque datum est cogi sed cogere mundo]
quis caelum posset nisi caeli munere nosse, 1
et reperire deum, nisi qui pars ipse deorum est ?

[95] et tum
[110] infidos
[115] munera nosset

---

[a] The influences of the Moon and Sun on the tides (*cf.* Pliny, *N.H.* 2. 215).

[b] *Cf.* Pliny, *N.H.* 2. 109, 221.

[c] The Moon.

[d] *Cf.* 4. 886-910.

Moon, now provoked by her retreat to the other side of the sky, and now attendant upon the Sun's yearly revolution [a]; thus that, submerged beneath the waves and in prison-shell [b] confined, animals adapt their forms to the motions of the Moon and copy your waning, Delia,[c] and your growth; thus you too return your features to your brother's car and a second time from it reseek them, and as much as he has grudged or lavished on you do you reflect, your star dependent on his. Thus lastly with the herds and dumb animals on earth: though they remain for ever ignorant of themselves and of law, nevertheless when nature summons them to the parent heaven, they lift up their minds and watch the sky and stars: they cleanse their bodies on beholding the horns of the nascent Moon, and note the storms about to come, the fine weather about to return. Who after this can doubt [d] that a link exists between heaven and man, *to whom, in its desire for earth to rise to the stars, gifts* outstanding did nature give and the power of speech and breadth of understanding and a wing-swift mind, and into whom alone indeed has God come down and dwells, and seeks himself in man's seeking of him? Pass over other [e] arts in which man is granted such enviable competence, gifts above our estate.[f] Who could know heaven save by heaven's gift and discover God save one who

[e] *I.e.* other than astrology.

[f] [111-114] interrupting the context (of man's affinity with heaven) with irrelevant doctrines: "I pass over the fact that nothing is given with impartial distribution, from which it is clear that the whole is the work of the creator, not of matter itself. I pass over the fact that fate is sure and inevitable and that it is given to matter to suffer, to heaven to exercise, compulsion."

quisve hanc convexi molem sine fine patentis
signorumque choros ac mundi flammea tecta,
aeternum et stellis adversus sidera bellum
[ac terras caeloque fretum subiectaque utrisque] 12
cernere et angusto sub pectore claudere posset,
ni sanctos animis oculos natura dedisset
cognatamque sibi mentem vertisset ad ipsam
et tantum dictasset opus, caeloque veniret 12
quod vocat in caelum sacra ad commercia rerum? 12
quis neget esse nefas invitum prendere mundum
et velut in semet captum deducere in orbem?
sed, ne circuitu longo manifesta probentur, 13
ipsa fides operi faciet pondusque fidemque;
nam neque decipitur ratio nec decipit umquam.
rite sequenda via est ac veris credita causis,
eventusque datur qualis praedicitur ante.
quod Fortuna ratum faciat, quis dicere falsum 13
audeat et tantae suffragia vincere sortis?

Haec ego divino cupiam cum ad sidera flatu
ferre, nec in turba nec turbae carmina condam,
sed solus, vacuo veluti vectatus in orbe
liber agam currus non occursantibus ullis 1
nec per iter socios commune regentibus actus,
sed caelo noscenda canam, mirantibus astris
et gaudente sui mundo per carmina vatis,
vel quibus illa sacros non invidere meatus
notitiamque sui, minima est quae turba per orbem.

[117] atque
[122] tantos
126: *see after* 270
[132] secunda
[137] in turbam
[139] ubera tam

---

[a] See Introduction p. xv.
[b] *I.e.* of the zodiac. Here follows the interpolated [120]:

shares himself in the divine?[a] Who could discern and compass in his narrow mind the vastness of this vaulted infinite, the dances of the stars, the blazing dome of heaven, and the planets' everlasting war against the signs,[b] had not nature endowed our minds with divine vision, had turned to herself a kindred intelligence, and had prescribed so great a science? Who, unless there came from heaven a power which calls us heavenward to the sacred fellowship of nature? Who could deny the sacrilege of grasping an unwilling heaven, enslaving it, as it were, in its own domain, and fetching it to earth? But, lest it take a long digression to vouch for what is plain to see, the faith it keeps[c] will create for our science authority and faith in it; for neither does our system deceive nor ever is deceived. Rightly and for true reasons trusted is the path that one must take, and the outcome follows even as before foretold. What Fortune has confirmed, what man would dare call false, gainsaying the casting of so great a vote?

136This is the theme I should wish with breath inspired to carry to the stars. Not in the crowd nor for the crowd shall I compose my song, but alone, as though borne round an empty circuit I were freely driving my car with none to cross my path or steer a course beside me over a common route, I shall sing it for the skies to hear, while the stars marvel and the firmament rejoices in the song of its bard, and for those to whom the stars have not grudged knowledge of themselves and their sacred motions, the smallest

". . . and land and sea beneath the sky, and what there is beneath each, . . ."

[c] *I.e.* by its true predictions (*cf.* 131-133).

illa frequens, quae divitias, quae diligit aurum, 14
imperia et fasces mollemque per otia luxum
et blandis diversa sonis dulcemque per aures
affectum, ut modico noscenda ad fata labore.
hoc quoque fatorum est, legem perdiscere fati.
 Et primum astrorum varia est natura notanda 15
carminibus per utrumque genus. nam mascula sex sunt,
diversi totidem generis sub principe Tauro :
cernis ut aversos redeundo surgat in artus.
alternant genus et vicibus variantur in orbem.
 Humanas etiam species in parte videbis, 1
nec mores distant : pecudum pars atque ferarum
ingenium facient. quaedam signanda sagaci
singula sunt animo, propria quae sorte feruntur :
nunc binis insiste ; dabunt geminata potentis
per socium effectus. multum comes addit et aufert, 1
ambiguisque valent, quis sunt collegia, fatis
ad meritum noxamque. duos per sidera Pisces
et totidem Geminos nudatis aspice membris.
his coniuncta manent alterno bracchia nexu,
dissimile est illis iter in contraria versis. 1
par numerus, sed enim dispar natura notanda est.
atque haec ex paribus toto gaudentia censu

145 frequens] fluit — 153 adversus . . . arcum
147 adversa — 157 quae iam

---

[a] Rendering *fasces*, the bundle of rods and axe carried by the lictors before the chief officers of state.

[b] *Cf.* 4. 118.

[c] Henceforth throughout the *Astronomica* "signs" will usually mean the signs of the zodiac.

[d] Thus masculine are ♈ ♊ ♌ ♎ ♐ ♒ ; feminine ♉ ♋ ♍ ♏ ♑ ♓.

[e] Human : ♊ ♍ ♒ ; bestial : ♈ ♉ ♋ ♌ ♏ ♑ ♓. Of the

society on earth. Vast is the crowd which worships wealth and gold, power and the trappings of office,[a] soft luxury amid ease, diversions of seductive music, and a happy feeling stealing through the ears, objects of slight labour compared with the understanding of fate. Yet this too is the gift of fate, the will to learn fate's laws.[b]

150 My song must first mark the differing nature of the signs [c] according to sex. For six are masculine, whilst as many, led by the Bull, are of the opposite sex: you see how he rises by his hind limbs when he reappears. They alternate their sex, changing one after another round the circle.[d]

155 You will also behold the human form in some, and the dispositions they bestow are not out of keeping; some will produce the nature of cattle and beasts.[e] Certain signs must with careful mind be noted as single, and these keep to an unshared estate.[f] Now turn to the double signs; being doubled they will exert influences the power of which is tempered by a partner. Much does a companion add and take away, and, when destiny hangs in the balance, the signs that are accompanied have power to decide for good or ill. Look among the constellations for the two Fishes and the Twins of like number with limbs unclad. The arms of the Twins are for ever linked in mutual embrace; but the Fishes face opposite ways and have different courses. Mark well that, though the two signs are alike in their duality, they are unlike in their nature. These are they that of the double signs go rejoicing in a full estate and

others ♎ is implied at 2. 528 to be human, while ♐ plainly combines both natures.

[f] Single: ♈ ♉ ♋ ♌ ♎ ♏ ♒.

signa meant, nihil exterius mirantur in ipsis
amissumve dolent, quaedam quod, parte recisa
atque ex diverso commissis corpore membris, 1
ut Capricornus et intentum qui derigit arcum
iunctus equo : pars huic hominis, sed nulla priori.
[hoc quoque servandum est alta discrimen in arte,
distat enim gemini duo sint duplane figura]
quin etiam Erigone binis numeratur in astris, 1
nec facies ratio duplex ; nam desinit aestas,
incipit autumnus media sub Virgine utrimque.
idcirco tropicis praecedunt omnibus astra
bina, ut Lanigero, Chelis Cancroque Caproque,
quod duplicis retinent conexo tempore vires. 1
ut, quos subsequitur Cancer per sidera fratres,
e geminis alter florentia tempora veris
sufficit, aestatem sitientem provehit alter ;
nudus uterque tamen, sentit quia uterque calorem,
ille senescentis veris, subeuntis at ille 1
aestatis : par est primae sors ultima parti.
quin etiam Arcitenens, qui te, Capricorne, sub ipso
promittit, duplici formatus imagine fertur :
mitior autumnus mollis sibi vindicat artus
materiamque hominis, fera tergo membra rigentem 1
excipiunt hiemem mutantque in tempora signum.
quosque Aries prae se mittit, duo tempora Pisces
bina dicant : hiemem hic claudit, ver incohat alter.
cum sol aequoreis revolans decurrit in astris,

185 et
187 quin etiam] nec iam
191 nunciamque
193 [hic]

[a] [173 f.] : "This distinction, too, must be preserved in our lofty art, for it makes a difference whether double signs are twin or of composite shape."

[b] She has two wings, and is so figured on the Farnese globe.

wonder at nothing foreign in themselves or mourn for aught lost, as do certain signs of amputated limb and members put together from unlike bodies, as Capricorn and he that joined to a horse takes aim with taut bow: the latter has part that is man, but the former none.[a] Erigone, too, is numbered among the double signs, but the duality in her appearance[b] is not the reason; for at the middle of the Virgin summer on one side ceases and autumn on the other begins.[c] Double signs precede all the tropic ones,[d] the Ram, the Claws, the Crab, and the Sea-goat, for the reason that, linking season with season, they possess double powers. Just as one of the twin brothers followed by the Crab round the zodiac imparts blossoming springtime, so the other brings on thirsting summer; yet each is unclad, for each feels the heat, the one of aging spring, the other of approaching summer: the last portion of the former is matched by the first degree of the latter. The Archer, too, who gives promise of Capricorn behind himself, comes shaped with twofold appearance: milder autumn claims the smooth limbs and body of his human allotment, whilst the animal portions in his rear prepare for frosty winter and change the sign to suit the change of season. And the two Fishes that the Ram sends before himself denote two seasons: one concludes winter, the other introduces spring. When the returning[e] Sun courses through

[c] Thus the double signs are ♊ ♍ ♐ ♑ ♓.

[d] The argument developed by the poet presupposes that the turning point of the seasons occurs in the *first* degree of the tropic signs (see Introduction p. lxxxi). Notice that the term is applied to Aries and Libra (equinoctial signs) as well as Cancer and Capricorn, the signs properly tropic.

[e] The Sun's annual orbit beginning in Aries.

hiberni coeunt cum vernis roribus imbres.
utraque sors umoris habet fluitantia signa.

Quin tria signa novem signis coniuncta repugnant
et quasi seditio caelum tenet. aspice Taurum
clunibus et Geminos pedibus, testudine Cancrum
surgere, cum rectis oriantur cetera membris;
ne mirere moras, cum sol aversa per astra
aestivum tardis attollat mensibus annum.

Nec te praetereat nocturna diurnaque signa
quae sint perspicere et propria deducere lege,
non tenebris aut luce suam peragentia sortem
(nam commune foret nullo discrimine nomen,
omnia quod certis vicibus per tempora fulgent
et nunc illa dies, nunc noctes illa sequuntur),
sed quibus illa parens mundi natura sacratas
temporis attribuit partes statione perenni.
namque Sagittari signum rabidique Leonis
et sua respiciens aurato vellere terga,
tum Pisces et Cancer et acri Scorpios ictu,
aut vicina loco, divisa aut partibus aequis,
omnia dicuntur simili sub sorte diurna.
cetera, vel numero consortia vel vice sedis,
interiecta locis totidem, nocturna feruntur.
quidam etiam sex continuis dixere diurnas
esse vices astris, quae sunt a principe signo
Lanigeri, sex a Libra nocturna videri.

[197] quod
[201] adversa
[216] nec . . . nec
[218] quin
[219] castris

---

[a] Pisces.

[b] Inverted: ♉ ♊ ♋ (all adjacent); upright: ♈ ♉ ♍♎♏ ♐ ♑♒♓.

[c] See Introduction pp. xxxviii f.

the watery stars,[a] then winter's rains mingle with showers of spring: each sort of moisture belongs in the double-sign that swims.

[197]Further, three adjacent signs are at variance with the other nine and a kind of dissension takes hold of heaven. Observe that the Bull rises by his hind quarters, the Twins by their feet, the Crab by his shell, whereas all the others rise in upright posture [b]; so wonder not at the delay when in tardy months [c] the Sun carries summertide aloft [d] through signs which rise hind-first.

[203]Fail not to perceive and from true rule deduce what signs are nocturnal, and what diurnal: they are not those that perform their function in darkness or daylight (the name would apply to all alike, since at regular intervals they shine at every hour, and now the nocturnal ones accompany the day, now the diurnal ones the night), but those on which nature, mighty parent of the universe, bestowed sacred portions of time in a permanent location. The signs of the Archer and the fierce Lion, he who looks round on the golden fleece of his back,[e] then the Fishes and Crab and the Scorpion of stinging lash, signs either adjacent or spaced at equal intervals, are all under like estate termed diurnal.[f] The others, identical in number and in the pattern of their spacing, for they are inserted into as many places, are called nocturnal. Some have also asserted that the diurnal stations belong to the six consecutive stars which begin with the Ram and that the six from the Balance count as

[d] Aloft to the tropic of Cancer.
[e] Aries.
[f] First classification: diurnal: ♈ ♋ ♌ ♏ ♐ ♓; nocturnal: ♉ ♊ ♍ ♎ ♑ ♒.

sunt quibus esse diurna placet quae mascula surgunt,
femineam sortem tutis gaudere tenebris.
Quin non nulla tibi nullo monstrante loquuntur
Neptuno debere genus, scopulosus in undis
Cancer et effuso gaudentes aequore Pisces.
at, quae terrena censentur sidera sorte,
princeps armenti Taurus regnoque superbus
lanigeri gregis est Aries pestisque duorum
praedatorque Leo et dumosis Scorpios arvis.
sunt etiam mediae legis communia signa,
ambiguus tergo Capricornus, Aquarius undis,
umida terrenis aequali foedere mixta.
Non licet a minimis animum deflectere curis,
nec quicquam rationis eget frustrave creatum.
fecundum est proprie Cancri genus, acer et ictu
Scorpios, et partu complentes aequora Pisces.
sed sterilis Virgo est, simili coniuncta Leoni,
nec capit, aut captos effundit, Aquarius ortus.
inter utrumque manet Capricornus corpore mixto
et qui Cretaeo fulget Centaurus in arcu,
communisque Aries aequantem tempora Libram
et Geminos Taurumque pari sub sorte recenset.
Nec tu nulla putes in eo commenta locasse

222 noctem
226 ut
228 positique
231 terrae
232: *see after* 4. 489
244 vocasse

---

[a] Second classification: diurnal: ♈ ♉ ♊ ♋ ♌ ♍; nocturnal: ♎ ♏ ♐ ♑ ♒ ♓.

[b] Third classification (*cf.* 2. 151-154): diurnal: ♈ ♉ ♌ ♎ ♐ ♒; nocturnal: ♉ ♋ ♍ ♏ ♑ ♓.

[c] The lion preys on herds and flocks, the scorpion is dangerous to them.

nocturnal.[a] There are those who fancy that the masculine signs are diurnal and that the feminine class rejoices in the safe cover of darkness.[b]

[223]Moreover some signs need no guide to tell you that they owe their origin to the Sea-god : the rock-clinging Crab in the shallows, and the Fishes that rejoice in expanse of water. Some stars are reckoned of earthly lot : they are the Bull, the chieftain of the herd ; the Ram proud of his lordship over the woolly flock ; the predator and bane respectively of both, the Lion and the Scorpion that haunts the thicket of the field.[c] Some signs there are betwixt and between, of a middle dispensation, since Capricorn's nature is compromised by its tail, the Waterman's by his stream : they are signs in which watery and earthly elements are mixed in an even compact.[d]

[234]You must not divert your attention from the smallest detail : nothing exists without reason or has been uselessly created. The Crab's kind is especially fertile, and fertile too are the sharp-stinging Scorpion and the Fishes that fill the ocean with their spawn. Barren is the Virgin, as also her neighbour the Lion ; and the Waterman fails to receive seed or, if he does so, spills it. Between the two groups fall Capricorn of compound form and the Centaur who glitters with his Cretan bow ; also intermediate is the Ram who counts of like estate the equinoctial Scales, the Twins, and the Bull.[e]

[244]Nor must you imagine that Nature has wrought

[d] Aquatic : ♋♓ ; terrene : ♈♉♌♏ (♊♍♎♐ belong here, but are not mentioned by Manilius) ; amphibious : ♑♒.

[e] Fertile : ♋♏♓ ; barren : ♌♍♒ ; intermediate : ♈♉♊♎♐♑.

naturam rerum, quod sunt currentia quaedam,
ut Leo et Arcitenens Ariesque in cornua tortus;
aut quae recta suis librantur stantia membris,
ut Virgo et Gemini, fundens et Aquarius undas;
vel quae fessa sedent pigras referentia mentes,
Taurus depositis collo sopitus aratris,
Libra sub emerito considens orbe laborum,
tuque tuos, Capricorne, gelu contractus in artus;
quaeve iacent, Cancer patulam distentus in alvum,
Scorpios incumbens plano sub pectore terrae,
in latus obliqui Pisces semperque iacentes.

Quod si sollerti circumspicis omnia cura,
fraudata invenies amissis sidera membris.
Scorpios in Libra consumit bracchia, Taurus
succidit incurvo claudus pede, lumina Cancro
desunt, Centauro superest et quaeritur unum.
sic nostros casus solatur mundus in astris
exemploque docet patienter damna subire,
omnis cum caelo fortunae pendeat ordo
ipsaque debilibus formentur sidera membris.

Temporibus quoque sunt propriis pollentia signa:
aestas a Geminis, autumnus Virgine surgit,
bruma Sagittifero, ver Piscibus incipit esse.
quattuor in partes scribuntur sidera terna;
hiberna aestivis, autumni verna repugnant.

[252] tuo . . . astris
[253] contra iacet
[259] cancri
[269] autumnis

---

[a] Running: ♈ ♌ ♐; standing: ♊ ♍ ♒; sitting: ♉ ♎ ♑; lying: ♋ ♏ ♓.

[b] Because it lacks bright stars (*cf.* 4. 530-534): there is an elegant pun here, since *lumina* can mean "bright stars" as well as "eyes."

a design of no purpose in that certain signs are running, as are the Lion, the Archer, and the Ram that ends in twisted horns ; or that some stand erect with their limbs perfectly poised, as the Virgin and Twins and Waterman pouring forth his stream ; or that some sit fatigued and reflect their weariness of mind, the Bull slumberous now the plough has left his shoulder, the Balance that sinks down after discharging its round of tasks, and you, Capricorn, whose limbs are shrivelled by the frost ; or that some lie flat, the Crab sprawling with distended belly, the Scorpion reposing on the ground beneath its smooth breast, the Fishes swimming sideways, for ever horizontal.[a]

[256]Now if you examine all the signs with keen attention, you will find some bereft of limbs which are lost. The Scorpion expends its arms on the Scales, the Bull sinks lame with leg doubled under it, the Crab lacks eyes,[b] whilst of the Centaur's one survives and one is missing.[c] Thus the sky assuages our misfortunes in its stars, and by example teaches us to undergo loss with fortitude, since the whole scheme of fortune depends upon heaven, and even constellations are fashioned with limbs deformed.[d]

[265]The signs also enjoy power in their special seasons : summer comes with the Twins, autumn with the Virgin ; winter begins with the Archer, spring with the Fishes. The four divisions of the year are each allotted three signs ; winter's are at war with summer's, the vernal with those of autumn.[e]

[c] Sagittarius is figured in profile.

[d] Deformed : ♉ ♋ ♏ ♐.

[e] Spring : ♓ ♈ ♉ ; summer : ♊ ♋ ♌ ; autumn : ♍ ♎ ♏ ; winter : ♐ ♑ ♒.

Nec satis est proprias signorum noscere formas
et privas quas dant leges nascentibus astra;
consensu quoque fata movent et foedere gaudent
atque aliis alia succedunt sorte locoque.
circulus ut dextro signorum clauditur orbe,
in tris aequalis discurrit linea ductus
inque vicem extremis iungit se finibus ipsa,
et, quaecumque ferit, dicuntur signa trigona,
in tria partitus quod ter cadit angulus astra
quae divisa manent ternis distantia signis.
Laniger ex paribus spatiis duo signa, Leonis
atque Sagittari, diverso conspicit ortu;
Virginis et Tauri Capricorno consonat astrum;
cetera sunt simili ratione triangula signa
per totidem sortes, desunt quae, condita mundo:
[sed discrimen erit dextris laevisque: sinistra
quae subeunt, quae praecedunt dextra esse feruntur;
dexter erit Tauro Capricornus, Virgo sinistra]
hoc satis exemplo est. at, quae divisa quaternis
partibus aequali laterum sunt condita ductu
quorum designat normalis virgula sedes,
haec quadrata ferunt. Libram Capricornus et illum
conspicit ante Aries atque ipsum a partibus aequis
Cancer et hunc laeva subeuntis sidera Librae.
semper enim in dextris censentur signa priora.
sic licet in totidem partes diducere cuncta

[126] primas
[272] aliae
[284] laevisque] seu eusa
[290] quadrata] ta
[293] [in]

[a] Rightward because it revolves to the right, *i.e.* clockwise in figures 3 ff. (pp. xlii ff.).
[b] *I.e.* from any point on the circumference.
[c] [284-286]: "but you will find a difference between right

[270]Nor is it enough to know the special shapes of the signs and the individual ordinances which the stars impose on men at their birth ; they also affect our destinies through their agreements with each other, for they rejoice in alliances and cooperate with one another according to their natures and locations. Where the circle of the zodiac's rightward wheel [a] is bounded,[b] a line runs out into three equal lengths and joins itself at limits which are mutually extreme ; the signs which the line strikes are called trigonal, because an angle is thrice formed and is assigned to three signs separated from each other by three intervening signs. The Ram beholds at equal distance the two signs of the Lion and the Archer rising on opposite sides of him ; the signs of the Virgin and the Bull are in harmony with Capricorn ; by the same principle the other trigonal signs which still remain are framed in the heavens in a like number of configurations [c] : this is example enough.[d] But the signs which are separated by quarters of the circle and are situated upon a figure with equal sides, the positions of which a square rule draws, these signs are called quadrate. Capricorn views Libra, whilst the Ram sees Capricorn ahead and is in turn beheld at an equal distance by the Crab ; and the Crab is perceived by Libra's leftward stars as it follows up : for preceding signs are reckoned as right signs. Even thus may one divide all the signs into groups of as many parts, and out of the twelve signs

signs and left : left signs are termed those which follow, right those which precede : to Taurus Capricorn is a right sign, Virgo a left."

[d] Trigons : (1) ♈♌♐ ; (2) ♉♍♑ ; (3) ♊♎♒ ; (4) ♋♏♓ : see Introduction p. xli and figure 3.

ternaque bis senis quadrata effingere signis,
quorum proposito reddentur in ordine vires.
Sed siquis contentus erit numerasse quadrata,
divisum ut signis mundum putet esse quaternis,
aut tribus ac binis signis ornare trigonum,
ut socias vires et amicos exigat ortus
foederaque inveniat mundi cognata per astra,
falsus erit. nam, quina licet sint undique signa,
qui tamen e trinis, quae quinto quoque feruntur
astra loco, fuerint nati, sentire trigoni
non poterunt vires : licet illud nomine servent,
amisere loco dotes numerisque repugnant.
nam, cum sint partes orbis per signa trecentae
et ter vicenae, quas Phoebi circuit ardor,
tertia pars eius numeri latus efficit unum
in tris perducti partes per signa trigoni.
hanc autem numeri non reddit linea summam,
si signum signo, non pars a parte notetur,
quod, quamvis duo sunt ternis dirimentibus astra,
si tamen extremam laevi primamque prioris
inter se conferre voles numerumque notare,
ter quinquagenas implebunt ordine partes ;
transibit numerus formam finesque sequentis
consumet ductus. licet ergo signa trigona

296 redduntur
298 [ut]
299 tria sub quinis
303 ex signis
312 a signo
313 [ternis] a

---

[a] Squares : (1) ♈♋♎♑ ; (2) ♉♌♏♒ ; (3) ♊♍♐♓ : see Introduction p. xli and figure 4.

[b] For example, take the trine signs Aries and Leo (which are separated by the three signs Taurus, Gemini, and Cancer) . . .

[c] . . . and join the 30th degree of Leo to the 1st degree of

create three squares, the powers of which shall be described in the order I have set before me.[a]

[297]But if anyone is content with reckoning squares in the belief that the sky is divided into arcs of four signs or with constructing a trigon with arcs of five signs, hoping thus to determine the signs' powers in alliance and the bonds of friendship between those born under them and discover the federations of heaven in kindred stars, that man will be deceived. For, even supposing that there were sets of five signs on every side, men who are born under such trine signs as are positioned at every fifth place will nevertheless be unable to feel the influences of the trigon; though the signs keep the name of trigon, they have forfeited its dowry by their position and are in revolt against the true numbers. For since in the zodiac there are three hundred and three score degrees traversed by the fiery Sun, a third of that number makes up one side of a trigon inscribed in the zodiac to divide it into three equal parts. However, the line does not give this numerical total, if sign is measured from sign and not degree from degree; for although you have two star-groups with three between,[b] yet if you care to join to each other the last point of the left sign and the first of the preceding and mark the total, they will duly measure thrice fifty degrees[c]; the number will be too great for the form of a trigon and will trespass upon the territory of the following line.[d] Thus, though such signs be

Aries, you will then have constructed an arc of 150 degrees, . . .

[d] . . . which will trespass (30 degrees) upon the territory of the following side of the trigon, which should have joined the 1st degree of Leo to the 1st of Sagittarius.

dicantur, partes non servant illa trigonas.
haec eadem species fallet per signa quadrata
(quod, cum totius numeri, qui construit orbem,
ter denae quadrum partes per sidera reddant,
evenit ut, prima signi de parte prioris
si partem ad summam ducatur virga sequentis,
bis sexagenas faciat ; sin summa prioris
et pars confertur subeuntis prima, duorum
signorum in medio numerum transisque referque,
triginta duplicat partes, pars tertia derit) ;
et, quamvis quartum a quarto quis computet astrum,
naufragium facient partes unius in ipsis.
non igitur satis est signis numerasse trigona
quadrative fidem quaeri per signa quaterna.
quadrati si forte voles effingere formam,
aut trinis paribus facies cum membra trigoni,
hic poscit quintam partem centesima summa,
illic amittit decimam. sic convenit ordo.
et, quiscumque quater iunctis favet angulus usque,
quaeque loca in triplici signarit linea ductu
cum sinuata viae linquet dispendia recta,
his natura dedit communi foedera lege
inque vicem affectus et mutua iura favoris.
quocirca non omnis habet genitura trigonis

322 denae] triginta
326 subiunctis
327 transitque refertque
332 caeli
337 cunctis
339 signata
343 f. = 318 f.

[a] For example, if from the 1st degree of Aries a line is drawn to the 30th of Cancer (these being quadrate signs), it makes 120 degrees ; . . .

[b] . . . if the 30th degree of Aries and the 1st of Cancer are connected, the intervening degrees (those of Taurus and

called trigonal, they do not preserve the correct degrees of a trigon. The same fallacy will cause deception in the case of quadrate signs: since, of the circle's total number, thrice ten degrees for every sign go to make up a square, it therefore follows that, if from the first point of a preceding sign a line is drawn to the last point of the quadrate sign behind, it makes twice sixty degrees [a]; but if the last point of the former and the first of that following are connected, should you go over and count the number shared by the two intervening signs, it will double thirty degrees, and a third part of the requisite total will be missing.[b] And though a man compute a fourth sign from a fourth, the degrees in themselves will cause the wreck of a whole sign. It is therefore not enough to count trigons by signs or to expect a true square from signs at intervals of four. If perchance you wish to construct the figure of a square, or when you are about to draw the sides of an equilateral triangle, the sum of a hundred degrees needs in the latter case increase of a fifth, in the former suffers loss of a tenth.[c] Thus does the count tally. And whatever points joined in a series of four the angle favours, and whatever point the straight line marks in its threefold track when it leaves the winding detour of the circumference, upon these has nature bestowed federation and common law, mutual goodwill and rights of friendship with each other. Therefore not every geniture enjoys the common

Gemini) will amount to 60; and of the required 90 degrees to be subtended by the side of a square, a third (30 degrees) will be missing.

[c] The side of a trigon subtends 100 plus $\frac{1}{5}$ of 100 = 120 degrees, of a square 100 minus $\frac{1}{10}$ of 100 = 90 degrees.

consensum signis, nec, cum sunt forte quadrata,
continuo inter se servant commercia rerum.
distat enim, partes consumat linea iustas
detrectetne modum numeri, quem circulus ambit,
nunc tris efficiens, nunc quattuor undique ductus,
quos in plura iubet ratio procedere signa
interdum, quam sunt numeris memorata per orbem.

Sed longe maior vis est per signa trigoni
quam quibus est titulus sub quarto quoque quadratis.
altior est horum summoto linea templo,
illa magis vicina meat caeloque recedit
et propius terras accedit visus eorum
aeraque infectum nostras demittit ad auras.

Debilia alternis data sunt commercia signis,
mutua nec magno consensu foedera servant,
invita angusto quod linea flectitur orbe.
nam, cum praeteriens formatur singula limes
sidera et alterno devertitur angulus astro
sexque per anfractus curvatur virgula in orbem,
a Tauro venit in Cancrum, tum Virgine tacta
Scorpion ingreditur, tum te, Capricorne, rigentem
et geminos a te Pisces aversaque Tauri
sidera contingens finit, qua coeperat, orbem.
alterius ductus locus est per transita signa,
utque ea praetereas quae sunt mihi singula dicta,
flexibus et totidem similis sit circulus illi.

346 continua
347 enim an
348 -etque
349 efficiet
358 deviaque
361 praedales formantur
365 tum] quo
366 adversaque tauro
370 illis

[a] It is implied that the quadrate relationship exists only in quadrate degrees, *e.g.* 1°♈, 1°♋, 1°♎, 1°♑, and so on.

[b] Hexagons: (1) ♈ ♊ ♌ ♎ ♐ ♒; (2) ♉ ♋ ♍ ♏ ♑ ♓: see Introduction p. xliii and figure 5.

feeling of its trigonal signs, nor do signs which happen to be quadrate necessarily preserve community of interests with each other.[a] For it makes a difference whether the line uses up the exact toll of its degrees or rejects the limit to the circumference's total by constructing on each side, in the one case three lines, in the other four, which the law of numbers sometimes forces to proceed into more signs than are specified by the number of degrees in the zodiac.

[352]But in the zodiac the trigon's power is far greater than that of those signs to which, reckoned as fourth from each other, is ascribed the title of quadrate. The side of the quadrate signs is higher, and their perimeter farther removed from us, whilst the side of trigons travels closer and draws away from heaven : their vision draws nearer to Earth and sends down to our atmosphere an air tempered by their influences.

[358]Weak is the connection between alternate signs, nor do they maintain with unfailing constancy the federation between each other, because the line is reluctant to bend in its circuit of short chords. A track is formed which passes by one constellation at a time, whilst the angle lodges at the alternate signs, the line making six changes of direction in its curve round the circle : it therefore leaves the Bull for the Crab, then after touching the Virgin enters Scorpion, thence reaches you, chill Capricorn, from you the two Fishes and the stars of the Bull averse, and finishes the circle where it began. The path of a second line lies through the signs that the first missed and is drawn in such a way that you pass by each of the signs I mentioned, so that its circle resembles the former in having as many chords.[b]

tertia convexo conduntur signa recessu ;
transversos igitur fugiunt subeuntia visus,
quod nimis inclinata iacent limisque videntur
vicinoque latent : ex recto certior ictus.
et, quia succedit convexo linea caelo,
singula circuitu quae tantum transeat astra,
visus eis procul est altoque vagatur Olympo
et tenuis vires ex longo mittit in orbem.
sed tamen est illis foedus sub lege propinqua,
quod non diversum genus est coeuntibus astris,
mascula sed maribus respondent, cetera sexus
feminei secum iungunt commercia mundi.
sic, quamquam alternis, par est natura figuris,
et cognata iacent generis sub legibus astra.

Iam vero nulla est haerentibus addita signis
gratia ; nam consensus hebet, quia visus ademptus.
in seducta ferunt animos, quae cernere possunt.
sunt etiam adversi generis conexa per orbem
mascula femineis semperque obsessa vicissim.
[disparibus non ulla datur concordia signis]

Sexta quoque in nullas numerantur commoda vires,
virgula per totum quod par non ducitur orbem

[374] converso
[372] iacent] cne
[377] vis eius
[380] quod euntibus
[382] se coniungunt
[383] quicquam . . . paret
[386] habet

Signs separated by one intervening sign are concealed in a curved recess; and so, though they follow the one in front, they escape its slanting gaze, since, lying at an angle too obtuse, they are only to be seen with sidelong glance and are hidden from their neighbour: sight strikes more surely straight ahead. And because their line comes close to the celestial vault, since in its circuit it only passes over one sign at a time, their sight of each other takes place at a great distance from us, ranging over the heights of heaven, and the influences it radiates from afar are faint when they reach the earth. However federation exists among them by virtue of their affinity, for the signs that are connected are not of unlike sex, but masculine respond to male, whilst the remainder are of female sex and form among themselves a heavenly fellowship. Thus is it that, though the signs of hexagons are alternate, their nature is similar, and as kindred stars they acknowledge the ties of sex.

[385]But no friendliness towards one another has been granted to adjacent signs: sympathy between them is blunted because the sight of each other is denied them. Their attentions are bestowed on distant signs which they can see. Again they are signs of opposite sex, linked male to female right round the circle, and each in turn is for ever beset by its neighbours.[a]

[391]Sixth signs,[b] as well, are not reckoned as capable of any influence, because their line is not traced through the whole circle in equal lengths, but strikes

[a] [390]: "To signs unlike no concord of any kind is given."

[b] *I.e.* signs separated by four signs.

sed duo signa ferit mediis summota quaternis,
tertius absumpto ductus non sufficit orbe.
At, quae diversis e partibus astra refulgent
per medium adverso mundum pendentia vultu
et toto divisa manent contraria caelo
septima quaeque, loco quamvis summota feruntur,
ex longo tamen illa valent viresque ministrant
vel bello vel pace suas, ut tempora poscunt,
nunc foedus stellis, nunc et dictantibus iras.
quod si forte libet, quae sunt contraria, signa
per titulos celebrare suos sedesque, memento
solstitium brumae, Capricornum opponere Cancro,
Lanigerum Librae (par nox in utroque diesque est),
Piscibus Erigonen, iuvenique urnaeque Leonem ;
Scorpios e summo cum fulget, Taurus in imo est,
et cadit Arcitenens Geminis orientibus orbi.
[hos servant inter sese contraria cursus]
sed, quamquam adversis fulgent contraria signis,
natura tamen interdum sociata feruntur,
et genere amplexis concordia mutua surgit :
mascula se paribus vel sic, diversa suorum
respondent generi. Pisces et Virginis artus
adversi volitant, sed amant communia iura,
et vincit natura locum ; sed vincitur ipsa
temporibus, Cancerque tibi, Capricorne, repugnat
femina femineo, quia brumae dissidet aestas.

393 ferunt
395 atque ea
401 phoebo
408 orbe
409 observant
412 genera exemplis
413 si . . . si
414 astra

[a] For example, such a line drawn from Aries will strike first Virgo and then Aquarius, but it cannot then rejoin Aries, since between Aquarius and Aries lies only one sign (Pisces).

two signs with four intervening, whilst a third side does not fit in since the circle is exhausted.[a]

[395]But the signs which shine from opposite quarters, poised with faces confronting each other across mid-sky, and remain at two extremes with all heaven between are seventh signs: though remote from each other by position, they nevertheless exercise influence from afar and furnish their powers in war and peace, at the bidding of the times, for now the planets proclaim alliance and now strife. Wherefore if perchance you wish to recount the names and abodes of the opposite signs, be sure to match mid-summer against winter, Capricorn against Cancer, the Ram against the Balance (night and day in each are equal), Erigone against the Fishes, the Lion against the youth with his urn; when Scorpion shines overhead, the Bull is in the depths, and the Archer sets as the Twins rise above the earth.[b] But though sign facing sign they shine opposed, yet because of their nature they are oft borne in alliance and a mutual sympathy springs up between them, linked as they are by the tie of sex: even in these circumstances[c] male calls to male at one with him, and the other signs to their own sex. The Fishes and the the limbs of the Virgin fly in opposition but cherish the bonds they share, and the tie of sex prevails over location; but over this tie in turn the seasons prevail: Cancer resists Capricorn, though females both, since summer conflicts with winter. The one

[b] Opposite signs: (1) ♈♎; (2) ♉♏; (3) ♊♐; (4) ♋♑; (5) ♌♒; (6) ♍♓: see Introduction p. xliv and figure 6. Here follows the interpolated [409]: "These are the courses kept by signs opposed to each other."

[c] *I.e.* in spite of being opposite signs.

hinc rigor et glacies nivibusque albentia rura,
hinc sitis et sudor nudusque in collibus orbis,
aestivosque dies aequat nox frigida brumae.
sic bellum natura gerit, discordat et annus,
ne mirere in ea pugnantia sidera parte.
at non Lanigeri signum Libraeque repugnant
in totum, quia ver autumno tempore differt
(fructibus hoc implet maturis, floribus illud)
sed ratione pari est, aequatis nocte diebus,
temporaque efficiunt simili concordia textu
permixtosque dies mediis hiemem inter et aestum
articulis uno servantia utrimque tenore
quo minus infesto decertent sidera bello.
talis erit ratio diversis addita signis.

His animadversis rebus quae proxima cura?
noscere tutelas adiectaque numina signis
et quae cuique deo rerum natura dicavit,
cum divina dedit magnis virtutibus ora,
condidit et varias sacro sub nomine vires,
pondus uti rebus persona imponere posset.
Lanigerum Pallas, Taurum Cytherea tuetur,
formosos Phoebus Geminos; Cyllenie, Cancrum,
Iuppiter, et cum matre deum regis ipse Leonem;
spicifera est Virgo Cereris fabricataque Libra
Vulcani; pugnax Mavorti Scorpios haeret;
venantem Diana virum, sed partis equinae,
atque angusta fovet Capricorni sidera Vesta;

424 repugnat 438 possit
430 tempore utrumque

[a] Cybele, the great mother-goddess, who is regularly attended by lions; she is mentioned only for rhetorical ornament, and the poet does not mean that she shares Jupiter's guardianship of Leo.

season brings frost and ice and a countryside white with snow, the other thirst and sweat and an earth with hillsides parched; and the chill wintry night rivals in length the summer's day. Thus nature wages warfare and the year is split in faction; so wonder not at the signs so situate doing battle. But the Ram's sign and Libra's do not engage in total combat: for though in point of season spring differs from autumn (the one fills the earth with fruit, the other with flowers), yet it enjoys a like principle, seeing that day is levelled with night in each. And the seasons, which are harmonious in their likeness of texture and, as links joining winter and summer, maintain with the same tenor on either side days of identical blend, bring it about that the constellations refrain from clashing in violent warfare. Such is the scheme to be found in confronting signs.

433What step must one take next, when so much has been learnt? It is to mark well the tutelary deities appointed to the signs and the signs which Nature assigned to each god, when she gave to the great virtues the persons of the gods and under sacred names established various powers, in order that a living presence might lend majesty to abstract qualities. Pallas is protectress of the Ram, the Cytherean of the Bull, and Phoebus of the comely Twins; you, Mercury, rule the Crab and you, Jupiter, as well as the Mother of the Gods,[a] the Lion; the Virgin with her sheaf belongs to Ceres, and the Balance to Vulcan who wrought it; bellicose Scorpion clings to Mars; Diana cherishes the hunter, a man to be sure, but a horse in his other half, and Vesta the cramped[b] stars of Capricorn; opposite

[b] *Cf.* 1. 271, note.

e Iovis adverso Iunonis Aquarius astrum est
agnoscitque suos Neptunus in aethere Pisces.
hinc quoque magna tibi venient momenta futuri,
cum ratio tua per stellas et sidera curret
argumenta petens omni de parte viasque
artis, ut ingenio divina potentia surgat
exaequentque fidem caelo mortalia corda.

Accipe divisas hominis per sidera partes
singulaque imperiis propriis parentia membra,
in quis praecipuas toto de corpore vires
exercent. Aries caput est ante omnia princeps
sortitus censusque sui pulcherrima colla
Taurus, et in Geminis aequali bracchia sorte
scribuntur conexa umeris, pectusque locatum
sub Cancro est, laterum regnum scapulaeque Leonis,
Virginis in propriam descendunt ilia sortem,
Libra regit clunes, et Scorpios inguine gaudet,
Centauro femina accedunt, Capricornus utrisque
imperitat genibus, crurum fundentis Aquari
arbitrium est, Piscesque pedum sibi iura reposcunt.

Quin etiam propriis inter se legibus astra
conveniunt, ut certa gerant commercia rerum,
inque vicem praestant visus atque auribus haerent
aut odium foedusve gerunt, conversaque quaedam
in semet proprio ducuntur plena favore.
idcirco adversis non numquam est gratia signis,

447 aequore
454 [periis] membra figuris
464 imperat et
471 [est]

[a] *I.e.* opposite Leo.
[b] *I.e.* you, when fully versed in astrology.
[c] *Cf.* 4. 704-709.
[d] *I.e.* one arm to each twin.

Jupiter [a] Juno has the sign of Aquarius, and Neptune acknowledges the Fishes as his own for all that they are in heaven. This scheme too will provide you with important means of determining the future when, seeking from every quarter proofs and methods of our art, your mind speeds among the planets and stars so that a divine power may arise in your spirit and mortal hearts [b] no less than heaven may win belief.

[453]Now learn how the parts of the human frame are distributed among the constellations, and how the limbs are subject each to a particular authority: over these limbs, out of all the parts of the body, the signs exercise special influence. The Ram [c] as chieftain of them all is allotted the head, and the Bull receives as of his estate the handsome neck; evenly bestowed,[d] the arms to shoulders joined are accounted to the Twins; the breast is put down to the Crab, the realm of the sides and the shoulder-blades are the Lion's, the belly comes down to the Maid as her rightful lot; the Balance governs the loins, and Scorpion takes pleasure in the groin; the thighs hie to the Centaur, Capricorn is tyrant of both knees, whilst the pouring Waterman has the lordship of the shanks, and over the feet the Fishes claim jurisdiction.

[466]Moreover, the stars have agreements among themselves according to special laws, and so enjoy fixed associations: upon each other they direct their gaze and to each other give ear; else they bear hatred or friendship; some are introverted and by the fullness of their self-esteem are drawn into themselves. And so sometimes goodwill exists between stars that are opposed and war is waged between

et bellum sociata gerunt ; alienaque sede
inter se generant coniunctos omne per aevum,
a triquetrisque orti pugnant fugiuntque vicissim ;
quod deus, in leges mundum cum conderet omnem,
affectus quoque divisit variantibus astris,
atque aliorum oculos, aliorum contulit aures,
iunxit amicitias horum sub foedere certo,
cernere ut inter se possent audireque quaedam,
diligerent alia et noxas bellumque moverent,
his etiam propriae foret indulgentia sortis,
ut se diligerent semper sibique ipsa placerent ;
sicut naturas hominum plerasque videmus
qui genus ex signis ducunt formantibus ortus.

Consilium ipse suum est Aries, ut principe dignum es
audit se Libramque videt, frustratur amando
Taurum ; Lanigero qui fraudem nectit et ultra
fulgentis geminos audit per sidera Pisces,
Virgine mens capitur visa. sic vexerat ante
Europam dorso retinentem cornua laeva
indutusque Iovi. Geminorum ducitur auris
ad iuvenem aeternas fundentem Piscibus undas
inque ipsos animus Pisces oculique Leonem.
Cancer et adverso Capricornus conditus astro
in semet vertunt oculos, in mutua tendunt
auribus, et Cancri captatur Aquarius astu.
at Leo cum Geminis aciem coniungit et aurem
Centauro ferus et Capricorni diligit astrum.
Erigone Taurum spectat sed Scorpion audit

474 utrique trisorti
488 geminos] videt atque
489 [visa] sic quondam
495 oculis
496 astro
498 centaurus geminos et

[a] *Cf.* 2. 616.

signs in alliance ; signs which share no ties of place beget men attached in life-long friendship, and men born of triangles fight and shun each other in turn. For, when God brought the whole universe under law, he also set the signs at variance by distributing the affections among them and connected the vision of some, the hearing of others. Thus he composed alliances of these signs in lasting federation, that some might see and hear each other, others might love or cause injury and war, and that others might cherish an addiction to their own natures, ever in love with themselves and finding favour in their own eyes : we see most men with dispositions such as these ; they owe their natures to the stars that gave them birth.

485 As befits a leader, the Ram is his own counsel ; he listens to himself and beholds the Balance, and his love for the Bull is thwarted. Against him the Bull weaves a net of deceit and hears the twin Fishes which shine beyond the Ram, whilst his heart is entranced at the sight of the Maid. Thus thrilled once long ago and serving as Jove's disguise, he had borne Europa on his back as with her left hand she clutched his horn. The hearing of the Twins is drawn to the youth who pours for the Fishes a never-ending stream, their hearts to the Fishes themselves, and their eyes to the Lion. The Crab and the Sea-goat's sign placed opposite turn their eyes upon themselves, at one another strain with their ears, whilst the Waterman is taken in by the wiles of the Crab. But the Lion joins his gaze to that of the Twins and his hearing—for he too is a fierce star [a]—to that of the Centaur and loves the sign of Capricorn. Erigone looks at the Bull but hearkens to the

atque Sagittifero conatur nectere fraudem.
Libra suos sequitur sensus solumque videndo
Lanigerum atque animo complexa est Scorpion infra.
ille videt Pisces, audit quae proxima Librae.
nec non Arcitenens magno parere Leoni
auribus atque oculis sinum fundentis Aquari
conspicere assuevit solamque ex omnibus astris
diligit Erigonen. contra Capricornus in ipsum
convertit visus (quid enim mirabitur ille
maius, in Augusti felix cum fulserit ortum ?)
auribus et summi captat fastigia Cancri.
at nudus Geminis intendit Aquarius aurem
sublimemque colit Cancrum spectatque reducta
tela Sagittiferi. Pisces ad Scorpion acrem
derexere aciem cupiuntque attendere Taurum.
has natura vices tribuit, cum sidera fixit.
his orti similis referunt per mutua sensus,
audire ut cupiant alios aliosque videre,
[horum odio nunc horum idem ducuntur amore]
illis insidias tendant, captentur ab illis.

Quin adversa meant alterna trigona trigonis,
alteraque in bellum diverso limite ducit
linea. sic veri per totum consonat ordo.
namque Aries, Leo et Arcitenens, sociata trigona
signa, negant Chelis foedus totique trigono

503 -que per omnia libram 520 alterna] etiam
516 census

[a] Virgo.

[b] [518] : ". . . the same people are one moment moved by hatred of these, the next by love of those, . . ."

[c] For a summary of these relationships see Introduction pp. xlvi ff. and figures 7, 8, and 9.

Scorpion and endeavours to plot deceit against the Archer. The Balance heeds his own counsel and has embraced with his gaze none but the Ram and with his heart the Scorpion beneath. The latter beholds the Fishes and listens to her who is Libra's neighbour.[a] The Archer moreover has grown used to waiting on the mighty Lion with his ears and beholding with his eyes the pot that the Waterman empties; he adores Erigone alone of all the stars. Capricorn on the other hand turns his gaze upon himself (what greater sign can he ever marvel at, since it was he that shone propitiously upon Augustus' birth?) and catches with his ears the height of topmost Cancer. And now to the Twins does the Waterman incline his ear, he naked too; he worships the Crab on high and contemplates the drawn bow of the Archer. The Fishes have fixed their glance upon the fierce Scorpion and crave to listen to the Bull. These are the reciprocal connections which Nature gave to the signs when she put them in their positions. Men born of these signs display the corresponding feelings towards each other: some they love to hear and others they love to behold,[b] they set traps for these and are ensnared by those.[c]

520 Now there are trigons[d] alternate to trigons which move in hostility to each other, and one[e] of the two diameters leads them to war along opposed paths. Thus the design of truth is consistent in every part. For indeed, the Ram, the Lion, and the Archer, allied trigonal signs, reject federation with

[d] The poet refers to the 1st and 3rd trigons (which are alternate to the 2nd and 4th): this is shown by 2. 523-535.
[e] That joining Aries and Libra.

quod Gemini excipiunt fundens et Aquarius undas.
idque duplex ratio cogit verum esse fateri,
quod tria signa tribus signis contraria fulgent,
quodque aeterna manent hominum bella atque ferarum.
[humana est facies Librae, diversa Leonis]
idcirco et cedunt pecudes, quod viribus amplis
consilium est maius. victus Leo fulget in astris,
aurea Lanigero concessit sidera pellis,
ipse suae parti Centaurus tergore cedit,
usque adeo est hominis virtus. quid mirer ab illis
nascentis Librae superari posse trigono?

Nec sola est ratio quae dat nascentibus arma
inque odium generat partus et mutua bella;
sed plerumque manent inimica tertia quaeque
lege, in transversum vultu defixa maligno,
quodque, manent quaecumque loco contraria signa
adversosque gerunt inter se septima visus,
tertia quaeque illis utriusque trigona feruntur;
ne sit mirandum si foedus non datur astris
quae sunt adversis signis cognata trigona.
quin etiam brevior ratio est per signa sequenda;
nam, quaecumque nitent humana condita forma
astra, manent illis inimica et victa ferarum.

534 victus quod
571 velle
574 quoque
577 phoebus

---

[a] The logic of 2. 528 requires *Chelae* ("Claws") to be here interpreted as a human Balance-holder, and so Libra is sometimes figured (*e.g.* Boll, *Sphaera* figure V).

[b] [529] interpolated to remind the reader that the Balance (see previous note) is male: "The form of the Balance is human, of other kind the Lion's."

[c] The twofold reason mentioned in 2. 526.

the Balance-holder and all his [a] trigon, a trigon completed by the Twins and the Waterman pouring his stream. And a twofold reason compels us to admit the truth of this : because the signs shine in diametric opposition, three against three, and because of the eternal state of war between man and beast.[b] And the beasts give way because intelligence prevails over brute strength. As one vanquished shines the Lion among the stars ; it was the theft of his golden fleece that gave the Ram his place in heaven ; to part of himself the Centaur gives way on account of his rear, to such an extent is manliness restricted to man. Why should I wonder then that men born under these signs are no match for Libra's trigon ?

570Nor is it [c] the only reason which imparts enmities to men at birth and begets a race for mutual hatred and hostility ; but alternate signs mostly abide in a state of antagonism, being rooted with evil sidelong stares and because signs [d] alternate to whatever signs are diametrically opposed and, separated by five, exchange confronting looks belong to the trigons of each ; so one must not wonder if signs deny peace to those stars which possess trigonal relationship with their confronting signs.[e] A simpler reason is also to be traced in the zodiac ; for to all the shining signs endowed with human form those of the beasts are ever hostile and ever by them vanquished. Nevertheless, there are individual signs

[d] . . . because (for example) ♉ ♓ and ♍ ♏, alternate to ♈ and ♎ respectively, belong to the trigons of ♎ and ♈ respectively.

[e] For example, Aries denies peace to Gemini and Aquarius, signs of the trigon of Libra, the confronting sign of Aries.

sed tamen in proprias secedunt singula mentes
et privata gerunt secretis hostibus arma. 54
Lanigero genitis bellum est cum Virgine natis
et Libra Geminisque et eis quos protulit Urna.
in partus Tauri sub Cancro nata feruntur
pectora et in Chelis et quae dant Scorpios acer
et Pisces. at, quos Geminorum sidera formant, 54
his cum Lanigero bellum est eiusque trigono.
in Cancro genitos Capricorni semina laedunt
et Librae partus et quos dat Virginis astrum
quique sub aversi numerantur sidere Tauri.
Lanigeri communis erit rabidique Leonis 55
hostis, et a totidem bellum subscribitur astris.
Erigone Cancrumque timet geminique sub arcu
Centauri et Pisces et te, Capricorne, rigentem.
maxima turba petit Libram, Capricornus et illi 5
adversus Cancer, Iuvenis quod utrimque quadratum est,
quaeque in Lanigeri numerantur signa trigonum.
Scorpios in totidem fecundus creditur hostes;
aequoreum iuvenem, Geminos, Taurum atque Leonem,
Erigonen Libramque fugit metuendus et ipse,
quique Sagittari veniunt de sidere partus. 5
hos Geminis nati Libraque et Virgine et Urna
depressisse volent. naturae lege iubente
haec eadem, Capricorne, tuis inimica feruntur.
at quos aeternis perfundit Aquarius undis,
in pugnam Nemeaeus agit totumque trigonum, 5
turba sub unius fugiens virtute ferarum.

542 gemini pisces pertulit unda
544 dat
551 et a] sed
552 taurumque geminumque
553 centaurum
555 iuvenis] chelis utrumque
566 fugiens] iuvenis

[a] Scorpio.

[b] Taurus and Scorpio.

which follow their own caprice and, having private foes, wage wars of their own.

[541]The Ram's children are at war with the offspring of the Virgin, the Balance, and the Twins, and with those whom the Urn has brought forth. Against the progeny of the Bull there advance men born under the Crab and under the Scales, and those produced by the fierce Scorpion and by the Fishes. But those created by the Twins' stars have war with the Ram and his trigon. The seed of Capricorn vexes those begotten under the Crab, and so do the children of the Balance, those produced by the Virgin's sign, and those that are mustered under the constellation of the Bull averse. The Ram shares his foes with the savage Lion ; and the same signs levy war on both. Erigone fears the Crab, him [a] that lurks under the bow of the two-formed Centaur, the Fishes, and you, chill Capricorn. The Balance is assailed by the largest host—Capricorn, and Cancer opposite him, the quadrate signs [b] on either side of the Youth, and those signs reckoned as of the Ram's trigon. The Scorpion is presumed prolific of as many foes ; it flees before the lad with his waters, the Twins, the Bull, the Lion, and (to whom it is itself an object of fear) the Maid and the Balance, and also the offspring that issues from the Archer's sign. And this offspring the children of the Twins, the Balance, Virgin, and the Urn will crave to vanquish. Bidden by the rule of their own nature, the self-same signs are borne in hostility against your sons, Capricorn. But those poured forth by the Waterman with his ever-flowing stream are provoked to battle by the Nemean and all his trigon, a host of beasts that flees before the manhood of one. The progeny of the

Piscibus exortos vicinus Aquarius urget
et gemini fratres et quos dat Virginis astrum
quique Sagittari descendunt sidere nati.
 Per tot signorum species contraria surgunt
corpora totque modis totiens inimica creantur.
idcirco nihil ex semet natura creavit
foedere amicitiae maius nec rarius umquam ;
perque tot aetates hominum, tot tempora et annos,
tot bella et varios etiam sub pace labores,
cum Fortuna fidem quaerat, vix invenit usquam.
unus erat Pylades, unus qui mallet Orestes
ipse mori ; lis una fuit per saecula mortis,
alter quod raperet fatum, non cederet alter.
[et duo, qui potuere sequi : vix noxia poenis,
optavitque reum sponsor non posse reverti,
sponsoremque reus timuit, ne solveret ipsum]
at quanta est scelerum moles per saecula cuncta,
quamque onus invidiae non excusabile terris !
venales ad fata patres matrumque sepulcra
non posuere modum sceleri, sed fraude nefanda
ipse deus Caesar cecidit, qua territus orbi
imposuit Phoebus noctem terrasque reliquit.
quid loquar eversas urbes et prodita templa
et varias pacis clades et mixta venena
insidiasque fori, caedes in moenibus ipsis

570-578 : *see after* 535 [582] pectore
[580] quotiens

[a] For a summary of these enmities see Introduction p. l.
[b] In the *Dulorestes* of Pacuvius King Thoas does not know which of his two captives is Orestes (who must die) : Pylades nobly insists that it is he, but is hotly contradicted by Orestes (*cf.* Cicero, *Lael.* 24). At this point the MSS. give [586-588], interpolated to bring in mention of Damon and Phintias

Fishes is attacked by the Waterman next door, by the twin brothers, by those brought forth of the Virgin's sign, and by those derived from the Archer's constellation.[a]

579From so many configurations of signs come beings opposed to each other, and thus variously and thus often is enmity created. For this reason nature has never created from herself anything more precious or less common than the bond of true friendship. And throughout the long history of mankind, ages and centuries so many, amid so many wars and the motley strife even of peace, when misfortune calls for loyal support, it scarce finds it anywhere. There was but one Pylades, but one Orestes, eager to die for his friend. Once only throughout the ages have men disputed for the prize of death, in that one snatched at a doom the other refused to yield.[b] Yet how great is the sum of villainy in every age! How impossible to relieve the earth of its burden of hate! Fathers sold to death and mothers murdered [c] *have not marked the limit of wickedness: even the god Caesar fell victim to deceit unspeakable, whereat in horror on the world* Phoebus brought darkness and forsook the earth.[d] Why tell of the sack of cities, the betrayal of temples, the many disasters suffered in time of peace, the mixing of poisons, treachery in the market-place, slaughter within the

(*cf.* Cicero, *De Off.* 3. 45): "And two there were that availed to follow their example: scarce could punishment find guilt to punish; guarantor prayed for the default of the one arraigned, whose fear was that his guarantor might gain him release."

[c] *I.e.* sold and murdered by their children. The reference is probably to the proscriptions (*cf.* Plutarch, *Sulla* 31. 6).

[d] *Cf.* Virgil, *Georg.* 1. 467 f.

et sub amicitiae grassantem nomine turbam ?
in populo scelus est et abundant cuncta furoris.
et fas atque nefas mixtum, legesque per ipsas
saevit nequities ; poenas iam noxia vincit.
scilicet, in multis quoniam discordia signis
corpora nascuntur, pax est sublata per orbem,
et fidei rarum foedus paucisque tributum,
utque sibi caelum sic tellus dissidet ipsa
atque hominum gentes inimica sorte feruntur.
Si tamen et cognata cupis dinoscere signa,
quae iungant animos et amica sorte ferantur,
Lanigeri partus cum toto iunge trigono.
simplicior tamen est Aries, meliusque Leone
prosequitur genitos et te, Centaure, creatos
quam colitur. namque est natura mitius astrum
expositumque suae noxae, nec fraudibus ullis,
nec minus ingenio molli quam corpore constans :
illis est feritas signis praedaeque cupido,
venalisque animus non numquam excedere cogit
commoditate fidem, nec longa est gratia facti ;
plus tamen in duplici numerandum est roboris esse,
cui commixtus homo est, quam te, Nemeaee, sub uno.
at, cum Lanigeri partus sub utroque *laborant*
vique urgente dolent amborum astuque, trigono
non parcit ; sed rara gerit pro tempore bella,
quae feritas utriusque magis prorumpere cogit.
idcirco et pax est signis et mixta querella.

[615] constant [621] quod . . . pro tempore

[a] These lines seem to allude to violence at Rome in the last days of the Republic.

very walls of Rome, and a conspiracy that skulks beneath friendship's name? [a] Crime is rife among the people, and all is filled with madness; the distinction between right and wrong is swept aside, and iniquity runs amok with the very laws as its weapons; guilt is now too great for punishment. Truly, since many are the signs in which men are born for discord, peace is banished throughout the world, and the bond of loyalty is rare and granted to few; and just as is heaven, so too is earth at war with itself, and the nations of mankind are subject to a destiny of strife.

608But should you also wish to learn which related signs exchange affection and are subject to a destiny of friendship, connect the offspring of the Ram with all his trigon. Yet the Ram is a simple creature and shows more respect for the children of the Lion and the Centaur's progeny than they for him. His is by nature a gentle sign, exposed to the harm that falls on gentleness; he is devoid of deceit, and his heart is as soft as his fleecy body. His fellow signs are marked by ferocity and a lust for spoil, and their covetous spirit oft impels them to break faith for their own ends; and their gratitude for a kindness is short-lived. However, in the double sign that has mixture of man we must reckon greater strength than is possessed by yours of single form, Nemean. Nevertheless, when the Ram's offspring by each *are harassed and smart under the aggressive violence and knavery of the pair*, then he loses patience with the trigon; even so, few are the wars he wages and only when occasion demands, and for the outbreak of these the fierceness of the others is rather to blame. And so, whilst the trigon enjoys peace, it is a peace

quin etiam Tauri Capricorno iungitur astrum,
nec magis illorum coeunt ad foedera mentes;
Virgineos etiam partus, quicumque creantur
Tauro, complecti cupiunt, sed saepe queruntur.
quos Geminique dabunt Chelaeque et Aquarius ortus
unum pectus habent fideique immobile vinclum,
magnus et in multos veniet successus amicos.
Scorpios et Cancer fraterna in nomina ducunt
ex semet genitos, nec non et Piscibus orti
concordant illis. saepe est et subdolus actus:
Scorpios aspergit noxas sub nomine amici;
at, quibus in lucem Pisces venientibus adsunt,
his non una manet semper sententia cordi,
commutant animos interdum et foedera rumpunt
ac repetunt, tectaeque lues sub fronte vagantur.
sic erit ex signis odium tibi paxque notanda.

Nec satis hoc, tantum solis insistere signis:
contemplare locum caeli sedemque vagarum.
parte genus variant et vires linea mutat.
nam sua quadratis veniunt, sua iura trigonis
et quae per senos decurrit virgula tractus
quaeque secat medium transverso limite caelum;
distat enim surgatne eadem subeatne cadatne.
hinc modo dat mundus vires, modo deterit idem,
quaeque illic sumunt iras, huc acta reponunt.
crebrius adversis odium est, cognata quadratis

[629] quosque dabunt chelae et quos dat aquarius   631: *not in the MSS.*   642: *see after* 706

---

[a] Here Bonincontrius, realizing the need for mention of Gemini, inserted a verse (631) of his own (*magnus erit Geminis amor et concordia duplex*: "The Twins will have a great love and twofold concord"), an attempt at emendation superseded by Housman's restoration of *Gemini* in 629.

wherewith is mingled wrangling. Next, the sign of the Bull is joined to Capricorn, but their temperaments do not blend in fellowship any better; those begotten of the Bull are eager to embrace the Virgin's children too, but often they quarrel. The issue of the Twins, the Scales, and the Waterman are of one heart and share a bond of loyalty which naught can sunder [a]; and they will achieve conspicuous success in winning many friends. The Scorpion and the Crab endow their sons with the name of brother, and the progeny of the Fishes is also united with them. But there are frequently sly dealings as well. The Scorpion sows trouble in the guise of friend. And those whom the Fishes father at birth never keep in their hearts for long a constant affection; ever and anon they change their sympathies and now forswear their ties and now renew them, and beneath a mask of blandness unseen hatreds come and go. As thus foretold by the signs must you mark war and peace.

[643]Nor is it enough merely to consider the signs in isolation: you must observe their place in heaven and the position of the planets. The signs vary their nature according to their quarter of the sky, and a line which joins them likewise suffers change of its strength. For special properties accrue to squares, to trigons, to the line which traverses six chords, and to that which cuts the midspace of the sky with diametric track; for it makes a difference whether the same line is rising, or is beneath the earth, or is setting. Hence the sky now adds strength and now takes it away, and the stars which conceive anger there discard it when driven here. Between opposite signs exists mostly hatred; and to quadrate signs

corpora censentur signis et amica trigonis.
nec ratio obscura est ; nam quartum quodque locavit
eiusdem generis signum natura per orbem.
quattuor aequali caelum discrimine signant
in quibus articulos anni deus ipse creavit,
ver Aries, Cererem Cancer Bacchumque ministrans
Libra, caper brumam genitusque ad frigora piscis.
nec non et duplici quae sunt conexa figura
quartum quemque locum retinent ; duo cernere Pisces
et geminos iuvenes duplicemque in Virgine formam
et duo Centauri licet uno corpora textu.
sic et simplicibus signis stat forma quadrata ;
nam neque Taurus habet comitem, nec iungitur ulli
horrendus Leo, nec metuit sine compare quemquam
Scorpios, atque uno censetur Aquarius astro.
sic quaecumque manent quadrato condita templo
signa parem referunt numeris aut tempore sortem
ac veluti cognata manent sub foedere tali.
idcirco affines signant gradibusque propinquis
accedunt unaque tenent sub imagine natos,
quotquot cardinibus, *prona vertigine mundi*
naturae vires propriae variante, moventur ;
quae, quamquam in partes divisi quattuor orbis
sidera quadrata efficiunt, non lege quadrati
censentur : minor est numeri quam cardinis usus.
longior in spatium porrecta est linea maius
quae tribus emensis signis facit astra trigona.
haec ad amicitias imitantis iura gradumque

[663] textum [666] corpore

---

[a] Literally "every fourth sign," the Romans counting at both ends, for example ♈♋♎♑, as specified in 658 f.

[b] Literally "the goat and fish," Capricorn being compounded of the two.

[c] *Cf.* 2. 176.

are mostly ascribed ties of kinship, ties of friendship to trigonal signs. And the reason is plain ; for in every third place [a] right round the zodiac nature has set a sign of similar character. Four at equal intervals mark out the heavens, and in them God himself has created joining-points of the year : the Ram brings spring, the Crab the summer's corn, the Balance the vintage, and the Goat-fish,[b] born to endure the cold, brings winter. Moreover those signs which possess a combination of two figures occupy every third place : thus may you see the Fishes twain, the two young men, the Maid of twofold aspect,[c] and the Centaur's two bodies wrought in a single woof. Likewise the single signs also enjoy the figure of a square ; for the Bull has no partner, nor has the fierce Lion another at his side ; the mateless Scorpion has no one to fear, and the Waterman is counted as of single sign. Consequently the signs situate in quadrate positions display a like condition of number or season and remain bound by such federation as though by ties of blood. For this reason the quadrate signs indicate connections by marriage, assist the near degrees of kinship, and stamp their children with identical features, when they move through the cardinal points, incurring the while variation of their natural *powers owing to the sky's forward rotation* ; although they divide the circle into four parts and form quadrate signs, they are not considered as falling under the law of the square : the numerical association counts for less than that of cardinal point. The line which passes by three signs and creates trigonal constellations is longer and covers a greater distance. These constellations conduct us to friendships that rival the connections and ties of blood and

sanguinis atque animis haerentia foedera ducunt,
utque ipsa ex longo coeunt summota recessu,
sic nos coniungunt maioribus intervallis.
haec meliora putant, mentes quae iungere possunt,
quam quae non numquam foedus sub sanguine fallunt.
proxima vicinis subscribunt, tertia quaeque
hospitibus. sic astrorum servabitur ordo.
adde suas partes signis, sua partibus astra ;
nam nihil in totum servit sibi : mixta feruntur,
ipsis dant in se partes capiuntque vicissim.
quae mihi mox certo digesta sub ordine surgent.
omnibus ex istis ratio est repetenda per artem,
pacata infestis signa ut discernere possis.

Perspice nunc tenuem visu rem, pondere magnam
et tantum Graio signari nomine passam,
dodecatemoria, in titulo signantia causas.
nam, cum tricenas per partes sidera constent,
rursus bis senis numerus diducitur omnis ;
ipsa igitur ratio binas in partibus esse
dimidiasque docet partes. his finibus ecce
dodecatemorium constans, bis senaque tanta
omnibus in signis ; quae mundi conditor ille
attribuit totidem numero fulgentibus astris,
ut sociata forent alterna sidera sorte,

676 pectora
683-686: *see after* 672
689 in se partes] fines astris
692 perdiscere
699 esse
700 constant . . . cuncta
702 numeros

[a] *I.e.* the distinctive powers of squares, trigons, and hexagons will be intensified or depressed, but not altered, by their passage through the cardinal points.

[b] *Cf.* 4. 294.

to alliances knit together by feelings of the heart; and just as the stars themselves come together notwithstanding the vast space between, so they unite us over distances greater than those which separate kin. These signs, which have power to join heart to heart, are considered superior to those which not uncommonly forswear their ties of blood. Adjacent signs aid neighbours, alternate signs aid guests. Thus the principle [a] of the signs will be preserved. You must apportion to the signs the divisions of other signs which belong to them, and to these divisions their proper signs; for no sign ministers exclusively to itself [b]; they intermingle as they revolve: they cede to each other parts in themselves and are ceded parts in turn. These matters will be treated by me next [c] and follow in due order. It is from all such details [d] that a true account is to be drawn in our science, if one is to distinguish peaceful from aggressive signs.

[693]Pray examine now a matter trivial in appearance, yet one of great moment, which does not permit description of itself save by a Greek word. I speak of the dodecatemories, of which the name proclaims the principle. The signs each consist of thirty degrees, and every total is further divided by twice six; the calculation therefore shows that in each division there are two and a half degrees. Here, then, is the dodecatemory, consisting of this territory; a dozen such divisions occur in all the signs. The great builder of heaven bestowed these divisions upon the equal number of shining signs, that the constellations might be associated in a system of giving and

[c] 2. 693 ff.

[d] Such as are mentioned in 645-651, 687-690.

et similis sibi mundus, et omnia in omnibus astra,
quorum mixturis regeret concordia corpus
et tutela foret communi mutua causa.
in terris geniti tali sub lege creantur ;
idcirco, quamquam signis nascantur eisdem,
diversos referunt mores inimicaque vota ;
et saepe in peius derrat natura, maremque
femina subsequitur : miscentur sidere partus,
singula divisis variant quod partibus astra,
dodecatemoriis proprias mutantia vires.

Nunc, quod sit cuiusque, canam, quove ordine constent,
ne vagus ignotis signorum partibus erres.
ipsa suo retinent primas in corpore partes
sidera, vicinae subeuntibus attribuuntur,
cetera pro numero ducunt ex ordine partes,
ultima et extremis ratio conceditur astris.
singula sic retinent binas in sidere quoque
dimidiasque eius partes, et summa repletur
sortibus exactis triginta sidere in omni.

Nec genus est unum, ratio nec prodita simplex,
pluribus inque modis verum natura locavit
diduxitque vias voluitque per omnia quaeri.
haec quoque comperta est ratio sub nomine eodem.
quacumque in parti nascentum tempore Luna

709 pecudes errat
710 sidera
716 vicinis
719 sidera
720 eius] sui
723 rerum

---

[a] *I.e.* by displaying in each sign the twelvefold division which marks itself.

[b] See Introduction p. li and figure 10.

[c] *I.e.* of that constellation's degrees.

[d] In fact what Manilius says in 725-737 is merely a roundabout way of expounding the doctrine of 715-721.

receiving, that heaven might everywhere display a likeness to itself,[a] and that all the signs might have place in each other : thus by this interchange would harmony prevail through the zodiac and its signs would protect one another in a common interest. Men born on earth are created under this law ; and so, although their nativities occur in the same signs, they exhibit different traits and conflicting desires ; and oft does nature stray from good to bad, and birth of female follows that of male : the genitures differ in a single constellation, because the individual signs vary on account of the distribution of their divisions and modulate their respective powers in the dodecatemories.

[713]Now shall I recount what dodecatemory belongs to each sign and in what order they are to be found,[b] lest you go astray and err because you know not the divisions of the signs. Within their own domain the constellations keep the first division for themselves, the next is bestowed upon the sign following, and the remaining signs according to their place in the sequence are allotted successive divisions, and the last assignment is made to the farthest sign away. Thus each sign occupies in every constellation two and a half of its degrees,[c] making a total of thirty degrees exacted from the whole zodiac.

[722]But there is more than one system of calculating dodecatemories, and the procedure handed down admits an alternative. Nature has placed the truth in more methods than one and has separated the paths which lead to herself, wishing the quest for her to be made along every path. The following procedure also [d] has been devised for these same dodecatemories. Multiply thrice four times what-

constiterit, numeris hanc ter dispone quaternis,
sublimi totidem quia fulgent sidera mundo.
inde suas illi signo, quo Luna refulsit,
quaeque hinc defuerant partes numerare memento.
proxima tricenas pariterque sequentia ducunt.
[hic ubi deficiet numerus, tunc summa relicta
in binas sortes adiecta parte locetur
dimidia, reliquis tribuantur ut ordine signis]
in quo destituent, eius tum Luna tenebit
dodecatemorium signi ; post cetera ducet
ordine quodque suo, sicut stant astra locata.

Haec quoque te ratio ne fallat, percipe paucis
(maior in effectu minor est) de partibus ipsis
dodecatemorii quota sit quod dicitur esse
dodecatemorium. namque id per quinque notatur
partes ; nam totidem praefulgent sidera caelo
quae vaga dicuntur, ducunt et singula sortes
dimidias, viresque in eis et iura capessunt.
in quo quaeque igitur stellae quandoque locatae
dodecatemorio fuerint spectare decebit ;
cuius enim stella in fines in sidere quoque

735 qua
737 quoque
738 perspice
739 de] quod
740 quota] quod
745 quocumque
747 stellae fine sint

---

[a] An ornate way of saying "allocate 30 degrees to the sign in which the Moon is situated, no matter how few or how many degrees of that sign the Moon has actually traversed."

[b] [732-734] (see Introduction p. liii): "When this number fails, then let the remainder be divided into portions of two

ever degree the Moon occupies at the moment of nativity, since so many are the constellations which shine in high heaven. Of this number be sure to allot to that sign in which the Moon shone both the degrees that she has traversed and those that were lacking to the count of thirty.[a] The next sign is allotted thirty and so are succeeding signs.[b] Whatever the sign in which the count gives out, that sign's dodecatemory shall the Moon be then occupying[c]; subsequently she will occupy the remaining dodecatemories, each one in its turn, according to the established order of the signs.[d]

[738]Now lest the following scheme too elude you, learn from a few words (for its importance is greater than its compass) what size portion of the dodecatemory is that which carries the name of dodecatemory.[e] Now the dodecatemory is divided into five parts; for so many are the stars called wanderers which with passing brightness shine in heaven[f]: they are each allotted half a degree wherein they assume power and authority. And so it will be right to observe in which dodecatemory each of the planets is at any moment positioned: for a planet will exert its influ-

and a half, so that these may be distributed in order among the remaining signs."

[c] Let the Moon be in 9° Aries: $9 \times 12 = 108$; of this number 30 are allotted to Aries, 30 to Taurus, 30 to Gemini, leaving 18, so that the count gives out in Cancer: thus 9° Aries occupies the dodecatemory of Cancer.

[d] This means no more than that, once the Moon's dodecatemory is fixed, the following dodecatemories are given to the following signs in regular order.

[e] Planetary dodecatemories: see Introduction p. liv and figure 11.

[f] Saturn, Jupiter, Mars, Venus, and Mercury (*i.e.* not including Sun or Moon).

inciderit, dabit effectus in viribus eius.
undique miscenda est ratio per quam omnia constant.
verum haec posterius proprio cuncta ordine reddam ; 7
nunc satis est docuisse suos ignota per usus,
ut, cum perceptis steterit fiducia membris,
sic totum corpus facili ratione notetur
et bene de summa veniat post singula carmen.
ut rudibus pueris monstratur littera primum
per faciem nomenque suum, tum ponitur usus,
tum coniuncta suis formatur syllaba nodis,
hinc verbi structura venit per membra legendi,
tunc rerum vires atque artis traditur usus
perque pedes proprios nascentia carmina surgunt,
singulaque in summam prodest didicisse priora
(quae nisi constiterint primis fundata elementis,
effluat in vanum rerum praeposterus ordo
versaque quae propere dederint praecepta magistri),
sic mihi per totum volitanti carmine mundum
erutaque abstrusa penitus caligine fata,
Pieridum numeris etiam modulata, canenti
quoque deus regnat revocanti numen in artem,
per partes ducenda fides et singula rerum
sunt gradibus tradenda suis, ut, cum omnia certa
notitia steterint, proprios revocentur ad usus.
ac, velut, in nudis cum surgunt montibus urbes,
conditor et vacuos muris circumdare colles

748 inciderint
756 tum] com-
758 verbis . . . verba
761 summa
768 arte
770 et
771 revocantur
772 consurgunt orbes

[a] For example, suppose Saturn (a malefic planet) happens to occupy a dodecatemory of Jupiter (a benefic), then its bad influences will be modified for the better.

ences as modified by the powers of that dodecatemory into whose territory in any given sign it has come.[a] From all quarters must be pieced together the design by which all things are ordered. However, all this in due order shall I explain hereafter[b]; now it is enough to teach new principles by demonstrating their uses, so that, when you have acquired confidence in your grasp of the elements, you will be thus able by simple reasoning to mark the complete pattern, and my poem can fittingly pass on from details and deal with the whole. Children who have not yet begun their lessons are first shown the shape and name of a letter, and then its value is explained: then a syllable is formed by the linking of letters; next comes the building up of a word by reading its component syllables; afterwards the meaning of expressions and the rules of grammar are taught, and then verses come into being and rise up on feet of their own. To reach the final goal it is important to have mastered each of the earlier steps, for unless these are firmly based upon underlying principles, instruction, as teachers too hurriedly expound their precepts out of order, will be ill-arranged and prove labour vainly spent. Thus as I wing my way in song throughout the whole universe and, entuning them to the Muses' rhythm, sing of fates drawn from deep-seated darkness and summon to my art the power by which God rules, I too must by degrees win credence and assign each matter to its correct step, so that, when all the parts are grasped with sure understanding, they may be applied to their proper uses. And as, when a city is being built on a bare mountain-side and its founder plans to encompass the empty

[b] A promise not fulfilled in the extant work of Manilius.

destinat, ante manus quam temptet scindere fossas
fervit opus (ruit ecce nemus, saltusque vetusti
procumbunt solemque novum, nova sidera cernunt,
pellitur omne loco volucrum genus atque ferarum,
antiquasque domos et nota cubilia linquunt,
ast alii silicem in muros et marmora templis
rimantur, ferrique rigor per pignora nota
quaeritur, hinc artes, hinc omnis convenit usus),
tum demum consurgit opus, cum cuncta supersunt,
ne medios rumpat cursus praepostera cura,
sic mihi conanti tantae succedere moli
materies primum rerum, ratione remota,
tradenda est, ratio sit ne post irrita neve
argumenta novis stupeant nascentia rebus.

Ergo age noscendis animum compone sagacem
cardinibus, qui per mundum sunt quattuor omnes
dispositi semper mutantque volantia signa:
unus ab exortu caeli nascentis in orbem,
qua primum terras aequali limite cernit,
alter ab adversa respondens aetheris ora,
unde fugit mundus praecepsque in Tartara tendit;
tertius excelsi signat fastigia caeli,
quo defessus equis Phoebus subsistit anhelis
reclinatque diem mediasque examinat umbras;
ima tenet quartus fundato nobilis orbe,
in quo principium est reditus finisque cadendi
sideribus, pariterque cccasus cernit et ortus.

775 vertit
779 templi
780 tempora
784 cunctanti
797 declinatque

[a] The cardines: see Introduction pp. lv f. and figure 12.

hill with walls, before his team attempt to cut trenches, doth work proceed apace ; and lo, a forest tumbles and ancient woodlands fall, beholding sun and stars unseen before ; all tribes of bird and beast are banished from the spot, leaving the immemorial homes and lairs they knew so well ; others, meanwhile, seek stone for walls and marble for temples and by means of sure clues search for sources of unbending iron ; from their different sides skill and experience of every kind combine to help ; and only when all materials are available in plenty does construction proceed, lest premature effort cause the project to break down in mid-course : so, as I strive to perform a mighty undertaking, must I first supply the matter of my theme, withholding explanation, lest hereafter explanation prove ineffectual and my arguments be silenced at the outset before some unanticipated fact.

788Come now, prepare an attentive mind for learning the cardinal points [a] : four in all, they have positions in the firmament permanently fixed and receive in succession the speeding signs. One looks out from the rising of the heavens as they are born into the world and has the first view of the Earth from the level horizon ; the second faces it from the opposite edge of the sky, the point from which the starry sphere retires and hurtles headlong into Tartarus ; a third marks the zenith of high heaven, where wearied Phoebus halts with panting steeds and rests the day and determines the mid-point of shadows ; the fourth occupies the nadir, and has the glory of forming the foundation of the sphere ; in it the stars complete their descent and commence their return, and at equal distances it beholds their risings

haec loca praecipuas vires summosque per artem
fatorum effectus referunt, quod totus in illis
nititur aeternis veluti compagibus orbis ;
quae nisi perpetuis alterna sorte volantem
cursibus excipiant nectantque in vincula, bina
per latera atque imum templi summumque cacumen,
dissociata fluat resoluto machina mundo.

Sed diversa tamen vis est in cardine quoque,
et pro sorte loci variant atque ordine distant.
primus erit, summi qui regnat culmine caeli
et medium tenui partitur limite mundum ;
quem capit excelsa sublimem Gloria sede
(scilicet haec tutela decet fastigia summa),
quidquid ut emineat sibi vindicet et decus omne
asserat et varios tribuendo regnet honores.
hinc favor et species atque omnis gratia vulgi,
reddere iura foro, componere legibus orbem
foederibusque suis externas iungere gentes
et pro sorte sua cuiusque extollere nomen.
proximus, est ima quamquam statione locatus,
sustinet aeternis nixum radicibus orbem,
effectu minor in specie sed maior in usu.
fundamenta tenet rerum censusque gubernat,
quam rata sint fossis, scrutatur, vota metallis
atque ex occulto quantum contingere possit.
tertius, aequali terris in parte nitentem
qui tenet exortum, qua primum sidera surgunt,
unde dies redit et tempus describit in horas,

803 aetheriis
814 deus omni
826 atque illi tollens

[a] MC (*medium caelum*).
[b] IMC (*imum caelum*).
[c] Horoscope.

and settings. These points are charged with exceptional powers, and the influence they exert on fate is the greatest known to our science, because the celestial circle is totally held in position by them as by eternal supports; did they not receive the circle, sign after sign in succession, flying in its perpetual revolution, and clamp it with fetters at the two sides and lowest and highest extremities of its compass, heaven would fly apart and its fabric disintegrate and perish.

[808]Each cardinal, however, enjoys a different influence; they vary according to their position, and they differ in rank. First place goes to the cardinal [a] which holds sway at the summit of the sky and divides heaven in two with imperceptible meridian; enthroned on high this post is occupied by Glory (truly a fit warden for heaven's supreme station), so that she may claim all that is pre-eminent, arrogate all distinction, and reign by awarding honours of every kind. Hence comes applause, splendour, and every form of popular favour; hence the power to dispense justice in the courts, to bring the world under the rule of law, to make alliances with foreign nations on one's own terms, and to win fame relative to one's station. The next point,[b] though situate in the lowest position, bears the world poised on its eternal base; in outward aspect its influence is less, but is greater in utility. It controls the foundations of things and governs wealth; it examines to what extent desires are accomplished by the mining of metal and what gain can issue from a hidden source. The third cardinal,[c] which on the same level as the Earth holds in position the shining dawn, where the stars first rise, where day returns and divides time

hinc inter Graias horoscopos editur urbes,
nec capit externum, proprio quia nomine gaudet.
hunc penes arbitrium vitae est, hic regula morum,
fortunamque dabit rebus, ducetque per artes,
qualiaque excipiant nascentis tempora prima,
quos capiant cultus, quali sint sede creati,
utcumque admixtis subscribent viribus astra.
ultimus, emenso qui condit sidera mundo
occasumque tenens summersum despicit orbem,
pertinet ad rerum summas finemque laborum,
coniugia atque epulas extremaque tempora vitae
otiaque et coetus hominum cultusque deorum.

Nec contentus eris percepto cardine quoque :
intervalla etiam memori sunt mente notanda
per maius dimensa suas reddentia vires.
quidquid ab exortu summum curvatur in orbem,
aetatem primam nascentisque asserit annos.
quod summo premitur devexum culmine mundi
donec ad occasus veniat, puerilibus annis
succedit teneramque regit sub sede iuventam.
quae pars occasus aufert imumque sub orbem
descendit, regit haec maturae tempora vitae
perpetua serie varioque exercita cursu.
at, qua perficitur cursus redeunte sub ortum,
tarda supinatum lassatis viribus arcum
ascendens, seros demum complectitur annos
labentemque diem vitae tremulamque senectam.

831 hunc tenet
[est] . . . morum est
845 aetatis primae
849 aufert] inter
851 perpetua] et propria
852 ortum] imo

[a] Occident.

into hours, is for this reason in the Greek world called the Horoscope, and it declines a foreign name, taking pleasure in its own. Within its domain lies the arbitrament of life and the formation of character; it will grant success to enterprises, open up the professions, and decide the early years that await men from their birth, the education they receive, and the station to which they are born, according as the planets approve and mingle their influences. The last point,[a] which puts the stars to rest after traversing heaven and, occupying the occident, looks down upon the submerged half of the sky, is concerned with the consummation of affairs and the conclusion of toil, marriages and banquets and the closing years of life, leisure and social intercourse and worship of the gods.

[841]Nor must you rest content with observing each cardinal point; you must note with a retentive mind the spaces between them,[b] which extend over a larger range and possess special powers. The curve which stretches from the orient to the topmost point of the circle claims the earliest age and infant years. The slope which sinks down from the summit of the sky till it reaches the occident succeeds to the years of childhood and includes in its province control of tender youth. The portion which appropriates the setting heaven and descends to the bottom of the circle rules the period of adult life, a period tested by incessant change and chequered fortunes. But the part by whose return to the orient heaven's course is done and which with enfeebled strength slowly ascends the backbent arc, this part embraces the final years, life's fading twilight, and palsied age.

[b] The quadrants: see Introduction p. lvi and figure 12.

Omne quidem signum sub qualicumque figura
partibus inficitur mundi ; locus imperat astris
et dotes noxamque facit ; vertuntur in orbem
singula et accipiunt vires caeloque remittunt.
vincit enim natura loci legesque ministrat
finibus in propriis et praetereuntia cogit
esse sui moris, vario nunc ditia honore,
nunc sterilis poenam referentia sidera sedis.
quae super exortum est a summo tertia caelo,
infelix regio rebusque inimica futuris
et vitio fecunda nimis ; nec sola, sed illi
par erit, adverso quae fulget sidere sedes
iuncta sub occasu. neu praestet, cardine mundi
utraque praetenta fertur deiecta ruina.
porta laboris erit : scandendum est atque cadendum.
nec melior super occasus contraque sub ortu
sors agitur mundi : praeceps haec, illa supina
pendens aut metuit vicino cardine finem
aut fraudata cadet. merito Typhonis habentur
horrendae sedes, quem Tellus saeva profudit,
cum bellum caelo peperit nec matre minores
exstiterunt partus. sed fulmine rursus in alvum
compulsi, montesque super rediere cadentes,

860 loci] genus
862 dives
863 -que [re]
870 orta
877 altum

---

[a] Here follows the poet's doctrine of the dodecatropos or twelve temples : see Introduction p. lvii and figure 13.

[b] Temple 12 (in Greek κακὸς δαίμων "Evil Spirit").

[c] Temple 6 (in Greek κακὴ τύχη "Evil Fortune").

[d] Temple 8.

[e] Temple 2.

[f] Temple 8 fears to fall into the hands of 7, temple 2 to be dropped by 1.

856In any geniture every sign is affected by the sky's division into temples [a]; position governs the stars, and endows them with power to benefit or harm; each of the signs, as it revolves, receives the influences of heaven and to heaven imparts its own. The nature of the position prevails, exercises jurisdiction within its province, and subjects to its own character the signs as they pass by, which now are enriched with distinction of every kind and now bear the penalty of a barren abode. The temple [b] that is immediately above the Horoscope and is the next but one to heaven's zenith is a temple of ill omen, hostile to future activity and all too fruitful of bane; nor that alone, but like unto it will prove the abode [c] which with confronting star shines below the occident and adjacent to it. And so that this temple should not outdo the former, each alike moves dejected from a cardinal point with the spectacle of ruin before its eyes. Each shall be a portal of toil: in one you are doomed to climb, in the other to fall. Not more fortunate is the portion [d] of heaven above the occident or that opposite [e] it below the orient; suspended, the former face downward, the latter on its back, they either fear destruction at the hands of the neighbouring cardinal or will fall if cheated of its support.[f] With justice are they held to be the dread abodes of Typhon, whom savage Earth brought forth when she gave birth to war against heaven and sons as massive as their mother appeared.[g] Even so, the thunderbolt hurled them back to the womb, the collapsing mountains recoiled upon them, and

[g] Typhon (or Typhoeus) is here (as in Horace) represented as being one of the Giants, though he was not born until after their defeat: his story is told by Apollodorus, *Bibl.* 1. 6. 3.

cessit et in tumulum belli vitaeque Typhoeus.
ipsa tremit mater flagrantem monte sub Aetna.
at, quae fulgentis sequitur fastigia caeli
proxima, neve ipsi cedat, cui iungitur, astro
spe melior, palmamque petens victrixque priorum
altius insurgit : summae comes addita finis,
in peiusque manent cursus nec vota supersunt.
quocirca minime mirum, si proxima summo
atque eadem integrior Fortunae sorte dicatur
cui titulus Felix. censum sic proxima Graiae
nostra subit linguae vertitque a nomine nomen.
Iuppiter hac habitat : venerandam crede regenti.
huic in perversum similis deiecta sub orbe
imaque summersi contingens culmina mundi,
adversa quae parte nitet, defessa peracta
militia rursusque novo devota labori
cardinis et subitura iugum sortemque potentem
nondum sentit onus mundi, iam sperat honorem.
Daemonien memorant Grai, Romana per ora
quaeritur inversus titulus. sub corde sagaci
conde locum numenque dei nomenque potentis,
quae tibi posterius magnos revocentur ad usus.
hic momenta manent nostrae plerumque salutis
bellaque morborum caecis pugnantia telis,

880 flagrantis
881 caelo
882 qui fungitur
887 interior
fortunae] veneranda
888 quod . . . si
890 venerandam] fortunae
896 opus
898 in versu
900 revocantur

[a] Temple 11.

[b] The Greek astrologers usually call temple 11 ἀγαθὸς δαίμων ("Good Spirit") and employ the term ἀγαθὴ τύχη ("Good Fortune") of temple 5.

Typhoeus was sent to the grave of his warfare and his life alike. Even his mother quakes as he blazes beneath Etna's mount. The temple [a] immediately behind the summit of bright heaven, and (not to be outdone by its neighbour) of braver hope, surges ever higher, being ambitious for the prize and triumphant over the earlier temples : consummation attends the topmost abode, and no movement save for the worse can it make, nor is aught left for it to aspire to. There is thus small cause for wonder, if the station nearest the zenith, and more secure than it, is blessed with the lot of Happy Fortune. So most closely does our language approach the richness of Greek and render name for name.[b] In this temple dwells Jupiter : let its ruler convince you that it is to be reverenced. Like this temple, but with an inverse likeness, is that [c] which is thrust below the world and adjoins the nadir of the submerged heaven, and which shines in the opposite region : wearied after completion of active service it is again marked out for a further term of toil, as it waits to shoulder the yoke of the cardinal temple and its role of power : not as yet does it feel the weight of the world, but already aspires to that honour. This seat the Greeks call Daemonie [d] : a rendering of the name in Roman speech is wanting. Lay carefully in your mind the region and the divinity and appellation of the puissant deity, so that hereafter the knowledge may be put to great use. Here largely abide the changes in our health and the warfare waged by the unseen weapons

[c] Temple 5.

[d] Housman can find only two occurrences of the term, each obscure and signifying temple 6 (normally called κακὴ τύχη "Evil Fortune").

viribus ambiguam geminis casusque deique
nunc huc nunc illuc sortem mutantis utraque.
sed medium post astra diem curvataque primum
culmina nutantis summo de vertice mundi
aethra Phoebus alit; sub quo quae corpora nostra
concipiunt vitia et fortunam, ex viribus eius
decernunt. Deus ille locus sub nomine Graio
dicitur. huic adversa nitens, quae prima resurgit
sedibus ex imis iterumque reducit Olympum,
pars mundi fratrumque vices mortesque gubernat;
et dominam agnoscit Phoeben, fraterna videntem
regna per adversas caeli fulgentia partes
fataque damnosis imitantem finibus oris.
huic parti Dea nomen erit Romana per ora,
Graecia voce sua titulum designat eundem.
arce sed in caeli, qua summa acclivia finem
inveniunt, qua principium declivia sumunt,
culminaque insurgunt occasus inter et ortus
suspenduntque suo libratum examine mundum,
asserit hanc Cytherea sibi per sidera sedem
et velut in facie mundi sua collocat ora,
per quae humana regit. propria est haec reddita parti
vis, ut conubia et thalamos taedasque gubernet:
haec tutela decet Venerem, sua tela movere.
nomen erit Fortuna loco, quod percipe mente,
ut brevia in longo compendia carmine praestem.

903 causasque
907 aethra] aethera
alit] aut
908 decernunt
909 concupiunt
910 huc
912 fulvumque nitet
923 veluti faciem
925 ut] et
gubernat
926 docet

---

[a] Temple 9.
[b] Θεός (the regular term).

of disease, wherein are engaged the two powers of chance and godhead affecting this region of uncertainty on either side, now for better, now for worse. The stars that follow midday, where the height of heaven first slopes downward and bows from the summit, these [a] Phoebus nourishes with his splendour; and it is by Phoebus's influence that they decree what ill or hap our bodies take beneath his rays. This region is called by the Greek word signifying God.[b] Shining face to face with it is that part [c] of heaven which rises first from the bottommost regions and brings back the sky once more: it controls the fortunes and fate of brothers; and it acknowledges the Moon for its mistress, who beholds her brother's realms shining on her from the other side of heaven and who reflects human mortality in the dying edges of her face. Goddess is the name in Roman speech to be given to this region, whilst the Greeks call it by the same word in their language.[d] But in the citadel of the sky, where the rising curve attains its consummation, and the downward slope makes its beginning, and the summit towers midway between orient and occident and holds the universe poised in its balance, here does the Cytherean claim her abode [e] among the stars, placing in the very face of heaven, as it were, her beauteous features, wherewith she rules the affairs of men. To the abode is fittingly given the power to govern wedlock, the bridal chamber, and the marriage torch; and this charge suits Venus, the charge of plying her own weapons. Fortune shall be this temple's name [f]; and mark it well, that I may take a short route in my

[c] Temple 3. [d] Θεά (the regular term).
[e] Temple 10. [f] The term is not found elsewhere.

at, qua subsidit converso cardine mundus
fundamenta tenens, aversum et suspicit orbem
ac media sub nocte iacet, Saturnus in illa
parte suas agitat vires, deiectus et ipse
imperio quondam mundi solioque deorum,
et pater in patrios exercet numina casus
fortunamque senum. | titulus, quem Graecia fecit,
Daemonium signat dignas pro nomine vires.
nunc age surgentem primo de cardine mundum
respice, qua solitos nascentia signa recursus
incipiunt, viridis gelidis et Phoebus ab undis
enatat et fulvo paulatim accenditur igni
asperum iter temptans, | Aries qua ducit Olympum.
haec tua templa ferunt, Maia Cyllenie nate,
pro facie signata nota, quod nomen et ipsi
auctores tibi dant. | una est tutela duorum :
[nascentum atque patrum, quae tali condita parte est]
in qua fortunam natorum condidit omnem
natura, ex illa suspendit vota parentum.
unus in occasu locus est super. ille ruentem
praecipitat mundum terris et sidera mergit,
tergaque prospectat Phoebi, qui viderat ora ;
ne mirere, nigri si Ditis ianua fertur
et finem vitae retinet*que repagula mortis.*
hic etiam ipse dies moritur, tellusque per orbem

930 adversum
938 signatque suas
940 solido . . . recursu
937a erit templum
945b artes
944 o facies
935b prima
947 illo
949 mersit
953 hinc
terrasque

[a] Temple 4.
[b] Another term not found elsewhere.
[c] Temple 1.

lengthy song. Where at the opposite pole the universe subsides, occupying the foundations, and from the depths of midnight gloom gazes up at the back of the Earth, in that region [a] Saturn exercises the powers that are his own: cast down himself in ages past from empire in the skies and the throne of heaven, he wields as a father power over the fortunes of fathers and the plight of the old. Daemonium is the name the Greeks have given it,[b] denoting influences fitting the name. Turn now your gaze upon heaven as it climbs up from the first cardinal point, where [c] the rising signs commence afresh their wonted courses, and a pale Sun swims upward from the icy waves and begins by slow degrees to blaze with golden flame as it attempts the rugged path where the Ram heads the procession of the skies. This temple, Mercury, son of Maia, men say is yours, marked for its bright aspect with a designation which writers also give you for name.[d] The one wardship is commissioned with two charges [e]; for in it nature has placed all fortunes of children and has made dependent on it the prayers of parents. There remains one region, that in the setting heaven.[f] It speeds the falling sky beneath the Earth and buries the stars. Now it looks forth on the back of the departing Sun, yet it once beheld his face; so wonder not if it is called the portal of sombre Pluto and keeps control over the end of life *and death's firm-bolted door*. Here dies even the very light of day, which the ground beneath steals away from the world and locks

[d] Alluding to the name Στίλβων ("Glistener"), a common astrological name for Mercury.

[e] [936]: "of sons and fathers, which is located in such a region."

[f] Temple 7.

surripit et noctis captum sub carcere claudit.
nec non et fidei tutelam vindicat ipsi
pectoris et pondus. tanta est in sede potestas
quae vocat et condit Phoebum recipitque refertque,
consummatque diem. tali sub lege notandae
templorum tibi sunt vires : quae pervolat omnis
astrorum series ducitque et commodat illis
ipsa suas leges, stellaeque ex ordine certo,
ut natura sinit, lustrant, variasque locorum
efficiunt vires, utcumque aliena capessunt
regna et in externis subsidunt hospita castris.
haec mihi sub certa stellarum parte canentur ;
nunc satis est caeli partes titulosque notasse
effectusque loci per se cuiusque deosque.
[cui parti nomen posuit, qui condidit artem,
octotropos ; per quod stellae diversa volantes
quos reddant motus, proprio venit ordine rerum]

[955] ipsam [959] quas

---

[a] A promise not fulfilled in the poet's extant work.

up captive in the dungeon of night. This temple also claims for itself the guardianship of good faith and constancy of heart. Such is the power that dwells in the abode which summons to itself and buries the Sun, thus surrendering that which it has received, and brings the day to its close. This is the system by which you must mark the powers of the temples : through them revolves the entire procession of the zodiac, which draws from them their laws and lends to them its own ; the planets, too, according as nature allows, traverse them in fixed order and modify the various influences of the temples whenever they occupy realms not their own and sojourn in an alien camp. Of these matters I shall treat at the place in my song appointed for the planets [a] ; it is enough for now to have recorded the temples of heaven and their names, the innate influences of each place, and the deities that dwell therein.[b]

[b] [968-970] : "The founder of astrology gave to this section the title Octotropos ; the motions of the planets, which fly through it in the opposite direction, shall follow at the proper place." For the implications of this passage, see Introduction pp. lxi f.

BOOK THREE

ALTHOUGH *not without fascination for virtuoso versifiers, for it is something of a tour de force, the third book is the least poetical of the five, exemplifying for the most part Manilius's skill in rendering numbers and arithmetical calculations in hexameters. The introduction takes the form of a priamel enumerating the themes of other poets and concluding with the problems of the astrological bard. Manilius resumes his task with an exposition of the twelve athla, which in turn leads to the main subject of the book: the determination of the degree of the ecliptic which is rising above the horizon at the moment of a nativity. This, the most important operation in the casting of a horoscope, causes today's astrologer no trouble, but before the advent of clocks, atlases, and standard time account had to be taken of the varying lengths of daylight and darkness in different latitudes and different times of the year. Having equipped us with the necessary rules for dealing with the problem, our instructor continues to render mathematical tables in verse-form by introducing us to chronocrators and to the length of life ordained for each of us by the stars. A final chapter, on tropic signs, if lacking in astrological significance, provides a more poetic note on which to close.*

LIBER TERTIUS

In nova surgentem maioraque viribus ausum
nec per inaccessos metuentem vadere saltus
ducite, Pierides. vestros extendere fines
conor et ignotos in carmina ducere census.
non ego in excidium caeli nascentia bella,
fulminis et flammis partus in matre sepultos,
non coniuratos reges Troiaque cadente
Hectora venalem cineri Priamumque ferentem,
Colchida nec referam vendentem regna parentis
et lacerum fratrem stupro, segetesque virorum
taurorumque trucis flammas vigilemque draconem
et reduces annos auroque incendia facta
et male conceptos partus peiusque necatos;
non annosa canam Messenes bella nocentis,
septenosve duces ereptaque fulmine flammis
moenia Thebarum et victam quia vicerat urbem,

[4] indignos . . . cantus [15] -ve] -que

---

[a] The war of the Giants, told by Hesiod, *Theog.* 664-735 (*cf.* 1. 421 ff., 5. 341 ff.).

[b] The Greek expedition against Troy, told by Homer in the *Iliad*, line 8 referring specially to Book 24 (*cf.* 2. 1 ff.).

[c] The story of Medea, told by Apollonius Rhodius in the *Argonautica*; he alludes to the hexameter poet rather than Euripides, although he gives a comprehensive survey (as at

BOOK 3

As I rise to fresh heights and venture a task beyond my strength, fearlessly entering untrodden glades, O Muses, be my guides. To widen your domains I strive, and to bring new treasure into song. I shall not tell of war conceived for heaven's destruction and offspring buried by the flames of the thunderbolt in its mother's womb [a]; or of the oath-bound kings [b] and how, in Troy's last hour, Hector's body was ransomed for his obsequies and fetched by Priam; or of the woman of Colchis [c] sacrificing to her guilty love father's realm and brother's mangled corpse, the crop of warriors, the bulls that breathed fierce flames, the unsleeping dragon, the restoration of the years of youth, the fires kindled by a gift of gold, and the children born in wickedness and yet more wickedly slain. I shall not sing of the agelong warfare for which Messene was to blame [d]; or of the chieftains seven, the walls of Thebes saved by a thunderbolt from the threatened flames, and the city

5. 465 ff., where he certainly alludes to tragedy): Medea's passion for Jason, betrayal of Aeetes, murder of Absyrtus; the feats of Jason; the rejuvenation of Aeson; Medea's gifts which killed Glauce and Creon, and her murder of her children.

[d] The Second Messenian War, which lasted for twenty years and was instigated by the Messenians, told by Rhianus in his epic poem *Messeniaca.*

germanosve patris referam matrisque nepotes,
natorumve epulas conversaque sidera retro
ereptumque diem, nec Persica bella profundo
indicta et magna pontum sub classe latentem
immissumque fretum terris, iter aequoris undis;
non regis magni spatio maiore canenda
quam sunt acta loquar. Romanae gentis origo,
quotque duces urbis tot bella atque otia, et omnis
in populi unius leges ut cesserit orbis,
differtur. facile est ventis dare vela secundis
fecundumque solum varias agitare per artes
auroque atque ebori decus addere, cum rudis ipsa
materies niteat. speciosis condere rebus
carmina vulgatum est, opus et componere simplex.
at mihi per numeros ignotaque nomina rerum
temporaque et varios casus momentaque mundi
signorumque vices partesque in partibus ipsis
luctandum est. quae nosse nimis, quid, dicere quantu
est?

[17,18] -que
[23] sint
[24] totque
[33] quorumque

[a] The story of the Seven against Thebes: Capaneus, who bore on his shield the words πρήσω πόλιν ("I shall fire the city"), was destroyed by the thunderbolt; the Epigoni, sons of the conquered Seven, ten years later marched against Thebes and took it (Apollodorus, *Bibl.* 3. 7. 2 ff.). Here and in the following two items Manilius seems to allude to dramatic compositions.

[b] The children of Oedipus, who married his mother Jocasta (*cf.* 5. 464).

[c] Atreus killed the sons of Thyestes, and served up their flesh to him at a banquet (*cf.* 5. 462).

[d] The story of Xerxes, who ordered the Hellespont to be

conquered because of its conquest [a]; I shall not tell of them that were brethren to their father, grandchildren to their mother,[b] or of sons served up at table, whereat the stars recoiled and took away the light of day.[c] Nor shall I tell of the Persian declaration of war upon the main, when a vast fleet hid the ocean, and a channel was let into the land, and a road laid on the waters of the sea.[d] Of the feats of the monarch styled the Great, taking longer to record than to achieve,[e] I shall not speak. The founding of the Roman nation, the periods of war and peace as numerous as the city's consuls, and the whole world's submission to a single people's rule,[f] this I put off. It is easy to spread sail before a favouring breeze, to work a fertile soil with different skills, and to add lustre to gold or ivory, since the raw material of itself shines bright. It is a hackneyed task to write poems on attractive themes and compose an uncomplicated work. But I must wrestle with numerals and names of things unheard of,[g] with the seasons, the changing fortunes and movements of the sky, with the signs' variations, and even with the portions of their portions. Ah, how great a task it is to put into words what passes understanding! Ah,

lashed and fettered, cut a canal through Athos, and bridged the Hellespont (*cf.* 5. 49), told by Choerilus of Samos in his epic poem *Persica*.

[e] Alexander: [Longinus], *De Subl.* 4. 2 quotes Timaeus as saying of Alexander that he conquered all Asia in less time than Isocrates took over his *Panegyric* urging war on Persia. Manilius is probably making a contemptuous allusion to the epic poem of Choerilus of Iasus (*cf.* Horace, *Epist.* 2. 1. 232 ff.).

[f] Presumably alluding to the *Annales* of Ennius.

[g] Like "dodecatemories" (2. 695).

carmine quid proprio ? pedibus quid iungere certis ?
huc ades, o quicumque meis advertere coeptis
aurem oculosque potes, veras et percipe voces.
impendas animum ; nec dulcia carmina quaeras :
ornari res ipsa negat contenta doceri.
et, siqua externa referentur nomina lingua,
hoc operis, non vatis erit : non omnia flecti
possunt, et propria melius sub voce notantur.

Nunc age subtili rem summam perspice cura,
quae tibi praecipuos usus monstrata ministret
et certas det in arte vias ad fata videnda,
si bene constiterit vigilanti condita sensu.
principium rerum et custos natura latentum
(cum tantas strueret moles per moenia mundi
et circum fusis orbem concluderet astris
undique pendentem in medium, diversaque membra
ordinibus certis sociaret corpus in unum,
aeraque et terras flammamque undamque natantem
mutua in alternum praebere alimenta iuberet,
ut tot pugnantis regeret concordia causas
staretque alterno religatus foedere mundus),
exceptum a summa nequid ratione maneret
et quod erat mundi mundo regeretur ab ipso,
fata quoque et vitas hominum suspendit ab astris,
quae summas operum partes, quae lucis honorem,
quae famam assererent, quae numquam fessa volarent.
quae, quasi, per mediam, mundi praecordia, partem
disposita, obtineant, Phoebum lunamque vagasque

[a] *Cf.* 1. 247-254, 2. 60-83, 4. 888-890.
[b] The signs of the zodiac.

how great to tell in fitting poetry, and this to yoke to a fixed metre ! Come hither, whoever is able to devote ear and eye to my emprise, and hearken to the truths I utter. Apply your mind to understand and seek not poetry that beguiles : my theme of itself precludes adornment, content but to be taught. And if any terms are spoken in a foreign tongue, blame this on subject, not on bard : some things defy translation, and are better expressed by the native word.

[43]Now regard with close attention a matter of prime importance : when explained, it will render you immense service and will furnish practitioners of our art with a sure road to seeing the future, provided that it is well understood and stored in an alert mind. First cause and guardian of all things hidden, nature erected mighty structures along the ramparts of the universe and so surrounded Earth, poised squarely in the centre, with a sphere of stars ; and by fixed laws she united separate limbs into a single body, ordaining that air and earth and fire and flowing water should each for the other provide mutual sustenance, in order that harmony might prevail over so many elements at variance and the universe stand firm in the bonds of a reciprocal federation.[a] Now in order that nothing should be excluded from the total scheme, and that what was born of heaven should be by heaven's own self controlled, nature also made men's lives and destinies dependent on the stars, so that in their unwearied revolution they should claim charge of the success of human activity, the boon of life, and fame. And to those stars [b] which, deployed about the central region, occupy the heart of the universe, as it were, and which outfly the Sun

evincunt stellas nec non vincuntur et ipsa,
his regimen natura dedit, propriasque sacravit
unicuique vices sanxitque per omnia summam,
undique uti fati ratio traheretur in unum.
nam, quodcumque genus rerum, quotcumque labores
quaeque opera atque artes, quicumque per omnia ca
humana in vita poterant contingere, sorte
complexa est, tot et in partes, quot et astra locarat,
disposuit, certasque vices, sua munera cuique
attribuit, totumque hominis per sidera censum
ordine sub certo duxit, pars semper ut eidem
confinis parti vicinis staret in astris.
horum operum sortes ad singula signa locavit,
non ut in aeterna caeli statione manerent
et cunctos hominum pariter traherentur in ortus
ex isdem repetita locis, sed tempore sedes
nascentum acciperent proprias signisque migrarent
atque alias alii sors quaeque accederet astro,
ut caperet genitura novam per sidera formam
nec tamen incerto confunderet omnia motu.
sed, cum pars operum quae prima condita sorte est
accepit propriam nascentis tempore sedem,
cetera succedunt signisque sequentibus haerent.
ordo ducem sequitur donec venit orbis in orbem.
has autem facies rerum per signa locatas,
in quibus omnis erit fortunae condita summa,

[71] res posuit
nomina

[80] alius [alii]
[83] parte est

and Moon and planets and are also themselves outflown, to these nature gave dominion: to each sign she devoted individual associations, and fixed in the zodiac for ever the total distribution, so that the influences upon destiny might be drawn from all quarters and concentrated into a single whole. The infinite variety of circumstances that might occur in human affairs, all the possible sufferings, the callings and the skills, and all the vicissitudes imaginable, these nature included in the allotment: she arranged them in as many portions as she had created signs, making a fixed dispensation, each portion having its special benefice; and by an unchanging system she so disposed the full reckoning of man's estate among the stars that any portion would always stand next to the same portion in adjacent signs. The lots of these activities she allocated each to a sign, not in such a way that they should remain in a permanent quarter of the sky and, always looked for in the same place, be drawn to influence all human activities alike, but so that they should receive their proper position according to the moment of birth and change from sign to sign, each lot at a different time moving to a different constellation, so that the nativity then meets with a new pattern in the zodiac, without however disturbing everything with irregular motions. But when the section of the activities which is assigned to the first lot receives its proper place at the moment of a nativity, the rest follow attached to the zodiacal signs in their usual sequence. The procession follows the leader until the circle of lots fills up the circle of signs. Now just as these aspects of human affairs, wherein the whole sum of fortune will be found committed, are when placed in the zodiac

ut cursu stellae septem laeduntve iuvantve
cardinibusve movet divina potentia mundum,
sic felix aut triste venit per singula fatum
talis et illius sors est speranda negoti.
haec mihi sollemni sunt ordine cuncta canenda
et titulis signanda suis rerumque figuris,
ut pateat positura operum nomenque genusque.

Fortunae sors prima data est. hoc illa per artem
censetur titulo, quia proxima continet in se
fundamenta domus domuique haerentia cuncta :
qui modus in servis, qui sit concessus in arvis
quamque datum magnas operum componere moles,
ut vaga fulgentis concordant sidera caeli.
post hinc militiae locus est, qua quidquid in armis
quodque peregrinas inter versantibus urbes
accidere assuevit titulo comprenditur uno.
tertia ad urbanos statio est numeranda labores
(hoc quoque militiae genus est, civilibus actis
compositum) fideique tenet parentia vincla ;
format amicitias et saepe cadentia frustra
officia, et cultus contingant praemia quanta
edocet, appositis cum mundus consonat astris.
iudiciorum opus in quarto natura locavit
fortunamque fori : fundentem verba patronum
pendentemque reum lingua nervisque loquentis
impositum, et populo nudantem condita iura
atque expensa sua solventem iurgia fronte,
cum iudex veri nihil amplius advocat ipso.

[89] ut cum [113] rostrisque
[90] mundi

[a] *Cf.* 2. 788 ff.
[b] *Cf.* Introduction pp. lxii ff. and figure 14.

favoured or frowned on by the seven planets in their courses, and just as a divine force rotates the skies through the cardinal points,[a] so an influence, propitious or malign, comes over each allotment, and even likewise must you expect the issue of such and such a business. I must sing of all these lots in their due order, and designate them with their respective titles and spheres of action, that the location and name and nature of the activities may be revealed.[b]

[96]The first lot has been assigned to Fortune. This is the name by which it is known in astrology, because it contains in itself the chief essentials of the home and all that attaches to the name of home : the limit accorded to the number of one's slaves and the amount of one's land, and the size of buildings it is given one to erect, according to the degree of harmony in the wandering stars of bright heaven. Thereafter follows the abode of warfare, where under a single title is comprised whatever is likely to happen in war or befall those that are busied in foreign cities. The third position is to be assigned to the business of the city (this, too, a kind of warfare, one made up of civil engagements) and contains the ties dependent on trust ; it shapes friendships and services oft vainly spent, and reveals the size of the rewards for devotion, when heaven is blessed with a harmonious disposition of the planets. In the fourth abode nature has placed the activity of the law-courts and the fortunes of the bar : the advocate pouring forth his case, and the accused no less hanging on the pleader's words and delivery than he is dependent upon them ; the judge revealing to laymen the secrets of the law and weighing up disputes which he settles with his usual gravity, when, arbiter of truth, he takes into

quidquid propositas inter facundia leges
efficit, hoc totum partem concessit in unam
atque, utcumque regunt dominantia sidera, paret.
quintus coniugio gradus est per signa dicatus
et socios tenet, et committens hospita iura
iungitur et similis coniungens foedus amicos.
in sexta dives numeratur copia sede
atque adiuncta salus rerum, quarum altera quanti
contingant usus monet, altera quam diuturni,
sidera ut inclinant vires et templa gubernant.
septima censetur saevis horrenda periclis,
si male subscribunt stellae per signa locatae.
nobilitas tenet octavam, qua constat honoris
condicio et famae modus et genus et specioso
gratia praetexto. nonus locus occupat omnem
natorum sortem dubiam patriosque timores
omniaque infantum mixta nutricia turba.
huic vicinus erit, vitae qui continet actum,
in quo sortimur mores, et qualibus omnis
formetur domus exemplis, quamque ordine certo
ad sua compositi discedant munera servi.
praecipua undecima pars est in sorte locata,
quae summam nostri semper viresque gubernat,
quaque valetudo constat, nunc libera morbis,
nunc oppressa, movent ut mundum sidera cumque.
non alia est sedes, tempusve genusve medendi
quae sibi deposcat vel cuius tempore praestet
auxilium et vitae sucos miscere salubris.

[121] committem hospitis una [140] quaeque
[136] quoque [141] et

---

[a] *I.e.* no matter what zodiacal sign the fifth station falls to: of course this is true of all the stations, and the poet only employs the words *per signa* here to fill out the verse.

account nothing but the truth. All that eloquence can achieve in debating proposed legislation is concentrated in a single part, and obeys the dictates of the dominating planets. Throughout the signs [a] the fifth station is dedicated to wedlock and controls alliances, whereto is joined the compact which unites the relationship of host and guest and binds like friend to like. In the sixth seat is reckoned abundance of means and therewith financial soundness : the one indicates the degree, the other the duration of one's resources, according as the planets modify and the temples [b] direct the influence of this seat. The seventh is accounted grim with danger dire, if the planets located in the signs ill accord. Social rank occupies the eighth ; in it are comprised privileged position, extent of fame, nobility of birth, and popularity with its showy cloak. The ninth position takes possession of all filial problems, parental worries, and the motley collection of all matters connected with the rearing of children. Next will be found the one that embraces the conduct of life ; herein our character is determined, likewise with what traditions every family is shaped, and in what appointed fashion servants discharge the tasks to which they have been severally deputed. The portion located in the eleventh lot is paramount : it has permanent control of our whole being and strength ; in it abides our health, now immune from and now yielding to sickness, according to the influence of the planets over the heavens. No other seat is there which claims the choice of remedy and the moment for administering it, or in whose hour therapy and the mixing of life-saving potions have

[b] Of the dodecatropos (2. 856 ff.).

ultimus et totam concludens ordine summam
rebus apiscendis labor est, qui continet omnis
votorum effectus, et, quae sibi quisque suisque
proponit studia atque artes, haec irrita ne sint.
seu ferat officium nutus blanditus in omnis,
aspera sive foro per litem iurgia temptet,
fortunamve petat pelago ventisque sequatur,
seu Cererem plena vincentem credita messe
aut repetat Bacchum per pinguia musta fluentem,
hac in parte dies atque hac momenta dabuntur,
si bene convenient stellae per signa sequentes ;
quarum ego posterius vires in utrumque valentis
ordine sub certo reddam, cum pandere earum
incipiam effectus. nunc, ne permixta legentem
confundant, nudis satis est insistere membris.

Et, quoniam certo digestos orbe labores
nominaque in numerum viresque exegimus omnis
(athla vocant Grai, quae cuncta negotia rerum
in genera et partes bis sex divisa coercent),
nunc, quibus accedant signis quandoque, canendum
perpetuas neque enim sedes eademve per omnis
sidera nascentis retinent, sed tempore mutant,
nunc huc nunc illuc signorum mota per orbem,
incolumis tamen ut maneat qui conditus ordo est.

157 earum] rerum   164 ascendant

---

[a] See Introduction p. lxiv.

[b] A promise not fulfilled in the poet's extant work.

[c] It is not known whom Manilius is following here. ἆθλα, translated by *labores* (the Greek is more properly ἆθλοι), is not a term found in any Greek work ; *sortes*, on the other hand, clearly renders κλῆροι, special degrees of the zodiac occupied by the planets.

greater efficacy.[a] Duly closing the total sum, the last activity deals with the attainment of our aims : it embraces all the issues of the vows we make and ensures that the skill and efforts expended for oneself and one's own fail not of success. Whether one is to offer one's service and submit to another's every beck and call, whether to embark on a bitter dispute by litigation in the courts, whether to seek fortune on the sea and pursue it with the winds, whether to put one's hopes on a crop that with huge harvest will exceed one's outlay or a vintage that will overflow with the rich must—this is the portion in which day and hour for decision shall be given,[a] if the planets as they move through the zodiac are favourably situated. Of the force they exert for good or ill I I shall later [b] tell in due order, when I come to rehearse their influences. Meanwhile it is enough to press on with the bare outlines, lest a complex presentation confuse my reader.

[160]Since I have expounded the fixed circle of activities arranged in order, with all their names and powers (athla [c] is the Greek name for the sections which contain all the affairs of life arranged in twice six classes and portions), now must my poem tell to which signs they are attached at any given moment. For they do not remain in permanent abodes or stay attached to the same stars for every person born, but they change them [d] according to time, moving now hither, now thither, through the circle of signs, yet in such a way that the fixed order of succession remains intact. Now, lest an incorrect figuration

[d] Abodes and stars. Thus the first lot does not always abide at the rising cardinal point (ascendant), nor is it always attached to the sign of Aries.

ergo age, ne falsa variet genitura figura,
si sua quemque voles revocare ad signa laborem,
Fortunae conquire locum per sidera cuncta,
quae primum est aerumnosis pars dicta sub athlis.
qui tibi cum fuerit certa ratione repertus,
cetera praedicto subeuntibus ordine signis
coniunges, teneant proprias ut singula sedes.
et, ne forte vagus Fortunae quaerere sedem
incipias, duplici certam ratione capesse.
cum tibi, nascentis percepto tempore, forma
constiterit caeli, stellis ad signa locatis,
transverso Phoebus si cardine celsior ibit
qui tenet exortum vel qui demergit in undas,
per tempus licet affirmes natum esse diei.
at, si subiectis senis fulgebit in astris
inferior dextra laevaque tenentibus orbem
cardinibus, noctis fuerit per tempora natus.
haec tibi cum fuerint certo discrimine nota,
tunc, si forte dies nascentem exceperit alma,
a sole ad lunam numerabis in ordine partes
signorum, ortivo totidem de cardine duces,
quem bene partitis memorant horoscopon astris.
in quodcumque igitur numerus pervenerit astrum
hoc da Fortunae. iunges tum cetera signis
athla suis, certo subeuntibus ordine cunctis.
at, cum obducta nigris nox orbem texerit alis,
siquis erit qui tum materna excesserit alvo,
verte vias, sicut naturae vertitur ordo.
consule tum Phoeben imitantem lumina fratris

[172] pars est numerosis
[177] certa
[183] sevis *GL*: sevus *M*

[a] Above the ascendant or above the occident.

mar your casting of a geniture, when you wish to apply each activity to its sign, look through the whole zodiac for the position of Fortune; this was the portion first described among the toilsome labours. When you have accurately determined this, then join the others in the prescribed order to the sequence of signs, so that each holds its appropriate abode. And lest perchance you err in essaying the quest for the seat of Fortune, learn its correct location by a twofold procedure. When you have noted the moment of nativity and have duly recorded the figuration of the heavens with the planets marked in their signs, then if the Sun is moving above that horizontal cardinal which holds it at its rising or that which plunges it into the waves,[a] you may be sure that the birth takes place in time of day. But if it is shining in the six submerged signs, beneath the cardinal points which hold the world in place on left and right, then the birth falls within the hours of night. When with sure judgement you have established this, then, if by chance bounteous day gains the nativity, number the degrees of the signs in order from the Sun to the Moon and count off as many from that ascendant point which is found after careful plotting of the zodiac and is called the Horoscope. Then whatever sign the number reaches is to be bestowed on Fortune. Next link the remaining activities to their signs, as they follow, all in their fixed sequence.[b] However, if a child leaves its mother's womb when night is enfolding the earth and mantling it with wings of darkness, reverse the process, even as nature's order is reversed. Seek guidance then from

[b] For an example of the diurnal calculation, see Introduction p. lxv and figure 15.

semper et in proprio regnantem tempore noctis ;
quotque ab ea Phoebus partes et signa recedit
tot numerare iubet fulgens horoscopos a se.
hunc Fortuna locum teneat subeuntibus athlis,
ordine naturae sicut sunt cuncta locata.

Forsitan et quaeras, agili rem corde notandam,
qua ratione queas, natalis tempore, nati
exprimere immerso surgentem horoscopon orbe.
quod nisi subtili visum ratione tenetur,
fundamenta ruunt artis nec consonat ordo ;
cardinibus quoniam falsis, qui cuncta gubernant,
mentitur faciem mundus nec constat origo
flexaque momento variantur sidera templi.
sed, quanta effectu, res est tam plena laboris
cursibus aeternis mundum per signa volantem,
ut totum lustret curvatis arcubus orbem,
exprimere et vultus eius componere certos
ac tantae molis minimum deprendere punctum :
quae pars exortum vel quae fastigia mundi
obtineat summi demersosve aequoris undis
auferat occasus aut imo sederit orbe.

Nec me vulgatae rationis praeterit ordo,
quae binas tribuit signis surgentibus horas
et paribus spatiis aequalia digerit astra,
ut parte ex illa, qua Phoebi coeperit orbis,
discedat numerus summamque accommodet astris,

[a] For an example of the nocturnal calculation, see Introduction pp. lxvi f. and figure 16.

the Moon, who ever reflects her brother's radiance and reigns in the hours of night, her own domain ; as many degrees and signs as is distant the Sun from her, so many the shining Horoscope bids you number from itself.[a] This place let Fortune hold with the other activities following, all positioned in their natural order.

[203]Perhaps you will ask—and the question requires a nimble brain's attention—how at the moment of birth it is possible to determine the native's horoscoping degree as it rises from the submerged bowl of the heavens. Unless this is understood and grasped with a sharp mind, the basis of our science collapses and its design is thrown into confusion, since, if the cardinal points, which control everything, are not true, the skies put on a false appearance, the figuration of the birth is unsound, and the stars are displaced, deflected by the shift of temple. But it is a matter as laborious as important to represent, as it revolves with the zodiac in ceaseless motion, the passage of heaven with its arched quadrants through the whole circle, to depict its exact features, and to determine a minute point of so vast a structure, be it the degree in which the ascendant or that in which the summit of the sky *on high is situated, or that by which, sinking into the waters of the deep*, the setting is appropriated, or that which rests on the circle's bottom.

[218]Nor am I unacquainted with a common method of calculating, which attributes two hours each to the risings of the signs and reckons the stars as identical with equal ascensions ; by this calculation, starting from the point where the Sun's orbit begins, the number of the hours is counted, with its total con-

donec perveniat nascentis tempus ad ipsum,
atque, ubi substiterit, signum dicatur oriri.
sed iacet obliquo signorum circulus orbe,
atque alia inflexis oriuntur sidera membris,
ast illis magis est rectus surgentibus ordo,
ut propius nodis aliquod vel longius astrum est.
vix finit luces Cancer, vix bruma reducit,
quam brevis ille iacet, tam longus circulus hic est ;
Libra Ariesque parem reddunt noctemque diemque.
sic media extremis pugnant extremaque summis.
nec nocturna minus variant quam tempora lucis,
sed tantum adversis idem stat mensibus ordo.
in tam dissimili spatio variisque dierum
umbrarumque modis quis credere possit in auras
omnia signa pari mundi sub lege meare ?
adde quod incerta est horae mensura neque ullam
altera par sequitur, sed, sicut summa dierum
vertitur, et partes surgunt rursusque recedunt ;
cum tamen, in quocumque dies deducitur astro,
sex habeat supra terras, sex signa sub illis.
quo fit ut in binas non possint omnia nasci,
cum spatium non sit sibi par pugnantibus horis,
si modo bis senae servantur luce sub omni,
quem numerum debet ratio sed non capit usus.
Nec tibi constabunt aliter vestigia veri,
ni, lucem noctemque paris dimensus in horas,

228 nobis
236 possit credere
237 manere
244 sibi par] tantum

---

[a] That is, the sign of Capricorn.
[b] Aries and Libra are at variance with Cancer and Capricorn ; Capricorn with Cancer.

verted into signs, until it reaches the precise time of birth: the sign wherein it halts is said to be the rising sign. But the circle of signs lies in an oblique belt, and some signs rise with slanting limbs, whilst others have on rising a more upright posture, according as a sign is either nearer to the equinoctial points or farther off. Cancer is loth to end the day, winter [a] to renew it; the Sun's heavenly orbit is as long in the former case as it is brief in the latter; Libra and Aries make equal night and day. Thus the middle signs are at variance with the end ones, and the lowest with the highest.[b] Nor do the hours of night alter any less than those of day but the same variation exists, though in opposite months. Who could believe that, with such uneven periods and such changeable limits of day and darkness, all the signs rise into the sky under a uniform law of heaven? Moreover, the length of an hour is a variable quantity, nor is any hour followed [c] by another of the same length, but just as the duration of days varies, so also their parts increase and again decrease; and yet, under whatever sign a day's course is run,[d] it has at any moment six signs above the earth and six below. It follows that the signs cannot all rise in a period of two hours, since owing to the discrepancy between the hours their duration is not uniform, if indeed twelve hours are contained between each dawn and dusk, a number which ordinary reckoning demands but strict application does not permit.

[247]Nor in any other way will you be set on the path of truth unless, apportioning daylight and

[c] That is, on the next day.

[d] No matter in what sign of the zodiac the Sun is situated.

in quantum vario pateant sub tempore noris,
regulaque exacta primum formetur in hora
quae surgens sidensque diem perpendat et umbras.
haec erit, in Libra cum lucem vincere noctes
incipiunt vel cum medio concedere vere.
tunc etenim solum bis senas tempora in horas
aequa patent, medio quod currit Phoebus Olympo.
is cum per gelidas hiemes summotus in austros
fulget in octava Capricorni parte biformis,
tunc angusta dies vernalis fertur in horas
dimidiam atque novem, sed nox oblita diei
bis septem apposita, numerus ne claudicet, hora
dimidia. sic in duodenas exit utrimque
et redit in solidum natura condita summa.
inde cadunt noctes surguntque in tempora luces,
nunc huc nunc illuc gradibus per sidera certis
impulsae, quorum ratio manifesta per artem
collecta est venietque suo per carmina textu,
donec ad ardentis pugnarunt sidera Cancri ;
atque ibi conversis vicibus mutantur in horas
brumalis, noctemque dies lucemque tenebrae
hibernam referunt, alternaque tempora vincunt.
atque haec est illas demum mensura per oras

[251] surgens sidensque diem] signemque diem sedem [269] quarum

[a] That is, in Aries.

[b] When the Sun is in Aries or Libra, *i.e.* at the celestial equator, it must in any latitude (outside the circumpolar regions) describe equal arcs above and below the horizon in twelve hours each, rising due east and setting due west (*cf.* Introduction p. lxix and figure 17).

[c] In subsequent discussions it is clear that Manilius (or rather his sources) reckoned solstices and equinoxes as taking place in the first degree : see Introduction pp. lxxxi ff.

[d] That is, at 443-482, where the general rule is given.

darkness into uniform hours, you ascertain the duration of daylight and darkness at different seasons of the year, and first create a fixed standard by accurately calculating an hour which with increase here and decrease there holds day and night evenly balanced. Such an hour will be found when the nights are about to prevail over daylight in Libra or yield to it in mid-spring.[a] For then and then only do the two periods of night and day extend alike to twice six hours, since the Sun is revolving at the heavenly equator.[b] And when, moved to the south in chill winter-tide, it shines in the eighth [c] degree of two-formed Capricorn, then the brief day extends to nine and a half equinoctial hours and night, forgetful of day, to hours twice seven, increased, lest the full number go short, by half an hour. Thus the sum of the unequal night and day established by nature breaks evenly into two separate amounts of twelve hours and comes together again to form the full number. From this point the nights diminish and the days increase in length, thrust now on this side of the mean in their course through the zodiac and now on that in fixed gradations (of which a clear account has been drawn up in our science and will follow in my poem at the proper place [d]), until their combat has reached the stars of scorching Cancer ; and there the division of day and night changes to the winter one, but with roles reversed : daylight brings back the length of winter's night, darkness the length of winter's day, and now it is the other period which has the upper hand. Such at least is the measurement in the latitude of the lands inundated by the Nile,[e]

[e] Incorrect: the measurement is that of the latitude of Rhodes.

quas rigat aestivis gravidus torrentibus amnis
Nilus et erumpens imitatur sidera mundi
per septem fauces atque ora fugantia pontum.
Nunc age, quot stadiis et quanto tempore surgant
sidera, quotque cadant, animo cognosce sagaci,
ne magna in brevibus lateant compendia dictis.
nobile Lanigeri sidus, quod cuncta sequuntur,
dena quater stadia exoriens duplicataque ducit
cum cadit, atque horam surgens eiusque trientem
occupat, occiduus geminat. tum cetera signa
octonis crescunt stadiis orientia in orbem
et totidem amittunt gelidas vergentia in umbras.
hora novo crescit per singula signa quadrante
tertiaque e quinta pars parte inducitur eius.
haec sunt ad Librae sidus surgentibus astris
incrementa : pari momento damna trahuntur
cum subeunt orbem. rursusque a sidere Librae
ordine mutato paribus per tempora versa
momentis redeunt. nam, per quot creverat astrum
Lanigeri stadia aut horas, tot Libra recedit ;
occiduusque Aries spatium tempusque cadendi
quod tenet, in tantum Chelae consurgere perstant.
excipiunt vicibus se signa sequentia versis.
haec ubi constiterint vigilanti condita mente

272 gradibus
277 careant
285 in quarta [pars] parte si ducitur
288 ad sidera
294 excipiunt vicibus] eius in exemplum

[a] A stade is—Manilius has forgotten to say, but its measurement is revealed by what follows—half a degree of the celestial equator, that portion of the celestial equator which takes two minutes to rise : thus one revolution com-

when its stream is swollen with summer flood, the Nile which, in sallying forth through seven channels and mouths which put to flight the current of the sea, resembles the planets of heaven.

[275]Now learn with a perceptive mind how many stades [a] and how much time the stars take to rise and set, lest great profit stay hidden in a few words. The Ram's distinguished [b] sign, which leads the rest, appropriates in rising forty stades, and twice as many when it sets; in ascension it occupies an hour and a third and doubles this at its descension. Then do the other signs in rising upon earth increase by eight stades each and as many lose when they sink into the chill darkness. Through each single sign the hour of ascension has an additional increase of a quarter, to which is added the third part of its fifth part. [c] These are the increments of the rising signs as far as the constellation of the Balance: decreases of like amount are subtracted, when the signs pass beneath the earth. And again from the constellation of the Balance the signs return with similar gradations, but backwards and through a reverse order of times. For in as many stades and hours the Ram's sign had risen, in so many does the Balance set; and the space and period of descension occupied by the Ram in setting are preserved by the Balance at its rising. The remaining signs succeed each other in an inverse progression. When this has been understood and stored in an alert mind, it is

prises 720 stades and takes 24 hours. For an arithmetical tabulation of this passage, see table 1 p. lxx.

[b] Distinguished for its Golden Fleece.

[c] Through each single sign the increment is a quarter of an hour (15 minutes) plus the fifteenth of that quarter of an hour (1 minute), *i.e.* the increment is 16 minutes.

iam facile est tibi, quod quandoque horoscopet astrum,
noscere, cum liceat certis surgentia signa
ducere temporibus propriasque ascribere in horas,
partibus ut ratio signo ducatur ab illo,
in quo Phoebus erit, quarum mihi reddita summa est.

Sed neque per terras omnis mensura dierum
umbrarumque eadem est, simili nec tempora summa
mutantur : modus est varius ratione sub una.
nam, qua Phrixei ducuntur vellera signi
Chelarumque fides iustaeque examina Librae,
omnia consurgunt binas ibi signa per horas,
quod medius recto praeciditur ordine mundus
aequalisque super transversum vertitur axem.
illic perpetua iunguntur pace diebus
obscurae noctes ; aequo stat foedere tempus ;
nec manifesta patet falsi fallacia mundi,
sed similis simili toto nox redditur aevo ;
omnibus autumnus signis, ver omnibus unum,
una quod aequali lustratur linea Phoebo.
nec refert illic quo sol decurrat in astro,
litoreumne coquat Cancrum contrane feratur,
[sideribus mediis an quae sint quattuor inter]
quod, quamquam per tris signorum circulus arcus
obliquus iaceat, recto tamen ordine zonae

296 tibi] si
299 signis . . . illis
300 quos
304 numquam
306 sibi
307 praecingitur
312 nec
315 illic] tunc
sol] Phoebus

[a] In terms of stades.
[b] In lands beneath the celestial equator.
[c] The impartial Claws and Libra's just balance are

easy for you to ascertain what sign is horoscoping at any moment, since it is then possible to compute ascensions of signs with exact periods and record them with the precise number of hours, so that you may make the calculation from the sign in which the Sun is situate in terms of the units [a] whose reckoning I have revealed.

[301]But the length of day and night is not the same throughout all lands, nor do the times of ascensions differ by the same increments: the amount of fluctuation varies, though the principle is the same. For in lands beneath the line where [b] revolve the fleece of the Phrixean sign and the impartial Claws and Libra's just balance,[c] all signs occupy two hours in rising,[d] since the sky is cut vertically in the middle and revolves evenly above the horizontal axis. In that latitude the murky nights are matched with the days in everlasting peace; the hours remain the same in a federation of equality; nor does the deceitfulness of the inconstant sky stand out, but throughout all time like night follows night; all the signs enjoy the equinox of autumn, all the equinox of spring, since impartial Phoebus pursues a consistent track. Nor in that latitude does it make any difference in what sign the Sun is coursing along, whether it roasts the shore-haunting Crab or is borne in the opposite sign,[e] because, although the zodiac lies aslant through the three tropical circles, nevertheless

identical, both phrases merely signifying the zodiacal sign Libra.

[d] This is only roughly true, for the ecliptic does not rise at right angles to the horizon (*cf.* Introduction pp. lxxi f. and figure 18).

[e] [317]: ". . . or in the two intermediate signs or in any between those four, . . ."

consurgunt supra caput in terrasque feruntur
et paribus spatiis per singula lustra resurgunt,
ac bene diviso mundus latet orbe patetque.
at, simul ex illa terrarum parte recedas,
quidquid ad extremos temet provexerit axes
per convexa gradus gressum fastigia terrae,
quam tereti natura solo decircinat orbem
in tumidum et mediam mundo suspendit ab omni,—
ergo ubi conscendes orbem scandensque rotundum
degrediere simul, fugiet pars altera terrae,
altera reddetur. sed, quantum inflexeris orbem,
tantum inclinabit caeli positura volantis,
et modo quae fuerant surgentia limite recto
sidera curvato ducentur in aethera tractu,
atque erit obliquo signorum balteus orbe
qui transversus erat, statio quando illius una est,
nostrae mutantur sedes. ergo ipsa moveri
tempora iam ratio cogit variosque referre
sub tali regione dies, cum sidera flexo
ordine conficiant cursus obliqua malignos,
longius atque aliis aliud propiusve recumbat.
pro spatio mora magna datur : quae proxima nobis
consurgunt, longos caeli visuntur in orbes ;
ultima quae fulgent, celeris merguntur in umbras.
et, quanto ad gelidas propius quis venerit Arctos,
tam magis effugiunt oculos brumalia signa,
vixque ortis occasus erit. si longius inde
procedat, totis condentur singula membris
tricenasque trahent conexo tempore noctes

320 -que caput [in]
324 praevecteris axe
330 inflexerit orbe
331 inclinavit
346 ortus

[a] *Cf.* Introduction p. lxxii and figure 19.
[b] Sagittarius and Capricorn.

these circles rise overhead vertically and sink vertically into the earth, and their risings through individual arcs take uniform periods, and due to the perfect bisection of the Earth the sky is half hidden and half visible.[a] Now when you move from this part of the world, the farther your footsteps take you towards the pole as you stride over the curved slopes of the Earth, which nature has shaped with even surface into a rounded sphere and suspended right in the middle of the universe—when, then, you climb the Earth's arc and, as you climb, at the same time move down, one part of the Earth will recede and another will take its place. But in proportion as you tilt the Earth, so will the axis of the rotating heavens incline, and the signs which had just now risen in a vertical arc will now be drawn into the air over an inclined track, and the baldric of signs which had just now moved from side to side across heaven will now move in an oblique orbit, since its location remains constant, whilst ours changes. It logically follows that in such a place the times of orbits above the horizon also fluctuate and cause days of variable duration, for in their tilted arrangement slanted star-groups describe niggardly courses, and one sign sets nearer, another farther away than others. The length of visibility is in inverse proportion to distance away: signs that rise nearest to us are seen traversing large arcs of the sky; those that shine farthest away swiftly sink into the darkness. And the nearer one approaches the chill Bears, so much the more do the winter signs[b] recede from sight, and they have scarcely risen when they set. If one proceed still farther, each one in all of its parts will be hidden and will bring thirty nights of consecutive duration and

et totidem luces adiment. sic parva dierum
efficitur mora et attritis consumitur horis
paulatimque perit, spatio fugientibus astris.
pluraque, per partes surrepto tempore, signa
quaerentur medio terrae celata tumore
abducentque simul Phoebum texentque tenebras,
mensibus ereptis donec sit debilis annus.
si vero natura sinat sub vertice caeli,
quem gelidus rigidis fulcit compagibus axis,
aeternas super ire nives orbemque rigentem
prona Lycaoniae spectantem membra puellae,
stantis erit caeli species, laterumque meatu
turbinis in morem recta vertigine curret.
inde tibi obliquo sex tantum signa patebunt
circuitu, nullos umquam fugientia visus
sed teretem acclini mundum comitantia spira.
hic erit una dies per senos undique menses
dimidiumque trahens contextis lucibus annum,
numquam erit occiduus quod tanto tempore Phoebus,
dum bis terna suis perlustrat cursibus astra,
sed circum volitans recto versabitur orbe.
at, simul e medio praeceps descenderit orbe
inferiora petens deiecto sidera curru
et dabit in pronum laxas effusus habenas,
per totidem menses iunget nox una tenebras
vertice sub caeli. nam quisquis spectat ab axe,
dimidium e toto mundi videt orbe rotundi,
pars latet inferior ; neque enim circumvenit illum
recta acies, mediaque tenus distinguitur alvo.

351 fulgentibus
364 semper
369 versetur ab
371 cursu
374 axe] omni

[a] Callisto, *i.e.* Ursa Major.

take away as many days. Thus but a brief stay is left for the days, and with the diminution of their hours it is soon spent and gradually disappears altogether, as the stars flee in space. And as by degrees the time is snatched away, more signs will become invisible, concealed by the intervening curve of the Earth: they will abduct the Sun as well and weave a web of unbroken darkness, until the year is weakened by the loss of whole months. But if nature were to let you walk beneath the zenith of heaven, which the cold pole sustains with inflexible support, over perpetual snow and a frozen earth which gazes up at the figure of Lycaon's daughter [a] facing down, then the spectacle of an erect heaven will be encountered, which with the revolution of its edges will rotate with upright spin like a top. At that spot six signs only, lying in an oblique semicircle, will you ever be able to see; their inclined curve will never disappear from sight but will rotate with the rounded sky. Everywhere here a single day will last six months, a day which will keep a half year in continuous light, for in all this time, during which his course will take him through six signs, Phoebus will never set but will rotate in circling flight around the erect heavens. But when he plunges beneath the equator, setting his car on a downward course for the southern signs, and speeding on the descent gives free rein to his steed, through as many months shall a single night prolong darkness beneath the zenith of heaven. For whoever looks out from the pole sees half of the whole sphere of the rounded heavens, whilst the lower is hidden; for the direct line of vision does not encompass the sky, but surveys it only as far as the extremity of its bulge.

effugit ergo oculos summo spectantis ab orbe
dum sex summersis vectatur Phoebus in astris,
abducitque simul luces tenebrasque relinquit
sideribus, donec totidem, quot mensibus actis
cesserat, inde redit geminasque ascendit ad Arctos.
hic locus in binas annum noctesque diesque
per duo partitae dirimit divortia terrae.

Et, quoniam quanto varientur tempora motu
et quibus e causis dictum est, nunc accipe, signa
quot surgant in quoque loco cedantque per horas,
partibus ut prendi possint orientia certis,
ne falsus dubia ratione horoscopos erret.
atque hoc in totum certa sub lege sequendum est,
singula quod nequeunt, per tot distantia motus,
temporibus numerisque suis exacta referri.
a me sumat iter positum, sibi quisque sequatur
perque suos tendat gressus, mihi debeat artem.
quacumque hoc parti terrarum quisque requiret,
deducat proprias noctemque diemque per horas
maxima sub Cancro minimis quae cingitur umbris ;
et sextam summae, fuerit quae forte, diurnae
vicino tribuat post Cancri templa Leoni ;
at quae nocturnis fuerit mensura tenebris
in totidem partes simili ratione secanda est,
ut, quantum una ferat, tantum tribuatur ad ortus
temporis averso nascenti sidere Tauro.

[385] variantur

---

[a] *Cf.* Introduction p. lxxiii and figure 20.

[b] For the arithmetic of the passage 385-442, see Introduction p. lxxiv and table 2.

[c] The exposition which follows presupposes that the solstice occurs at the beginning of Cancer. As an example, let the longest day be 15 hours, the night 9 hours . . .

[d] . . . then the rising of Leo will take 2½ hours . . .

Thus, while he rides through the six submerged signs, Phoebus shuns the gaze of the observer on top of the Earth and at one and the same time steals daylight from the signs and leaves them darkness, until, as many months spent as before his departure, he comes back and rises towards the two Bears. Such a place twice separates the year into night and day, once in each hemisphere of the divided Earth.[a]

385 And [b] since I have described the fluctuation with which the times of day and night vary and the reasons for the fluctuation, learn now how many hours the signs take to rise and set in any given place, in order that you can compute the precise moment when each of a sign's degrees is rising. Then the horoscope will not be in error, falsified by inaccurate reckoning. The information is in all cases to be found by using a general principle, since, inasmuch as the individual signs differ with such great fluctuation, they cannot be recorded with every conceivable calculation of their periods and degrees. Let each one take the path laid down by me, and, following it for himself, trace that path with his own footsteps; let him owe the method to me. Whatever be the latitude of the Earth in which anyone desires this information, let him determine to their true number of hours the periods of light and darkness on the longest day which at Cancer's sign is interposed between the least portions of darkness [c]; and let him attribute a sixth of the longest day's hours, whatever be the sum, to the Lion, next neighbour to the temple of the Crab [d]; and the measurement of night's darkness is likewise to be divided into as many parts, so that the time of one portion is to be assigned to the Bull's rising when his hind-foremost constellation is

has inter quasque accipiet Nemeeius horas
quod discrimen erit, per tris id divide partes, 4
tertia ut accedat Geminis, qua tempora Tauri
vincant, atque eadem Cancro similisque Leoni,
sed certa sub lege, prioris semper ut astri 4
incolumem servent summam crescantque novando.
sic erit ad summam ratio perducta priorem 4
quam modo divisis Nemeaeus duxerit horis.
inde pari Virgo procedat temporis auctu. 4
his usque ad Chelas horarum partibus aucta 4
per totidem e Libra decrescent sidera partes.
et, quantis in utrumque moris tollentur ad ortus, 4
diversam in sortem tantis mergentur ad umbras.
haec erit horarum ratio ducenda per orbem
signorum: nunc in noscenda pone laborem
illa, quot stadiis oriantur quaeque cadantque.
quae quater et cum ter centum vicenaque constent,
detrahitur summae tota pars, quota demitur usque 4
omnibus ex horis aestivae nomine noctis,
solstitium summo peragit cum Phoebus Olympo.
quodque his exsuperat demptis id ducito in aequas

406 tauro
407 vincat
412 servet crescensque
409 quomodo
413 denas auctis
414 [e] librae
415 modis
419 [quater et] centum numerus v.
420 utque

---

[a] . . . and that of Taurus $1\frac{1}{2}$ hours ; . . .

[b] . . . one third of $2\frac{1}{2} - 1\frac{1}{2}$ is $\frac{1}{3}$, which will be the differential between the signs ; . . .

[c] . . . thus the risings of Gemini, Cancer, and Leo will take respectively $1\frac{5}{6}$, $2\frac{1}{6}$, and $2\frac{1}{2}$ hours, . . .

[d] . . . as was found above (p. 192, note *d*) ; . . .

[e] . . . Virgo will rise in $2\frac{5}{6}$ hours, and so does Libra ; . . .

born.[a] Divide the difference between these hours and those received by the Nemean into three parts [b]; let one of these three parts be given to the Twins, so that they may exceed by it the hours of the Bull, and the same amount to the Crab and the like to the Lion, but under the strict condition that they ever preserve intact the sum of the former sign and grow by accumulating.[c] Your reckoning will thus arrive at the former total which just now the Nemean secured upon partition of the hours.[d] Thence let the Virgin go ahead with equal increment of time. With these fractions of hours the signs increase as far as the Claws [e]; with the same gradations they will decrease after Libra.[f] And the periods, greater or smaller, in which the stars will rise at their ascensions will be occupied by them in reverse magnitude when they plunge to the shades.[g] This is the scheme of hours to be completed right round the circle *of signs: now direct your efforts to learning* the other scheme, namely the totals of stades they each occupy in rising and setting. The stades amount in all to four times and to thrice a hundred, plus twenty,[h] and from the total is subtracted a fraction proportionate to that which every year is subtracted from the day's full hours for the credit of midsummer night, when the Sun accomplishes his solstice in the heights of heaven.[i] When this subtraction has been carried out, divide the remainder into six equal parts,

[f] . . . Scorpio will rise in $2\frac{1}{2}$ hours, Sagittarius in $2\frac{1}{6}$ hours, and so on to Pisces (and Aries), who will take $1\frac{1}{6}$ hours ; . . .

[g] . . . thus Virgo will set in $1\frac{1}{6}$ hours, Leo in $1\frac{1}{2}$ hours, and so on to Aries, who will take $2\frac{5}{6}$ hours.

[h] 720.

[i] Continuing the above example (p. 192, note *c*): from 720 subtract a number which is to it as 9 is to 24, viz. 270 ; . . .

sex partes, sextamque ardenti trade Leoni.
rursus qui steterit numerus sub nomine noctis
eius erit signo Tauri pars illa dicanda.
quodque hanc exsuperat partem, superatur ab illa,
distinguitque duas medio discrimine summas,
tertia pars eius, numero super addita Tauri,
tradetur Geminis. simili tum cetera lucro
procedent numeros semper tutata priores
augebuntque novo vicinas munere summas,
donec perveniant ad iustae sidera Librae :
ex illa totidem per partes sic breviantur
Lanigeri ad fines ; conversaque omnia lege
accipiunt perduntque paris cedentia sortes.
haec via monstrabit stadiorum ponere summas
et numerare suos ortus per sidera cuncta.
quod bene cum propriis simul acceptaveris horis,
in nulla fallet regione horoscopos umquam,
cum poterunt certis numerari singula signa
temporibus parte ex illa quam Phoebus habebit.

Nunc, quibus hiberni momentis surgere menses
incipiant (neque enim paribus per sidera cuncta
procedunt gradibus, nivei dum vellera signi
contingant aequum luces cogentia et umbras

[430] traditur [432] novem . . . munero

[a] . . . then Leo will in rising occupy one sixth of 720−270 (= 450), viz. 75 stades ; . . .

[b] . . . and Taurus one sixth of 270, viz. 45 stades ; . . .

[c] . . . one third of 75−45 is 10, which will be the differential between the signs ; . . .

[d] . . . thus Gemini in rising will occupy 10 plus Taurus's 45, viz. 55 stades, . . .

[e] . . . and Virgo (and Libra) 85 stades ; . . .

[f] . . . Scorpio in rising will occupy 75 stades, Sagittarius 65

and hand over a sixth to the blazing Lion.[a] Again, a like fraction of the number of stades which stands to the credit of night is to be accounted to the sign of the Bull.[b] And of the amount which exceeds the latter portion and is held in excess by the former, and which separates the two totals with an intervening difference, a third portion,[c] added on top of the Bull's toll, shall be handed to the Twins.[d] Then the other signs will go ahead with similar increment, ever preserving the preceding amounts, and will augment their neighbour's total with a further increase until they arrive at the constellation of the impartial Balance [e]: from the Balance they likewise decrease in the same gradation as far as the territory of the Ram [f]; and by a reverse principle all signs at their settings receive or lose as many amounts of stades.[g] This method will teach you how to arrive at the totals of stades and to calculate the ascension of each of the signs. As soon as you have firmly grasped this, together with the correct numbers of hours, in no latitude will the Horoscope ever elude you, since you will be able to calculate from accurate times the degree each sign is distant from the one occupied by the Sun.

443 Now [h] let me explain with what increments the days of the winter months begin to grow; for not with equal gradation in every sign do they increase in their journey to the spotless fleece of the sign which forces day and darkness to shoulder the same

stades, and so on to Pisces (and Aries), who will occupy 35 stades; . . .

[g] . . . thus Virgo in setting will occupy 35 stades, Leo 45 stades, and so on to Aries, who will occupy 85 stades.

[h] For the arithmetic of the passage 443-482, see Introduction pp. lxxv f. and table 3.

ferre iugum), magna est ratio breviterque docenda.
principio capienda tibi est mensura diei
quam minimam Capricornus agit, noctisque per horas
quam summam ; quodque a iusto superaverit umbris,
perdiderint luces, eius pars tertia signo
tradenda est medio semper, qua sorte retenta
dimidio vincat primum, vincatur et ipsum
extremo : totum in partes ita digere tempus.
his opibus tria signa valent ; sed summa prioris
ac medii numeri coniuncta sequentibus astris
cesserit, ut, senis fuerit si longior horis
brumali nox forte die, Capricornus in horam
dimidiam attollat luces, et Aquarius horam
ipse suam proprie ducat summaeque priori
adiungat, Pisces tantum sibi temporis ipsi
constituant, quantum accipiant de sorte prioris,
et tribus expletis horis noctemque diemque
Lanigero tradant aequandam tempore veris.
incipit a sexta tempus procedere parte
dividuum ; triplicant vires haerentia signa

450 quoque ad . . . umbras
451 et trepident
456 ac medii] accedit
457 ternis
458 hora

[a] The exposition which follows presupposes that the solstice occurs at the very beginning of Capricorn : anticipating the poet's illustration (457 ff.), let the longest night be 15 hours, the shortest day 9 hours ; . . .

[b] . . . the mean is 12, and one third of 15 − 12 or 12 − 9 is 1, . . .

[c] . . . which (denoting the hourly increase in the length of days) is to be given to the middle sign of Capricorn, Aquarius and Pisces, viz. to Aquarius ; . . .

[d] . . . Aquarius receives 1 hour, which is one half more than Capricorn's increase (= ½ hour) and one half less than Pisces' increase (= 1½ hours) : . . .

[e] . . . these are the individual increases, . . .

[f] . . . not the accumulated ones, for the accumulated

yoke: the principle is important, yet may be briefly taught. First you must measure in hours the shortest day and the longest night spent by Capricorn [a]; and a third of the amount which darkness has secured in excess of the mean and daylight has lost of it [b] must in any latitude be surrendered to the middle of the winter signs [c]; retaining this portion it surpasses the first sign by a half of the increment it received, and by the same fraction it is itself surpassed by the last sign [d]: into these amounts must you apportion the whole time. These are the increases enjoyed by the three signs [e]; but you will find that the sum of the first and middle numbers is combined and given to the constellation which follows.[f] Thus, if perchance midwinter night is six hours longer than midwinter day, Capricorn increases his days by half an hour, Aquarius draws his own private increase of an hour, which he joins to the half hour of the preceding sign,[g] and the Fishes obtain as much increment of time on their own account as they receive from the allotment of the preceding sign,[h] and having accumulated three hours they hand over night and day to the Ram to be levelled at springtide. The increments of time [i] begin their advance with a sixth [j]; the following constellation trebles this

increase of the last constellation, Pisces, is ½ hour (Capricorn) plus 1 hour (Aquarius) plus its individual increase, 1½ hours, viz. 3 hours: thus the length of day at the beginning of Capricorn was 9 hours, at the end of Pisces (= beginning of Aries) it is 12 hours.

[g] Thus the accumulated increase of Aquarius is 1½ hours.

[h] The individual increase of Pisces is 1½ hours, which it adds to the 1½ hours it inherits from Aquarius.

[i] The accumulated increases.

[j] One sixth of the difference between the extreme (15 hours, or whatever it is) and the mean of 12 hours.

ultimaque acceptas duplicant. ita summa diebus
redditur, aequatae solvuntur faenore noctes
rursus et incipiunt propria de sorte diebus
cedere diversa labentia tempora lege.
namque Aries totidem deducit noctibus horas
quot prius abstulerant proprio sub nomine Pisces,
hora datur Tauro, cumuletque ut damna priora
dimidiam adiungunt Gemini. sic ultima primis
respondent, pariterque, illis quae proxima fulgent,
et media aequatis censentur viribus astra.
[praecipuosque gerunt varianda ad tempora motus]
hac vice descendunt noctes a sidere brumae
tolluntur que dies, annique invertitur orbis,
solstitium tardi dum fit sub sidere Cancri ;
tumque diem brumae nox aequat, tempora noctis
longa dies, similique redit, quam creverat, actu.

Illa etiam poterit nascens via ducere ad astrum
quod quandoque vadis emissum redditur orbi.
nam quota sit lucis, si luce requiritur, hora
aspicies, atque hunc numerum revocabis in ipsum
multiplicans decies, adiectis insuper eidem
quinque tamen summis, quia qualicumque sub hora

478 discedunt . . . ad sidera

---

[a] Thus the accumulated increases are ½, 1½, and 3 hours respectively.

[b] At the beginning of Aries, where day and night are equal.

[c] 1½.

[d] The increases to the hours of day are thus from 12 to 13½ in Aries, from 13½ to 14½ in Taurus, and from 14½ to 15 in Gemini.

[e] Gemini corresponds to Capricorn in increasing the length of day by ½ hour.

amount and the last doubles what it receives.[a] Thus the mean sum is restored to the days, and the nights, reduced to parity,[b] are cleared of debt and in turn start lending to the days from their own capital sums of time which decrease in the opposite sequence. For the Ram deducts as many hours from the nights as the Fishes had previously taken away on their own account [c]; an hour is given to the Bull, and the Twins add a further half so as to amass the debits of the preceding signs.[d] Thus the last sign corresponds to the first [e]; and likewise assessed with equal wealth are the signs that shine next to these,[f] and so also the two middle ones.[g] This is the sequence in which the nights diminish and the days increase from the constellation of midwinter,[h] and in which the year's cycle revolves until the solstice takes place at the sign of the lingering [i] Crab; then the night matches the winter's day, then the long day the space of winter's night, and returns over a course similar to that over which it had increased.

483The following method will also be able to guide you to the rising sign that at any given moment is released from ocean and given to the earth. For you must ascertain what hour of the day it is, if the Horoscope is sought by day, and repeatedly heap this number on itself by multiplying it ten times, adding to the amount, however, a further five times,

[f] Taurus and Aquarius, which both increase the length of day by 1 hour.

[g] Aries and Pisces, which both increase the length of day by $1\frac{1}{2}$ hours. Here has been inserted [477]: "And they provide the greatest increments to produce the inequality of the times."

[h] From the beginning of Capricorn.

[i] Because in Cancer the Sun lingers longest.

ter quinas mundi se tollunt sidera partes.
hic ubi constiterit numerus, coniungere et illas,
quae superent Phoebo partes per signa, memento.
ex hac tricenas summa per sidera partes
distribues, primamque vicem, quo Phoebus in astro
fulserit, inde aliis, solem quaecumque sequentur.
tum quo subsistet numerus consumptus in astro
quave in parte suam summam nomenque relinquet
haec erit exoriens et pars et forma per ignes.

*   *   *   *   *

contineat partes. ubi summam feceris unam,
tricenas dabis ex illa per singula signa,
donec deficiat numerus ; quaque ille sub astri
parte cadet, credas illam cum corpore natam
esse hominis pariterque orbem vidisse per ignes.
sic erit ipse tibi rapidis quaerendus in astris
natalis mundi certoque horoscopos ortu,

[496] numerique relinquit

---

[a] Let the moment of birth be 4 hours after sunrise when the Sun is in the 10th degree of Gemini : 4 times 15 is 60 ; . . .

[b] . . . add the 10 degrees the Sun is left with in Gemini : that makes 70 ; . . .

[c] . . . from 70 give 30 to Gemini, and 30 to Cancer : . . .

[d] the number is thus used up in Leo, and so the Horoscope will be found in the 10th degree of Leo. At this point in the text several verses have been lost, but their contents clearly dealt with a nocturnal procedure (*cf.* 485) and are easily reconstructed.

because in any hour of the day the signs ascend through thrice five degrees.[a] When this number is found, be sure to join to it the degrees which the Sun is left with in the zodiac.[b] From this total you must distribute thirty degrees to each sign, the first portion to the sign in which the Sun is shining, then to the others as they follow in the wake of the Sun.[c] Then the sign in which the number is used up and comes to a stop, and the degree in which it abandons its total and the name of number, will be the degree and starry figure which will be rising with its fires.[d] *If the Horoscope is sought by night, the procedure is the same, for once more you are to count back from the sign in which the Sun is situate, but the numerical calculation is more laborious. Ascertain the hour of night and multiply it by fifteen* [e]*; add thereto the degrees the Sun has completed in his sign* [f]*; add also one hundred and eighty so that your reckoning will day's* degrees contain.[g] When you have arrived at a grand total, give thirty degrees of it to each sign, until the number fails [h]; and in whatever degree of a constellation it runs out, this,[i] you can be confident, was born together with the native's body and, at the same time as the native, beheld the earth with its fires. In this way you must seek among the fast-rotating stars the rising point of the heavens and the

[e] Let the moment of birth be 4 hours after sunset when the Sun is in the 10th degree of Gemini: 4 times 15 is 60; . . .

[f] . . . add the 10 degrees the Sun has completed in Gemini: that makes 70; . . .

[g] . . . add 180: that makes 250; . . .

[h] . . . give 30 each to Gemini, Cancer, Leo, Virgo, Libra, Scorpio, Sagittarius, and Capricorn: that leaves 10; . . .

[i] . . . thus the Horoscope will be found in the 10th degree of Aquarius.

ut, cum exacta fides steterit sub cardine primo,
fallere non possint summi fastigia caeli,
non celeres obitus, stent fundamenta sub imo,
[stent veri stellarum obitus verique subortus]
sideraque in proprias vires sortesque recedant.
Nunc sua reddentur generatim tempora signis,
quae divisa etiam proprios ducuntur in annos
et menses lucesque suas horasque dierum,
per quae praecipuas ostendunt singula vires.
primus erit signi, quo Sol effulserit, annus,
annua quod lustrans consumit tempora mundum ;
proximus atque alii subeuntia signa sequuntur.
Luna dabit menses, peragit quod menstrua cursum.
tutelaeque suae primas horoscopos horas
asserit atque dies, traditque sequentibus astris.
sic annum mensesque suos natura diesque
atque ipsas voluit numerari signa per horas,
omnia ut omne foret divisum tempus in astra
perque alterna suos variaret sidera motus,
ut cuiusque vices ageret redeuntis in orbem.
idcirco tanta est rerum discordia in aevo
et subtexta malis bona sunt lacrimaeque sequuntur
vota nec inconstans servat fortuna tenorem ;
usque adeo permixta fluit nec permanet usquam,
amisitque fidem variando cuncta per omnis.

505 -acta] hac
507 non celeres] nosceret *M*: non feret *GL*
508 obitus] ortus
509 sidera] dena
513 quos
515 annum
527 in cunctos

[a] The epithet "swift" applies no more to settings than to risings but, like "fast-rotating" at the beginning of the sentence, has been chosen to emphasize that the rapid motion of the celestial sphere is a stumbling block in the accurate casting of a horoscope.

Horoscope with its fixed ascension, so that, when precise accuracy attaches to the location of the first cardinal point, the zenith of heaven on high will not be able to elude you, nor the swift[a] setting, and the foundations will be correctly fixed at the nadir,[b] and the signs will duly light upon their proper portions and influences.

510 Now I shall assign their special periods of life in classes to the signs; for the signs are also allotted to their own particular years and months and days and hours of days; and during these periods they each exercise especial influence. The first year of life will belong to that sign in which at birth the Sun has shone, since the Sun takes a year's duration to traverse the firmament; the next and subsequent years are consecutively bestowed upon the signs in their order. The Moon shall denote the months, since in a month it completes its course. The Horoscope brings under its regency the first days and the first hours, and hands the others to the following signs. Thus did nature wish year and months and days and even hours to be duly counted out through the signs, that every period of time might be distributed over every sign of the zodiac and vary its influences through the sequence of signs, according as it made a change to each one as it came round in the circle. This is why such extremes of experience are found in the passage of time, and good is linked to bad, sorrow attends success, and in its inconstancy fortune maintains no steady course: to such an extent is it varied and changing, nowhere remaining the same; and by its commutation of everything in

[b] [508]: "... correctly fixed the true settings and true risings of the planets, ..."

non annis anni nec menses mensibus usquam
conveniunt, seque ipsa dies alia usque requirit
horaque non ulli similis producitur horae,
tempora quod distant, propriis parentia signis,
per numeros omnis aevi divisa volantis,
talisque efficiunt vitas casusque animantum,
qualia sunt, quorum vicibus tum vertimur, astra.

Sunt quibus et caeli placeat nascentis ab orae
sidere, quem memorant horoscopon inventores,
parte quod ex illa describitur hora diebus,
omne genus rationis agi per tempora et astra
et capite ex uno menses annosque diesque
incipere atque horas tradique sequentibus astris;
et, quamquam socia nascantur origine cuncta,
diversas tamen esse vices, quod tardius illa,
haec citius peragant orbem. semel omnia ad astra
hora die, bis mense dies venit, unus in anno
mensis et exactis bis sex iam solibus annus.
difficile est in idem tempus concurrere cuncta,
unius ut signi pariter sit mensis et annus
atque dies atque hora simul: sibi discrepat ordo.
saepe fit ut, mitis tulerint qui sideris annum,
asperioris agant mensem; si mensis in astrum

531 aliumque
533 sistant
535 vitas] menses minantur
537 horae
538 inventuros
543 nascuntur
545 venit omnis ad astrum
550 agam

[a] Manilius appears to be talking about daylight hours and ignoring the hours of night. Certainly he is explicit enough elsewhere, at 3. 245, for example. But it is conceivable that he is here following a source in which the day (νυχθήμερον) was divided into, not twenty-four hours, but twelve; these "double-hours" were a regular Babylonian period of time

the lives of us all it has forfeited our trust. Nowhere does year bear resemblance to year or month to month; the very day, ever different, looks in vain for the replica of itself, and no hour's space is fully run in the likeness of another, because the periods are mutually at variance, being, inasmuch as they are divided into all the sections of fleeting time, subject to their own special signs, and hence bring it about that the lives and fortunes of the living are a reflection of the stars, by whose changes we ourselves then change.

537There are some who approve of an alternative scheme: from the sign at the edge of the rising heavens, which the founders of astrology call the Horoscope, since from that point are measured the hours of day, every class of calculation is made both in the divisions of time and in their distribution to signs; the years and months and days and hours start from the same source and are passed on to the following signs; however, even though all the periods proceed from a common origin, the sequences are dissimilar, because some periods are slower, some quicker to complete the circle. To every sign there comes an hour just once a day,[a] a day twice in the month, a month once in the year, and a year once in twelve annual courses of the Sun. It would be difficult for all the periods to revolve concurrently in the same time, so that together in a single sign the same year and month *and day and hour coincide*: *the cycle of time is at variance with itself. It oft happens that those who have drawn the year of a gentle sign* are passing the month of a harsher one; oft that, if the

and are elsewhere encountered in astrology (*cf.* Boll, *Sphaera* 311 ff.).

laetius inciderit, signum sit triste diei ;
si fortuna diem foveat, sit durior hora.
idcirco nihil in totum sibi credere fas est,
non annos signis, menses vertentibus annis,
mensibus aut luces, aut omnis lucibus horas,
quod nunc illa nimis properant, nunc illa morantur,
et modo dest aliis, modo adest, vicibusque recedit
aut redit atque alio mutatur tempore tempus
interpellatum variata sorte dierum.

Et, quoniam docui, per singula tempora, vitae
quod quandoque genus veniat, cuiusque sit astri
quisque annus, cuius mensis, simul hora diesque,
altera nunc ratio, quae summam continet aevi,
reddenda est, quot quaeque annos dare signa ferantur.
quae tibi, cum finem vitae per sidera quaeris,
respicienda manet ratio numerisque notanda.
bis quinos annos Aries unumque triente
fraudatum dabit. appositis tu, Taure, duobus
vincis, sed totidem Geminorum vinceris astro,
tuque bis octonos, Cancer, binosque trientes,
bisque novem, Nemeaee, dabis bessemque sub illis.
Erigone geminatque decem geminatque trientem,
nec plures fuerint Librae quam Virginis anni.

[557] alius [571] nemeae tribues

[a] Since there are twelve signs of the zodiac and twelve hours (see previous note) and twelve months, the cycles of hours and months will always remain constant, that is, the first hour of any day and the first month of any year will always fall to the same sign, and similarly with the second, third, and so on. But there being, in the Roman calendar at any rate, sequences of 30, 31, and 28 (29) days in a month, the cycle of days soon gets out of step. For example, if Jan. 1 is allotted to Aries, so will be Jan. 13, Jan. 25, Feb. 6, Feb. 18, Mar. 2, etc., with Feb. 1 falling to Scorpio, and Mar. 1 to Pisces, etc., thus producing an irregular cycle.

month falls to a benign constellation, the sign of the day will be a sign of woe ; oft that, if fortune smiles on the day, the hour will be a sombre one. It is not ordained that anything should place total reliance on itself : the years may not rely on their signs alone, nor the months on those of the revolving years, nor the days on those of the months, nor any hour on the sign of the day. The reason is that now these speed ahead, these lag behind ; one period now deserts the others, and now is of their company ; it successively moves away or comes back, and undergoes the changing influences of another, always being disturbed by the irregular cycle of the days.[a]

[560]Having taught what kind of life lies in store for us at any given moment, dealing with the separate periods of time, and to what star each year and month and therewith hour and day belong, I must now tell of another computation, which deals with the span of man's lifetime and reveals the number of years each sign is held to confer. When seeking to foretell length of life by the stars, you must keep this scheme in mind and mark well its amounts.[b] Twice five years shall the Ram bestow, plus one cheated of a third. You, Bull, exceed this with the addition of two, but by as many are exceeded by the sign of the Twins. Twice eight, together with two thirds, shall you, Cancer, bestow, and you, Nemean beast, twice nine and two thirds joined thereto. Erigone both doubles ten and doubles a third, nor shall the Balance have more years than the Maid. Scorpion will equal

[b] The amounts (in years) are as follows : Aries, Pisces $10\frac{2}{3}$; Taurus, Aquarius $12\frac{2}{3}$; Gemini, Capricorn $14\frac{2}{3}$; Cancer, Sagittarius $16\frac{2}{3}$; Leo, Scorpio $18\frac{2}{3}$; Virgo, Libra $20\frac{2}{3}$ (*cf.* Introduction p. lxxix and figure 21).

Scorpios aequabit tribuentem dona Leonem.
Centauri fuerint eadem quae munera Cancri.
ter quinos, Capricorne, dares, si quattuor essent
appositi menses. triplicabit Aquarius annos
quattuor et menses vitam producet in octo.
Piscibus est Aries et sorte et finibus haerens :
lustra decem tribuent solis cum mensibus octo.

Nec satis est annos signorum noscere certos,
ne lateat ratio finem quaerentibus aevi :
templa quoque et partes caeli sua munera norunt
et proprias tribuunt certo discrimine summas,
cum bene constiterit stellarum conditus ordo.
sed mihi templorum tantum nunc iura canentur ;
mox veniet mixtura suis cum viribus omnis,
cum bene materies steterit percognita rerum
non interpositis turbatarum undique membris.
si bene constiterit primo sub cardine Luna,
qua redit in terras mundus, nascensque tenebit
exortum, octo tenor decies ducetur in annos
si duo decedant. at, cum sub culmine summo
consistet, tribus hic numerus fraudabitur annis.
bis quadragenos occasus dives in actus
solis erat, numero nisi desset olympias una.
imaque tricenos bis fundamenta per annos
censentur bis sex adiectis messibus aevo.

589 turbatur
591 quo
592 octonos
598 mensibus

[a] This seems to look forward to an account of planetary influences, which, however, is not found in the poet's extant work.

[b] Temple 1 = 80−2 = 78. *Cf.* Introduction pp. lxxx f. and figure 22.

[c] Temple 10 = 80−3 = 77.

[d] Literally "olympiad," which though properly a four-

the Lion's bounty of gifts. The award of the Centaur will be the same as the Crab's. You, Capricorn, would donate thrice five years, were four months added. The Waterman will treble four years and prolong life a further eight months. As in domain, so in dispensation is the Ram an adherent of the Fishes: they shall impart ten revolutions of the Sun together with eight months.

581Nor is it enough to learn the total of years allotted by the signs, lest the whole scheme be hidden from us when we seek the measure of men's lives: the temples and portions of the sky also have gifts of their own to bestow, and impart in a precise gradation their respective amounts, when the configuration formed by the planets proves auspicious. But I now shall sing only of the ordinances of the temples; later the whole complex design will appear with its full force,[a] when the constituent parts of the universe are firmly grasped and not made confusing with portions scattered everywhere. If the Moon is auspiciously placed in the temple of the first cardinal point, where the heavens return to earth, and its rising occupies the ascendant, the course of life shall be prolonged to eighty years, if two withdraw.[b] But when it stands at the summit of the sky, the former number is cheated of three years.[c] The setting were rich in twice forty circuits of the Sun, were not one lustrum [d] lacking from the count.[e] The nethermost base is assessed at years twice thirty with twice six harvests added to the

year period is often used by the Latin poets (as a result of the Roman principle of inclusive reckoning) to denote a lustrum or five-year period (Ovid, *Ex Ponto* 4. 6. 5 f.).

[e] Temple 7 = 80 − 5 = 75.

quodque prius natum fuerit dextrumque trigonum
hoc sexagenos tribuit duplicatque quaternos.
quod fuerit laevum praelataque signa sequetur
tricenos annos duplicat, tris insuper addit.
quaeque super signum nascens a cardine primum
tertia sors manet et summo iam proxima caelo
haec ter vicenos geminat, tris abstrahit annos.
quaeque infra veniet spatio divisa sub aequo
per quinquagenas complet sua munera brumas.
quemque locum superat nascens horoscopos, ille
dena quater revocat vertentis tempora solis
accumulatque duos cursus iuvenemque relinquit.
at qui praecedit surgentis cardinis oram
vicenos ternosque dabit nascentibus annos
vix degustatam rapiens sub flore iuventam.
quod super occasus templum est ter dena remittit
annorum spatia et decimam tribus applicat auctis.
inferius puerum interimet, bis sexque peracti
immatura trahent natales corpora morti.

Sed tamen in primis memori sunt mente notanda
partibus adversis quae surgunt condita signa
divisumque tenent aequo discrimine caelum ;
quae tropica appellant, quod in illis quattuor anni
tempora vertuntur signis nodosque resolvunt

[604] sors manet] forma [611] horam

---

[a] Temple 4 = 60+12 = 72.
[b] Temple 9 = 60+8 = 68.
[c] Temple 5 = 60+3 = 63.
[d] Literally "the third temple from . . ." (by the Roman principle of inclusive reckoning).
[e] The Horoscope.
[f] Temple 11 = 60−3 = 57.
[g] Temple 3 = 50.
[h] Temple 2 = 40+2 = 42.
[i] Temple 12 = 23.

toll.[a] And the trigon of the Horoscope which rose first and is on the right bestows sixty and the double of four.[b] The trigon on the left and following in the wake of the preceding signs doubles thirty years and adds three over and above.[c] And the temple which is separated by one intervening sign[d] from the first sign rising at the cardinal point[e] and which is now next to heaven's peak, this multiplies a score by three and takes three years away.[f] The temple which comes below, separated from the cardinal point by an equal space, completes its endowment over fifty winters.[g] That place, above which stands the rising Horoscope, four times repeats ten seasons of the revolving Sun and heaps thereon two courses, and leaves one still a young man.[h] That which precedes the border of the rising quarter shall give years three and twenty to those born under it, plucking ere its bloom their youth scarce tasted.[i] The temple above the setting allows thrice ten circuits of years and adds thereto a tithe in augmenting them by three.[j] The lower one will despatch its wards in childhood, the completion of twice six birthdays consigning to death their undeveloped bodies.[k]

[618]But above all one must mark with a retentive mind the signs which rise from their places in opposite parts of the sky and mark its division into equal portions. They are called tropic signs,[l] since in them turn the four seasons of the year and untie the bonds which fasten them together; they bring change to

[j] Temple 8 = 30+3 = 33.
[k] Temple 6 = 12.
[l] Cancer, Capricorn, Aries, Libra (*i.e.* the term tropic includes the equinoctial signs: *cf.* 2. 178 and note).

totumque emutant converso cardine mundum
inducuntque novas operum rerumque figuras.
  Cancer ad aestivae fulget fastigia zonae
extenditque diem summum parvoque recessu
destruit et, quanto fraudavit tempore luces,
in tantum noctes auget: stat summa per omnis.
tum Cererem fragili properant destringere culmo,
Campus et in varias destringit membra palaestras,
et tepidas pelagus iactatum languet in undas.
tunc et bella fero tractantur Marte cruenta
nec Scythiam defendit hiems; Germania sicca
iam tellure fugit Nilusque tumescit in arva.
hic rerum status est, Cancri cum sidere Phoebus
solstitium facit et summo versatur Olympo.
  Parte ex adversa brumam Capricornus inertem
per minimas cogit luces et maxima noctis
tempora, producitque diem tenebrasque resolvit,
inque vicem nunc damna legit, nunc tempora supplet.
tunc riget omnis ager, clausum mare, condita castra,
nec tolerant medias hiemes sudantia saxa,
statque uno natura loco paulumque quiescit.
  Proxima in effectum et similis referentia motus
esse ferunt luces aequantia signa tenebris.
namque Aries Phoebum repetentem sidera Cancri
inter principium reditus finemque coercet
tempora diviso iungens concordia mundo,
convertitque vices victumque a sidere Librae

630 destringunt
631 trepidum
645 diebus
649 brumae

[a] Before the advancing Romans, no longer protected by its swamps as it was in the winter (*cf.* Tacitus, *Ann.* 2. 5. 3).

[b] The Sun's return journey is from Capricorn, and the sign of the Ram marks the halfway stage.

the whole sky as it revolves on its axis, giving a new look to the works of man and the face of nature.

625 Gleaming in summer's topmost circle, Cancer prolongs the day to its greatest length and then shortens it in retreating by small degrees, increasing night by the amount which it stole from day: for the sum of day and night remains constant. Then men make haste to strip the grain from the brittle stalk, while the Field of Mars strips their bodies for a variety of exercise, and after its storms the sea sinks back to rest upon warm waters. Then, too, the savage War-god wages bloody battles, from which not now does winter protect the Scythians; Germany retreats,[a] its land now dry; and the Nile rises to flood its fields. This is the state of things when the sun makes solstice in Cancer's sign and is found at heaven's summit.

637 On the other side Capricorn forces sluggish winter through the shortest day and longest duration of night, and begins to lengthen daylight and dispel darkness; by turns it now steals time from day and now repairs day's loss. Then every field is frost-bound; the sea is closed to ships; the camp is stationary; rocks covered with rime are unable to endure midwinter's cold: nature stands motionless and for a while is still.

644 Close to these in their influence and displaying variation similar to each other are, they say, the signs which level the hours of light and darkness. For the Ram arrests the Sun, as it reseeks Cancer's stars, between the start and end of its return journey,[b] and dividing heaven in half, matches night and day in harmony; and he reverses the sequence, bidding day, overpowered by the length of night

exsuperare diem iubet et succumbere noctes,
aestivi donec veniant ad sidera Cancri.
tum primum miti pelagus consternitur unda
et varios audet flores emittere tellus;
tum pecudum volucrumque genus per pabula laeta
in Venerem partumque ruit, totumque canora
voce nemus loquitur frondemque virescit in omnem.
viribus in tantum signi natura movetur.

Huic ex adverso simili cum sorte refulget
Libra diem noctemque pari cum foedere ducens,
tantum quod victas usque ad se vincere noctes
ex ipsa iubet, ad brumae dum tempora surgant.
tum Liber gravida descendit plenus ab ulmo
pinguiaque impressis despumant musta racemis;
mandant et sulcis Cererem, dum terra tepore
autumni resoluta patet, dum semina ducit.

Quattuor haec et in arte valent, ut tempora vertunt
sic hos aut illos rerum flectentia casus
nec quicquam in prima patientia sede manere.
sed non per totas aequa est versura figuras,
annua nec plenis flectuntur tempora signis.
una dies sub utroque aequat sibi tempore noctem,
dum Libra atque Aries autumnum verque figurant;
una dies toto Cancri longissima signo,
cui nox aequalis Capricorni sidere fertur:
cetera nunc urgent vicibus, nunc tempora cedunt.
una ergo in tropicis pars est cernenda figuris,

661 tum . . . vincat   670 annua] omnia

since leaving Libra's sign, now be the master, and bidding the nights succumb, till Cancer's summer sign is reached. Then first the sea is calmed with tranquil wave, and the earth dares to send forth flowers in all their variety ; then amid happy pastures the tribes of bird and beast hasten to mate and breed, and the whole woodland speaks with melodious voice and grows green to full foliage. So deeply is nature stirred by the potency of the sign.

[658]Opposite the Ram the Balance shines back at it : its lot is like the Ram's, for it treats day and night on equal terms, save that it bids the night, outweighed before, increase thenceforth until reaching its peak at winter-tide. This is the season when the Wine-god comes down in full strength from the laden elm, and the rich must pours foaming from pressed bunches of grapes ; and this the season when men commit the corn to the furrows, whilst the soil, relaxed by autumn's warmth, opens to clasp the seed.

[666]These four signs have great power in the art of which I tell, for as they mark the changing seasons, so do they alter this issue of affairs or that, suffering naught to persist in its initial state. Yet the change is not uniform throughout the figures of the signs, nor do the seasons of the year undergo change in every part of the sign. One day alone at either season makes equal to itself the night, while the Scales and Ram are shaping autumn and spring ; one day alone in the whole of the Crab's sign is longest, matched by an equal night in the Sea-goat's star : the other days and nights now wax and now wane in turn. Therefore in the tropic signs you must look for but one degree to alter the face of the

quae moveat mundum, quae rerum tempora mutet,
facta novet, consulta alios declinet in usus,
omnia in aversum flectat contraque revolvat.
has quidam vires octava in parte reponunt ;
sunt quibus esse placet decimae ; nec defuit auctor
qui primae momenta daret frenosque dierum.

681 decimas

[a] This is surprisingly phrased, for Manilius himself mostly

heavens, to change the seasons of nature, to undo what is done, to deflect to other ends what has been projected, to turn all things in the opposite direction and reverse their movement. Some ascribe these powers to the eighth degree ; some hold that they belong to the tenth ; nor was an authority lacking to give to the first degree the decisive influence and the control of the days.[a]

treats the first as the tropic degree : for further discussion of the matter and its connection with the precession of the equinoxes, see Introduction pp. lxxxi ff.

BOOK FOUR

It *is in his fourth book that Manilius spreads his wings most widely, for here every aspect of his teaching and every facet of his genius are attractively illustrated. At the beginning, as at the end, we hear the indignant voice of the apostle of Stoicism. There follows, with a change to the poet's lighter style, an exposition of the characteristics and influences of the signs of the zodiac (a foretaste of that treatment of the extra-zodiacal signs which will occupy most of the last book). The next two chapters, on decans and injurious degrees, present us with another glimpse of the versifier of astrological and mathematical tables. A somewhat cursory interlude on the rising of individual degrees of the ecliptic forms the transition to the chief subject of the latter portion of the book, a geographical survey and an account of the dominion of the zodiac over the various parts of the world. Ecliptic signs restore to us the philosophical poet, who takes for his peroration the theme of man as microcosm and image of the divine.*

LIBER QUARTUS

Quid tam sollicitis vitam consumimus annis
torquemurque metu caecaque cupidine rerum
aeternisque senes curis, dum quaerimus, aevum
perdimus et nullo votorum fine beati
victuros agimus semper nec vivimus umquam,
pauperiorque bonis quisque est, quia plura requirit
nec quod habet numerat, tantum quod non habet optat
cumque sibi parvos usus natura reposcat
materiam struimus magnae per vota ruinae
luxuriamque lucris emimus luxuque rapinas,
et summum census pretium est effundere censum ?
solvite, mortales, animos curasque levate
totque supervacuis vitam deplete querellis.
fata regunt orbem, certa stant omnia lege
longaque per certos signantur tempora casus.
nascentes morimur, finisque ab origine pendet.
hinc et opes et regna fluunt et, saepius orta,
paupertas, artesque datae moresque creatis
et vitia et laudes, damna et compendia rerum.
nemo carere dato poterit nec habere negatum
fortunamve suis invitam prendere votis

[6] qui
[8] sui
[19] clades
[20] caret damno

---

[a] Line 16 also appears in the verse inscription *CIL* 2. 4426.

BOOK 4

Oh, why do we spend the years of our lives in worry, tormenting ourselves with fears and senseless desires; grown old before our time with anxieties which never end; forfeiting length of days by our very quest for it; setting no limit to our wishes, so that their fulfilment leaves us still unblest, but ever playing the part of men who mean to live yet never do? Everyone is the poorer for his possessions because he looks for more: none counts his blessings, but only lusts for what he lacks. Though nature needs only modest requirements, we build higher and higher the peak from which to fall, and purchase luxury with our gains, and with love of luxury the fear of dispossession, until the greatest boon that wealth can confer is the squandering of itself. Set free your minds, O mortals, banish your cares, and rid your lives of all this vain complaint! Fate rules the world, all things stand fixed by its immutable laws, and the long ages are assigned a predestined course of events. At birth our death is sealed, and our end is consequent upon our beginning.[a] Fate is the source of riches and kingdoms and the more frequent poverty; by fate are men at birth given their skills and characters, their merits and defects, their losses and gains. None can renounce what is bestowed or possess what is denied; no man by prayer may seize fortune if it demur, or escape if it

aut fugere instantem : sors est sua cuique ferenda.
an, nisi fata darent leges vitaeque necisque,
fugissent ignes Aenean, Troia sub uno
non eversa viro fatis vicisset in ipsis ?
aut lupa proiectos nutrisset Martia fratres,
Roma casis enata foret, pecudumque magistri
in Capitolinos duxissent fulmina montes,
includive sua potuisset Iuppiter arce,
captus et a captis orbis foret ? igne sepulto
vulneribus victor repetisset Mucius urbem,
solus et oppositis clausisset Horatius armis
pontem urbemque simul, rupisset foedera virgo,
tresque sub unius fratres virtute iacerent ?
nulla acies tantum vicit : pendebat ab uno
Roma viro regnumque orbis sortita iacebat.
quid referam Cannas admotaque moenibus arma
Varronemque fuga magnum | Fabiumque morando
postque tuos, Trasimenne, lacus, | cum vincere posset,
accepisse iugum victae Carthaginis arces,
seque ratum Hannibalem nostris cecidisse catenis
exitium generis furtiva morte luisse ?
adde etiam Latias acies Romamque suismet
pugnantem membris, adice et civilia bella
et Cimbrum in Mario Mariumque in carcere victum.

[23] aut
[38a] pugum
[39b] morantem
[38b] quam
[41] seque] spe
[42] exiliumque regi
[43] italas

---

[a] Virgil, *Aen.* 2. 632 f.
[b] Romulus and Remus.
[c] Referring to the establishment by shepherds on the Capitol of the cult of Jupiter the Thunderer.
[d] Gaius Mucius Scaevola (*cf.* 1. 779, Livy 2. 12).
[e] Horatius Cocles (*cf.* 1. 781, Livy 2. 10).

draw nigh: each one must bear his appointed lot. For were not fate the arbiter of life and death, would the flames have receded before Aeneas,[a] and Troy, saved from destruction by one man's survival, have emerged triumphant in the very day of ruin? Would the she-wolf of Mars have suckled the brothers[b] exposed to die, Rome have grown out of cottages, shepherds have brought thunder to the Capitol's slopes, or Jupiter consented to confinement in his citadel[c] and the world been vanquished by a vanquished people? Would Mucius[d] have quenched the fire with the blood from his wounds and returned victorious to Rome, Horatius single-handed have shut both bridge and city to the assailing host,[e] a maiden have annulled a treaty,[f] and three brothers have fallen to the valour of one?[g] It was no army that won that great victory: Rome then depended on a single warrior and, though fated to rule the world, despaired. What need have I to tell of Cannae and enemy arms brought to the city walls, of the heroism of Varro's flight and Fabius's delays? What need to tell how after the battle at your lake, Trasimene,[h] when victory lay in her grasp, the towers of humbled Carthage bowed to the yoke and Hannibal, judging he had fallen into our clutches, expiated in an inglorious death the destruction of his race? Add to these the battles in Latium[i] and Rome in arms against her own person; add too the civil wars, a Cimbrian helpless at the sight of Marius, himself in

[f] Cloelia (*cf.* 1. 780, Livy 2. 13).

[g] The Curiatii to one of the Horatii (*cf.* 1. 778 f., Livy 1. 25).

[h] As at 566 the poet wrongly dates the battle of Lake Trasimene after Cannae.

[i] The Social War.

quod, consul totiens, exul, quod de exule consul
adiacuit Libycis compar iactura ruinis
eque crepidinibus cepit Carthaginis urbem,
hoc, nisi fata darent, numquam fortuna tulisset.
quis te Niliaco periturum litore, Magne,
post victas Mithridatis opes pelagusque receptum
et tris emenso meritos ex orbe triumphos,
cum te iam posses alium componere Magnum,
crederet, ut corpus sepeliret naufragus ignis
eiectaeque rogum facerent fragmenta carinae ?
quis tantum mutare potest sine numine fati ?
ille etiam caelo genitus caeloque receptus,
cum bene compositis victor civilibus armis
iura togae regeret, totiens praedicta cavere
vulnera non potuit : toto spectante senatu,
indicium dextra retinens nomenque, cruore
delevit proprio, possent ut vincere fata.
quid numerem eversas urbes regumque ruinas,
inque rogo Croesum Priamique in litore truncum,
cui nec Troia rogus ? quid Xerxen, maius et ipso
naufragium pelago ? quid capto sanguine regem
Romanis positum, raptosque ex ignibus ignes
cedentemque viro flammam quae templa ferebat ?

[46] [de]
[53] te iam] iam etiam
[64] priamumque

---

[a] A Cimbrian slave was sent to kill Marius at the nadir of his fortunes but lost his nerve in the overwhelming presence of the man (Valerius Maximus 2. 10. 6).

[b] *Cf.* 1. 793. Pompey's triumphs were gained in Africa, Europe, and Asia (Velleius 2. 40. 4).

[c] As another Alexander.

[d] Julius Caesar, descended from Venus and deified after his death.

prison helpless.[a] An exile after being so oft consul, and consul again after being an exile, he rivalled in his fallen fortunes the Punic ruins in which he took refuge, and then from the rubble of Carthage emerged to capture Rome : never had chance permitted this to happen save by ordinance of fate. Who, Great Pompey, after your victory over the forces of Mithridates, your recovery of the seas from piracy, and your three triumphs gained from campaigns which traversed the earth,[b] would have believed that, when you could now represent yourself as another styled the Great,[c] you were destined to perish on Egyptian shores with but fire of shipwrecked wood to burn your corpse and remnants of an upcast barque to make your pyre ? Who can experience such change of fortune except by fate's decree ? When after his victory and happy settlement of civil strife he was administering the laws of peace, even he who was born of heaven and was to heaven restored[d] could not escape the violence so oft foretold : before the eyes of the assembled senate he obliterated with his own blood the evidence of the plot and the list of the conspirators which he held in his hand : all this so that fate could prevail. Need I recount the cities destroyed, the kings overthrown, Croesus on the pyre,[e] the headless corpse of Priam on the shore,[f] denied Troy even as a pyre ? Or Xerxes, whose shipwreck was greater than sea could contrive ? Need I tell of a man of servile stock placed over the Romans as their king,[g] or of fire rescued from fire and flames which were ravaging a temple but yielded to a

[e] *Cf.* Herodotus 1. 86 f.
[f] *Cf.* Virgil, *Aen.* 2. 557.
[g] Servius Tullius.

quot subitae veniunt validorum in corpora mortes
seque ipsae rursus fugiunt errantque per ignes !
ex ipsis quidam elati rediere sepulcris,
atque his vita duplex, illis vix contigit una.
ecce levis perimit morbus graviorque remittit ;
succumbunt artes, rationis vincitur usus,
cura nocet, cessare iuvat, mora saepe malorum
dat pausas ; laeduntque cibi parcuntque venena.
degenerant nati patribus vincuntque parentes
ingeniumque suum retinent ; transitque per illum,
ex illo fortuna venit. furit alter amore
et pontum tranare potest et vertere Troiam,
alterius frons est scribendis legibus apta.
ecce patrem nati perimunt natosque parentes
mutuaque armati coeunt in vulnera fratres.
non hominum hoc bellum est ; coguntur tanta moveri
inque suas ferri poenas lacerandaque membra.
quod Decios non omne tulit, non omne Camillos
tempus et invicta devictum mente Catonem,
materies in rem superat sed lege repugnat.
et neque paupertas breviores excipit annos
nec sunt immensis opibus venalia fata,
sed rapit ex tecto funus Fortuna superbo
indicitque rogum summis statuitque sepulcrum.
quantum est hoc regnum, quod regibus imperat ipsis !
quin etiam infelix virtus et noxia felix,

[76] causas [87] morte
[84] hominum] nostrum

[a] Lucius Caecilius Metellus, who rescued the sacred fire from the Temple of Vesta when it was burned (Livy, *Perioch.* 19).
[b] Like Leander. [c] Like Paris.
[d] Like Oedipus. [e] Like Medea.

mortal?[a] How oft without warning does death visit the bodies of the strong, how oft again does death flee its very self and wander forth through the funeral fires! Some carried out to burial have returned even from the grave, gifted with a twofold life whilst others have scarcely one. Why, a slight ailment kills, whilst one more serious allows of a reprieve; therapy fails, rational practice is baffled, and attention harms, whilst negligence proves beneficial and delay oft brings a cessation of complaints; and nourishment injures, whilst poison spares the patient. Sons fall short of their fathers' merits or rise above their parentage and keep a nature that is their own. The fortunes of a dynasty rise with one member of a house and are eclipsed with another. This one is a frenzied lover and can swim the straits[b] or ruin Troy,[c] that one's gravity is suited to the framing of law. See, sons slay a father,[d] and parents their children,[e] and brothers meet armed to shed each other's blood.[f] This violence is no work of men's: a destiny drives them to such awful passions and to suffer such punishment and mutilation of limb. Not every age has produced men like the Decii[g] and Camillus,[h] or a Cato with a spirit unconquered in defeat[i]: more than enough material exists to accomplish such an end, but resists through fate's decree. Again, the poor do not fall heir to fewer years of life, nor may one's span be purchased with vast wealth. But Fortune carries off a corpse from stately halls, appoints a pyre and assigns a tomb for men of highest station. What wonderful kingship is this, whose bidding even kings obey! Moreover, it happens that virtue fares ill, and guilt fares well;

[f] Like Eteocles and Polynices. [g] *Cf.* 1. 789.
[h] *Cf.* 1. 784. [i] Cato of Utica (*cf.* 1. 797).

et male consultis pretium est, prudentia fallit;
nec Fortuna probat causas, sequiturque merentis,
sed vaga per cunctos nullo discrimine fertur.
scilicet est aliud, quod nos cogatque regatque,
maius, et in proprias ducat mortalia leges
attribuatque suos ex se nascentibus annos
fortunaeque vices. permiscet saepe ferarum
corpora cum membris hominum: non seminis ille
partus erit; quid enim nobis commune ferisque,
quisve in portenti noxam peccarit adulter?
astra novant formas caelumque interserit ora.
denique, si non est, fati cur traditur ordo,
cunctaque temporibus certis ventura canuntur?
nec tamen haec ratio facinus defendere pergit
virtutemve suis fraudare in praemia donis.
nam neque mortiferas quisquam minus oderit herbas
quod non arbitrio veniunt sed semine certo,
gratia nec levior tribuetur dulcibus escis
quod natura dedit fruges, non ulla voluntas.
sic hominum meritis tanto sit gloria maior
quod caelo laudem debent, rursusque nocentis
oderimus magis in culpam poenasque creatos.
nec refert scelus unde cadat, scelus esse fatendum.
hoc quoque fatale est, sic ipsum expendere fatum.
[quod quoniam docui, superest nunc ordine certo
caelestis fabricare gradus, qui ducere flexo
tramite pendentem valeant ad sidera vatem]

110 minus] magis   115 gaudente venit

---

[a] For such portents *cf.* Livy 27. 11. 5; 31. 12. 8.
[b] *Cf.* 2. 149. In the MSS. follows the interpolation [119-121]:

poorly conceived plans are rewarded, whilst foresight fails; nor does Fortune examine the merits of a case and attend the deserving, but moves capriciously through the lives of all without distinction. Clearly, there is another and greater power to constrain and rule us, and to subject mortal affairs to laws of its own: it gives birth to men and at their birth determines the number of their years and the changes of their fortunes. Oft it conjoins body of beast with limb of man [a]: yet that will be found no product of the seed. What has our human nature in common with a beast's? When has the sin of adultery been punished with a monstrous birth? It is the stars which fashion strange shapes; it is heaven which intrudes these hybrid features. After all, if a pattern of fate does not exist, how comes it that it is represented as existing and all that will come to pass at particular moments prophesied? Not that this reasoning goes so far as to defend crime or deprive virtue of the gifts that are its reward. For no one will hate poisonous plants the less because they grow from their appointed seed and not by their own choice, nor will tasty food find less favour because the crops are nature's bounty and not given us by any decision of their own. Let man's merits, therefore, possess glory all the greater, seeing that they owe their excellence to heaven; and, again, let us hate the wicked all the more, because they were born for guilt and punishment. Crime, whencesoever sprung, must still be reckoned crime. This, too, is sanctioned by fate, that I should thus expound the rule of fate.[b]

"After this exposition, it now remains to construct in the correct order steps mounting to heaven, that they may avail to guide the hesitating seer by a winding path to the stars."

Nunc tibi signorum mores summumque colorem
et studia et varias artes ex ordine reddam.
Dives fecundis Aries in vellera lanis
exutusque novis rursus spem semper habebit,
naufragiumque inter subitum censusque beatos
crescendo cadet et votis in damna feretur,
in vulgumque dabit fructus et mille per artes
vellera diversos ex se parientia quaestus :
nunc glomerare rudis nunc rursus solvere lanas,
nunc tenuare levi filo nunc ducere telas,
nunc emere et varias in quaestum vendere vestes,
quis sine non poterant ullae subsistere gentes
vel sine luxuria. tantum est opus, ipsa suismet
asseruit Pallas manibus dignumque putavit,
seque in Arachnaeo magnam putat esse triumpho.
haec studia et similis dicet nascentibus artes,
et dubia in trepido praecordia pectore finget
seque sua semper cupientia vendere laude.
Taurus simplicibus dotabit rura colonis
pacatisque labor veniet ; nec praemia laudis
sed terrae tribuet partus. summittit in astris
colla iugumque suis poscit cervicibus ipse.
ille suis Phoebi portat cum cornibus orbem
militiam indicit terris et segnia rura
in veteres revocat cultus, dux ipse laboris,
nec iacet in sulcis solvitque in pulvere pectus.
Serranos Curiosque tulit fascesque per arva

[123] studii [et] [142] antris
[136] triumphum

---

[a] Ovid's account (*Met.* 6. 1-145) hardly justifies this self-esteem.

[b] Gaius Atilius Regulus Serranus (*cf.* Pliny, *N.H.* 18. 20).

[c] Manius Curius Dentatus (*cf.* 1. 787).

[122]Now shall I declare to you in due order the characters, the predominant quality, the pursuits, and the different skills which the signs impart.

[124]The Ram, who is rich with an abundance of fleecy wool and, when shorn of this, with a fresh supply, will ever cherish hopes; he will rise from the sudden shipwreck of his affairs to abundant wealth only to meet with a fall, and his desires will lead him to disaster; he will yield his produce for the common benefit, the fleece which by a thousand crafts gives birth to different forms of gain: now workers pile into heaps the undressed wool, now card it, now draw it into a tenuous thread, now weave the threads to form webs, and now they buy and sell for gain garments of every kind; no nation could dispense with these, even without indulgence in luxury. So important is this work that Pallas herself has claimed it for her own hands, of which she has judged it worthy, and deems her victory over Arachne a token of her greatness.[a] These are the callings and allied crafts that the Ram will decree for those born under his sign: in an anxious breast he will fashion a diffident heart that ever yearns to sell itself for praise.

[140]The Bull will dower the countryside with honest farmers and will come as a source of toil into their peaceful lives; it will bestow, not gifts of glory, but the fruits of the earth. It bows its neck amid the stars and of itself demands a yoke for its shoulders. When it carries the sun's orb on its horns, it bids battle with the soil begin and rouses the fallow land to its former cultivation, itself leading the work, for it neither pauses in the furrows nor relaxes its breast in the dust. The sign of the Bull has produced a Serranus [b] and a Curius,[c] has carried the rods of office through the

tradidit, eque suo dictator venit aratro.
laudis amor tacitae ; mentes et corpora tarda
mole valent, habitatque puer sub fronte Cupido.
Mollius e Geminis studium est et mitior aetas
per varios cantus modulataque vocibus ora
et gracilis calamos et nervis insita verba
ingenitumque sonum : labor est etiam ipse voluptas.
arma procul lituosque volunt tristemque senectam,
otia et aeternam peragunt in amore iuventam.
inveniunt et in astra vias numerisque modisque
consummant orbem postque ipsos sidera linquunt :
natura ingenio minor est perque omnia servit.
in tot fecundi Gemini commenta feruntur.
Cancer ad ardentem fulgens in cardine metam,
quam Phoebus summis revocatus cursibus ambit,
articulum mundi retinet lucesque reflectit.
ille tenax animi nullosque effusus in usus
attribuit varios quaestus artemque lucrorum :
merce peregrina fortunam ferre per urbes
et gravia annonae speculantem incendia ventis
credere opes orbisque orbi bona vendere posse
totque per ignotas commercia iungere terras
atque alio sub sole novas exquirere praedas
et rerum pretio subitos componere census.

159 postquam 163 curribus
162 victam

---

[a] Like Lucius Quinctius Cincinnatus (*cf.* Cicero, *Cato Maior* 56).

[b] By their ability to predict the stars' movements.

[c] As the Sun moves through Cancer, the days become

fields, and has left its plough to become a dictator.[a] Its sons have the love of unsung excellence ; their hearts and bodies derive strength from a massiveness that is slow to move, whilst in their faces dwells the boy-god Love.

[152]From the Twins come less laborious callings and a more agreeable way of life, provided by varied song and voices of harmonious tone, slender pipes, the melodies inborn in strings and the words fitted thereto : those so endowed find even work a pleasure. They would banish the arms of war, the trumpet's call, and the gloom of old age : theirs is a life of ease and unfading youth spent in the arms of love. They also discover paths to the skies, complete a survey of the heavens with numbers and measurements, and outstrip the flight of the stars [b] : nature yields to their genius, which it serves in all things. So many are the accomplishments of which the Twins are fruitful.

[162]Shining at the hinge of the year by the blazing turning-point which when recalled the Sun rounds in his course on high, the Crab occupies a joint of heaven and bends back the length of day.[c] Of a grasping spirit and unwilling to give itself in service the Crab distributes many kinds of gain, and skill in making profits ; he enables a man to carry his investment of foreign merchandise from city to city and, with an eye on steep rises in the price of corn, to risk his money upon sea-winds ; to sell the world's produce to the world, to establish commercial ties between so many unknown lands, to search out under foreign skies fresh sources of gain, and from the high price of his goods to amass sudden wealth. With heaven's

longer, until the solstice is reached, and are then "bent back," *i.e.* become shorter.

ignava et, celeris optando sortibus annos,
dulcibus usuris aequo Iove tempora vendit.
ingenium sollers suaque in compendia pugnax.

Quis dubitet, vasti quae sit natura Leonis
quasque suo dictet signo nascentibus artes?
ille novas semper pugnas, nova bella ferarum
apparat, et spolio vivit pecorumque rapinis;
hos habet hoc studium, postes ornare superbos
pellibus et captas domibus praefigere praedas
et pacare metu silvas et vivere rapto.
sunt quorum similis animos nec moenia frenent,
sed pecudum mandris media grassentur in urbe
et laceros artus suspendant fronte tabernae
luxuriaeque parent caedem mortesque lucrentur.
ingenium ad subitas iras facilisque recessus
aequale et puro sententia pectore simplex.

At quibus Erigone dixit nascentibus aevum
ad studium ducet mores et pectora doctis
artibus instituet, nec tam compendia census
quam causas viresque dabit perquirere rerum.
illa decus linguae faciet regnumque loquendi
atque oculos mentis, qui possint cernere cuncta
quamvis occultis naturae condita causis.
hinc et scriptor erit velox, cui littera verbum est
quique notis linguam superet cursimque loquentis
excipiat longas nova per compendia voces.
in vitio bona sunt: teneros pudor impedit annos,

[173] navigat
[174] Iove] que
[180] hos] hoc
[182] rapto] victor
[184] membris
190: *see after* 201
[195] possit
[200] bonas ut

favour he also sells seasons of idleness at rates of interest to his liking, wishing the swift passage of time to add to the principal. His is a shrewd nature, and he is ready to fight for his profits.

[176]Who can doubt the nature of the monstrous Lion, and the pursuits he prescribes for those born beneath his sign? The lion ever devises fresh fights and fresh warfare on animals, and lives on spoil and pillaging of flocks. The sons of the Lion are filled with the urge to adorn their proud portals with pelts and to hang up on their walls the captured prey, to bring the peace of terror to the woods, and to live upon plunder. There are those whose like bent is not checked by the city-gates, but they swagger about in the heart of the capital with droves of beasts; they display mangled limbs at the shop-front, slaughter to meet the demands of luxury, and count it gain to kill. Their temper is equally prone to fitful wrath and ready withdrawal, and guileless are the sentiments of their honest hearts.

[189]The temperaments of those whose span of life she pronounces at their birth Erigone will direct to study, and she will train their minds in the learned arts. She will give not so much abundance of wealth as the impulse to investigate the causes and effects of things. On them she will confer a tongue which charms, the mastery of words, and that mental vision which can discern all things, however concealed they be by the mysterious workings of nature. From the Virgin will also come the stenographer: his letter represents a word, and by means of his symbols he can keep ahead of utterance and record in novel notation the long speech of a rapid speaker. But with the good there comes a flaw: bashfulness handicaps

magnaque naturae cohibendo munera frenat
ora magisterio nodisque coercita Virgo.
nec fecundus erit (quid mirum in virgine ?) partus.
  Librantes noctem Chelae cum tempore lucis
per nova maturi post annum munera Bacchi
mensurae tribuent usus ac pondera rerum
et Palamedeis certantem viribus ortum,
qui primus numeros rebus, qui nomina summis
imposuit certumque modum propriasque figuras.
hic etiam legum tabulas et condita iura
noverit atque notis levibus pendentia verba,
et licitum sciet, et vetitum quae poena sequatur,
perpetuus populi privato in limine praetor.
non alio potius genitus sit Servius astro,
qui leges proprias posuit, cum iura retexit.
denique, in ambiguo fuerit quodcumque locatum
et rectoris egens, diriment examina Librae.
  Scorpios armata violenta cuspide cauda,
qua, sua cum Phoebi currum per sidera ducit,
rimatur terras et sulcis semina miscet,
in bellum ardentis animos et Martia castra
efficit et multo gaudentem sanguine mentem
nec praeda quam caede magis. quin ipsa sub armis
pax agitur : capiunt saltus silvasque peragrant,

190 nudosque   221 civem
214 proprias] potius

[a] The words denoting numbers, *i.e.* ἕν, δύο, τρία (as it might be with us "one, two, three" etc.).

[b] The figures denoting numbers, *i.e.* ΑΒΓ (as it might be with us "1, 2, 3" etc.).

[c] Abbreviations, specifically those used in the legal code: among examples Isidore, *Orig.* 1. 23, cites BF *bonum*

the early years of such persons, for the Maid, by holding back their great natural gifts, puts a bridle on their lips and restrains them by the curb of authority. And (small wonder in a virgin) her offspring is not fruitful.

[203]Balancing night with the length of day when after a year's space we enjoy the new vintage of the ripened grape, the Scales will bestow the employment of weights and measures and a son to emulate the talents of Palamedes, who first assigned numbers to things, and to these numbers names,[a] fixed magnitudes, and individual symbols.[b] He will also be acquainted with the tables of law, abstruse legal points, and words denoted by compendious signs [c]; he will know what is permissible and the penalties incurred by doing what is forbidden; in his own house he is a people's magistrate holding lifelong office. Under no other sign would Servius [d] more fittingly have been born, who in interpreting the law framed legislation of his own. Indeed, whatever stands in dispute and needs a ruling the pointer of the Balance will determine.

[217]By virtue of his tail armed with its powerful sting, wherewith, when conducting the Sun's chariot through his sign, he cleaves the soil and sows seed in the furrow, the Scorpion creates natures ardent for war and active service, and a spirit which rejoices in plenteous bloodshed and in carnage more than in plunder. Why, these men spend even peace under arms: they fill the glades and scour the woods; they

factum, SC *senatus consultum*, RP *respublica*, PR *populus Romanus*.

[d] Servius Sulpicius Rufus, extolled as the greatest of jurists by Cicero, *Brutus* 151-153.

nunc hominum, nunc bella gerunt violenta ferarum,
nunc caput in mortem vendunt et funus harenae,
atque hostem sibi quisque parat, cum bella quiescunt.
sunt quibus et simulacra placent et ludus in armis
(tantus amor pugnae), discuntque per otia bellum
et quodcumque pari studium producitur arte.

At, quibus in bifero Centauri corpore sors est
nascendi concessa, libet subiungere currus,
ardentis et equos ad mollia ducere frena
et totis armenta sequi pascentia campis,
quadrupedum omne genus positis domitare magistris,
exorare tigres rabiemque auferre leoni
cumque elephante loqui tantamque aptare loquendo
artibus humanis varia ad spectacula molem.
quippe ferae mixtum est hominis per sidera corpus
impositumque manet, quocirca regnat in illas.
quodque intenta gerit curvato spicula cornu,
et nervos tribuit membris et acumina cordi
et celeris motus nec delassabile pectus.

Vesta tuos, Capricorne, fovet penetralibus ignes :
hinc artes studiumque trahis. nam quidquid in usus
ignis eget poscitque novas ad munera flammas
sub te censendum est. scrutari caeca metalla,
depositas et opes terrarum exurere venis,
materiamque manu certa duplicare erit a te,
quidquid et argento fabricetur, quidquid et auro.
quod ferrum calidi solvant atque aera camini
consummentque foci Cererem, tua munera surgent.
addis et in vestes studium mercemque fugantem
frigora, brumalem servans per saecula sortem,

[247] terramque
[248] duplicari et arte
[251] consumentque
[252] fugacem

wage fierce warfare now against man, now against beast, and now they sell their persons to provide the spectacle of death and to perish in the arena, when, warfare in abeyance, they each find themselves foes to attack. There are those, too, who enjoy mock-fights and jousts in arms (such is their love of fighting) and devote their leisure to the study of war and every pursuit which arises from the art of war.

[230]But they whose lot it is to be born under the Centaur of double form delight in yoking a team, in bringing a fiery horse to obey the pliant reins, in following herds which graze all over the grasslands, and in imposing a master on every kind of quadruped and taming them : they soften tigers, rid the lion of his fierceness, speak to the elephant and through speech adapt its huge bulk to human skills in a variety of displays. Indeed, in the stars of this constellation the human form is blended with a beast's and placed above it ; wherefore it has lordship over beasts. And because it carries a shaft poised on drawn bow, it imparts strength to limb and keenness to the intellect, swiftness of movement, and an indefatigable spirit.

[243]In her shrine Vesta tends your fires, Capricorn : and from her you derive your skills and callings. For whatever needs fire to function and demands a renewal of flame for its work must be counted as of your domain. To pry for hidden metals, to smelt out riches deposited in the veins of the earth, to fold sure-handed the malleable mass—these skills will come from you, as will aught which is fashioned of silver or gold. That hot furnaces melt iron and bronze, and ovens give to the wheat its final form, will come as gifts from you. You also give a fondness for clothes and wares which dispel the cold, since your lot falls for all

qua retrahis ductas summa ad fastigia noctes
nascentemque facis revocatis lucibus annum.
hinc et mobilitas rerum, mutataque saepe
mens natat ; et | Veneri mixto cum crimine servit
pars prior, at | melior iuncto sub pisce senecta est.
 Ille quoque, inflexa fontem qui proicit urna,
cognatas tribuit iuvenalis Aquarius artes :
cernere sub terris undas, inducere terris,
ipsaque conversis aspergere fluctibus astra
litoribusque novis per luxum illudere ponto
et varios fabricare lacus et flumina ficta
et peregrinantis domibus suspendere rivos.
mille sub hoc habitant artes, quas temperat unda.
quippe etiam mundi faciem sedesque movebit
sidereas caelumque novum versabit in orbem.
tempore non ullo subolem taedebit Aquari,
quae per aquas veniunt, operum, fontesque sequuntur.
mite genus dulcesque fluunt a sidere partus,
pectora nec sordent ; faciles in damna feruntur ;
nec dest nec superest census. sic profluit urna.
 Ultima quos gemini producunt sidera Pisces,
his erit in pontum studium, vitamque profundo
credent et puppes aut puppibus arma parabunt,
quidquid et in proprios pelagus desiderat usus.
innumerae veniunt artes : vix nomina rebus

---

[a] *I.e.* reservoirs and aqueducts.
[b] *Cf.* Vitruvius 8. 1. 1-5 and Flores 95 ff.
[c] Referring to a model of the celestial system operated by water, such as that of Ctesibius mentioned by Pappus of Alexandria, 8. 2. Archimedes' famous orrery was driven by air.

time in winter's season, wherein you shorten the nights you have brought to their greatest length and give birth to a new year by enlarging the daylight hours. Hence comes a restless quality in their lives and a mind which is often changed and floats this way and that; the first half of the sign is the slave of Venus, and that with guilt involved, but a more virtuous old age is promised by the conjoined fish below.

[259]The youthful Waterman, who from upturned pot pours forth his stream, likewise bestows skills which have affinity with himself: how to divine springs under the ground and conduct them above, to transform the flow of water so as to spray the very stars, to mock the sea with man-made shores at the bidding of luxury, to construct different types of artificial lakes and rivers,[a] and to support aloft for domestic use streams that come from afar.[b] Beneath this sign there dwell a thousand crafts regulated by water. Why, water will even set in motion the face of heaven and the starry habitations, and will cause the skies to move in a novel rotation.[c] *Never will the sons of Aquarius grow tired* of the works which come in the wake of water and follow springs. They who issue from this sign are a gentle sort and a lovable breed, and no meanness of heart is theirs; they are prone to suffer losses; and of riches they have neither need nor surfeit. Even thus doth the urn's stream flow.

[273]The folk engendered by the two Fishes, the last of the signs, will possess a love of the sea; they will entrust their lives to the deep, will provide ships or gear for ships and everything that the sea requires for activity connected with it. The consequent skills are numberless: so many are the components of even

sufficiunt, tot sunt parvae quoque membra carinae.
adde gubernandi studium, quod venit in astra
et pontum caelo vincit. bene noverit orbem
fluminaque et portus, mundum ventosque, necesse est
iamque huc atque illuc agilem convertere clavum
et frenare ratem fluctusque effundere rector,
iam remis agere et lentas inflectere tonsas.
quin placidum ductis everrere retibus aequor
litoribusque suis populos exponere captos
aut uncos celare cibis aut carcere fraudem,
navalis etiam pugnas, pendentia bella,
attribuunt pelagique infectos sanguine fluctus.
fecundum genus est natis et amica voluntas
et celeres motus mutataque cuncta per aevum.

Hos tribuunt mores atque has nascentibus artes
bis sex materia propria pollentia signa.

Sed nihil in semet totum valet : omnia vires
cum certis sociant signis sub partibus aequis
et velut hospitio mundi commercia iungunt
conceduntque suas partes retinentibus astris.
quam partem Graiae dixere decanica gentes.
a numero nomen positum est, quod partibus astra
condita tricenis triplici sub sorte feruntur
et tribuunt denas in se coeuntibus astris
inque vicem ternis habitantur sidera signis.

279 quod] per
280 bene] et
283 rectos
284 iam] aut
285 quid
298 Graiae] deganae

[a] *Cf.* 2. 688.
[b] *Cf.* Introduction p. lxxxv and figure 23.

a small ship that there are scarcely enough names for things. There is also the art of navigation, which has reached out to the stars and binds the sea to heaven. The pilot must have sound knowledge of the earth, its rivers and havens, its climate and winds; how on the one hand to ply the mobile helm this way and that, and brake the ship and spread apart the waves, and how on the other to drive the ship by rowing and to feather the lingering blades. The Fishes further impart to their son the desire to sweep tranquil waters with dragnets and to display on shores which are their own the captive peoples of the deep, either by hiding the hook within the bait or the guile within the weel. Naval warfare too is of their gift, battles afloat, and blood-stained waves at sea. The children of this sign are endowed with fertile offspring, a friendly disposition, swiftness of movement, and lives in which everything is ever apt to change.

[292]These are the characters and these the skills which the twelve signs, by virtue of the powers its substance gives to each, bestow on men born under them.

[294]But no sign has exclusive control over itself[a]: all share their powers with certain signs in equal portions, and in a spirit of hospitality, as it were, they form a heavenly fellowship and surrender the parts of which they are composed to the keeping of other signs. This department of our art the Greeks have termed the system of decans.[b] The name is derived from the numeral, since the signs, which consist of thirty degrees, enjoy a tripartite arrangement and allot ten degrees to each of the signs associating with themselves, the constellations one after the other providing a home for three signs each. Thus nature

sic altis natura manet consaepta tenebris
et verum in caeco est multaque ambagine rerum ;
nec brevis est usus nec amat compendia caelum, 3
verum aliis alia opposita est et fallit imago
mentiturque suas vires et munera celat.
quae tibi non oculis, alta sed mente fuganda est
caligo, penitusque deus, non fronte, notandus. 3
Nunc quae sint coniuncta quibus quove ordine redda
ne lateant aliae vires aliena per astra.
namque Aries primam partem sibi vindicat ipsi,
altera sors Tauro, Geminis pars tertia cedit.
sidera sic inter divisum dicitur astrum
totque dabit vires dominos quotcumque recepit. 3
diversa in Tauro ratio est, nec parte sub ulla
censetur : Cancro primam mediamque Leoni,
extremam Erigonae tribuit. natura per astrum
stat tamen et proprias miscet per singula vires.
Libra decem partes Geminorum prima capessit, 3
Scorpios adiunctas ; Centauri tertia pars est,
nec quicquam numero discernitur, ordine cedit.
Cancer in adversum Capricorni derigit astrum
bis quinas primum partes, dignatus in illo
temporis articulo sub quo censetur et ipse, 3
quod facit aequalis luces brumalibus umbris
cognatamque gerit diverso in cardine legem ;
alterius partis perfundit Aquarius ignes,

303 conspecta
307 nomina
314 [sidera] sic inter denas

[a] Housman interprets "not from your eyes but from the depths of your mind," but see Bühler 478 f.

[b] The winter solstice, which will fall in the first decan, whether the tropic occurs in the 1st, 8th, or 10th degree (the alternatives specified in 3. 680-682).

is ever hedged about with deep darkness, and truth is hidden and wrapped in much complexity. Heaven is reached by no brief effort and does not favour short cuts; but the shape of one sign is placed in front of and conceals others, and dissimulates and hides its real influences and gifts. This obscurity must be dispelled not by the eyes but by profundity of intellect,[a] for it is in its interior and not in its outward appearance that the divine is to be apprehended.

[310] Now shall I tell which signs are joined to which, and in what sequence, lest the influences which others exercise in signs foreign to them go unperceived. The Ram claims his first decan for himself, the second portion falls to the Bull, the third decan to the Twins. Thus the constellation is said to have been distributed among the signs, and it will exert the influences of as many masters as it has received. The scheme works differently in the sign of the Bull, for he is not reckoned under any of his decans: he bestows the first on the Crab, the middle on the Lion, and the last on the Maid. Yet the Bull's nature persists throughout his sign and blends his peculiar influences in every section. The Balance is first to appropriate ten degrees of the Twins; Scorpio claims the next ten; the third decan is the Centaur's, who is in no way treated differently in the number of degrees, but yields to the others in precedence. The Crab proffers twice five degrees to Capricorn's opposite sign first, for in the latter he is deemed worthy of the seasonal juncture [b] with which Capricorn himself is assessed because in winter he makes the daylight equal to the darkness and at his opposite cardinal-point observes a kindred law; the fires of the second decan are bedewed by the Waterman, who is followed by the

quem subeunt Pisces extremo sidere Cancri.
at Leo consortis meminit sub lege trigoni
Lanigerumque ducem recipit Taurumque quadrato
coniunctum sibi ; sub Geminis pars tertia fertur :
hos quoque contingit per senos linea flexus.
praecipuum Erigone Cancro concedit honorem
cui primam tribuit partem ; vicina relicta est
vicino, Nemeaee, tibi ; pars ipsius una est
quae fastidito concessa est iure potiri.
sed Libra exemplo gaudet, pariterque regentem
noctes atque dies diverso in tempore secum
Lanigerum sequitur : veris iuga temperat ille,
haec autumnalis componit lucibus umbras:
nulli concedit primam, traditque sequenti
vicinam partem ; Centauri tertia summa est.
Scorpios in prima Capricornum parte locavit,
alterius dominum fecit cui nomen ab undis,
extremas voluit partes sub Piscibus esse.
at qui contento minitatur spicula nervo
Lanigero primas tradit sub iure trigoni
et medias Tauro partes Geminisque supremas.
nec manet ingrati Capricornus crimine turpis
sed munus reddit Cancro recipitque receptus
principiumque sui donat ; coniuncta Leonis
regna ferunt, summas partes et Virginis esse.
fontibus aeternis gaudens urnaque fluenti
iura sui Librae permittit prima regenda,

341 horas
343 f. *after* 348
350 turpi
352 leoni

[a] *Cf.* 2. 279 f.
[b] *I.e.* in his first decan.
[c] *Cf.* 2. 664-666.
[d] *Cf.* 2. 368-370.
[e] By the Crab and the Lion.
[f] To Scorpio.

Fishes in the Crab's remotest stars. But the Lion, mindful of his partner under the trigon's law,[a] welcomes the Ram as leader,[b] and next the Bull, who is united to him by square [c]; the third decan passes under the dominion of the Twins: by his hexagonal line the Lion reaches these too.[d] Erigone bestows especial honour on the Crab, to whom she gives first decan; the neighbouring one she leaves to you, her neighbour, O Lion of Nemea; one decan belongs to herself, which, since tenure of it was disdained by the others,[e] was granted to her to possess. The Balance has the pleasure of a precedent, and imitates the Ram, who though at the opposite season agrees with it in fixing equal hours of day and night: the Ram regulates the scales of spring, while the Balance matches autumn's nights with autumn's days. It yields its first decan to none, and hands the adjoining one to the sign which follows it [f]; the third factor belongs to the Centaur. The Scorpion has installed Capricorn in the first decan; has made the man named after water the lord of the second; and has willed that its last degrees be subject to the Fishes. Now he who with bow-string tensed ever threatens to let fly his shaft surrenders his first degrees to the Ram in accordance with the trigon's pact, the midmost to the Bull, and the last to the Twins. Capricorn risks not the charge of base ingratitude but discharges his obligation to the Crab: welcomed himself to the Crab's domain, he welcomes the Crab to his, and presents him with his initial portion; the bordering realm is reckoned as the Lion's, and the endmost degrees as the Maid's. He that rejoices in the never-ending stream that pours from his urn transfers control of the first rights over himself to the Balance,

haerentisque decem partes Nepa vindicat ipsi ;
summas Centaurus retinet iuvenale per astrum.
iam superant gemini Pisces, qui sidera claudunt.
Lanigero primos tradunt in finibus usus,
perque decem medias partes tu, Taure, receptus ;
quod superest, ipsi sumunt, utque orbe feruntur
extremo sic et sortis pars ultima cedit.

Haec ratio retegit latitantis robora mundi
in plurisque modos repetitaque nomina caelum
dividit et melius sociat, quo saepius, orbem.
nec tua sub titulis fallantur pectora notis :
dissimulant, non ostendunt mortalibus astra.
altius est acies animi mittenda sagacis
inque alio quaerendum aliud iunctisque sequendum
viribus ; et, cuius signi quis parte creatur,
eius habet mores atque illo nascitur astro.
talis per denas sortes natura feretur.
testis erit varius sub eodem sidere fetus,
quodque in tam multis animantum milibus, uno
quae veniunt signo, tot sunt, quot corpora, mores,
et genus externum referunt aliena per astra,
confusique fluunt partus hominum atque ferarum.
scilicet in partes iunguntur condita pluris
diversasque ferunt proprio sub nomine leges.

[356] nota [369] quaerendo mali quod

[a] An error of the poet's. According to the principle of his system the decans here conferred upon Aries and Taurus belong to Capricorn and Aquarius respectively.

[b] Those of the decans.

whilst the adjoining ten degrees are claimed by the Scorpion; the last degrees in the young man's sign are occupied by the Centaur. There now remain the Fishes twain, which close the circle of signs. They bestow on the Ram [a] enjoyment of the first parts of their territory; and to the middle ten degrees are you admitted, O Bull; the remainder they keep for themselves, and even as they are situate at the extremity of the zodiac's circle, so does the last decan of their allotment fall to them.

[363]This is the scheme which unveils the powers of the mysterious universe: it divides the heavens in more ways than one so that the names of the signs recur; the more frequent the recurrence, the closer is the association of the signs. Do not suffer your mind to be deceived by their familiar names: they mask other signs,[b] which they conceal from mortals. The knowing mind's keen edge must cut more deeply: one sign must be sought in another, and in our inquiry we must take account of the powers of each, for a man also receives the characteristics of the sign dominating the decan of his nativity and is born under the influence of that sign too. Such is the nature of the forces to be found in the decans. Evidence of this will be found in the diversity of those born under the same constellation, and the fact that among so many thousands of living beings sprung from a single sign there are as many dispositions as there are individuals, that born under signs alien to their natures they display the qualities of other stars, and that men and beasts are born without distinction in human and bestial signs alike. Naturally, since they consist of several components, the signs form associations with them, and each sign passes various decrees under its

nec tantum lanas Aries nec Taurus aratra
nec Gemini Musas nec merces Cancer amabit,
nec Leo venator veniet nec Virgo magistra,
mensuris aut Libra potens aut Scorpios armis
Centaurusque feris, igni Capricornus et undis
ipse suis Iuvenis geminique per aequora Pisces ;
mixta sed in pluris sociantur sidera vires.

"Multum" inquis "tenuemque iubes me ferre labo
rursus et in magna mergis caligine mentem,
cernere cum facili lucem ratione viderer."
quod quaeris, deus est : conaris scandere caelum
fataque fatali genitus cognoscere lege
et transire tuum pectus mundoque potiri.
pro pretio labor est nec sunt immunia tanta,
ne mirere viae flexus rerumque catenas.
admitti potuisse sat est : sint cetera nostra.
at nisi perfossis fugiet te montibus aurum,
obstabitque suis opibus super addita tellus.
ut veniant gemmae, totus transibitur orbis,
nec lapidum pretio pelagus cepisse pigebit.
annua solliciti consument vota coloni,
et quantae mercedis erunt fallacia rura !
quaeremus lucrum ventis Martemque sequemur
in praedas. pudeat tanto bona velle caduca.
luxuriae quoque militia est, vigilatque ruinis

388 *after* 396 402 ventis] naves

own name. Not only in wool will the Ram take pleasure, the Bull in the plough, the Twins in the Muses, or the Crab in trade ; nor will the Lion come forth as a hunter only, or as a teacher the Maid ; not only over measures will the Balance preside, the Scorpion over arms, the Centaur over beasts, Capricorn over fire, the Youth over his own waters, or over the seas the Fishes twain : but the signs are mixers and acquire further influences from their fellowships.

[387] "But," you say, "the task you bid me undertake is great and subtle, and you are plunging my mind back into deep darkness just when I thought a simple principle was enabling me to see light." The object of your quest is God : you are seeking to scale the skies and, though born beneath the rule of fate, to gain knowledge of that fate ; you are seeking to pass beyond your understanding and make yourself master of the universe. The toil involved matches the reward to be won, nor are such high attainments secured without a price ; so wonder not at the winding route and the intricacy of things. It is enough that we have been given power to make the search : let the rest be left to us. Unless you mine mountains, gold will elude your grasp, and the earth that is heaped above will bar access to the wealth it hides. Men will traverse the entire globe to make jewels available, and will not shrink from occupying the sea to gain the precious pearl. Each year the anxious farmer will utter every prayer he knows, and yet how small is the yield of the treacherous countryside ! We shall face the perils of sea-winds in our search for gain and follow the god of war in hope of booty. Ah, shame on those willing to pay so high a price for perishable goods ! Luxury too entails a kind of military service : the

venter, et, ut pereant, suspirant saepe nepotes.
quid caelo dabimus ? quantum est, quo veneat omne?
impendendus homo est, deus esse ut possit in ipso.

Hac tibi nascentum mores sunt lege notandi.
nec satis est signis dominantia discere signa
per denos numeros et quae sint insita cuique ;
sed proprias partes ipsas spectare memento
vel glacie rigidas vel quas exusserit ignis,
et sterilis sine utroque tamen, quas largior umor
quasve minor iusto vitiat. namque omnia mixtis
viribus et vario consurgunt sidera textu.
est aequale nihil. terrenos aspice tractus
et maris et variis fugientia flumina ripis :
crimen ubique frequens et laudi noxia iuncta est.
sic sterilis tellus laetis intervenit arvis
ac subito rumpit parvo discrimine foedus ;
et modo portus erat pelagi iam vasta charybdis,
laudatique cadit post paulum gratia ponti ;
et nunc per scopulos, nunc campis labitur amnis,
et, faciens iter aut quaerens, curritve reditve.
sic etiam caeli partes variantur in astris :
ut signum signo, sic a se discrepat ipsum
momentoque negat vires usumque salubrem,
quodque per has geritur partes sine fruge creatur
aut cadit aut multis sentit bona mixta querellis.

413 [sine]
414 quaque minoribus tovit
417 variis] partis
419 tellus] terris
424 et] aut
uritve

[a] The gourmet's tastes are expensive.
[b] The principle of the decans explained in 294-386.
[c] *Cf.* Introduction p. lxxxvii and table 4.

glutton keeps sleepless watch over that which proves his ruin,[a] and profligates oft pant for their own undoing. What then shall we give for heaven? What is the worth of that, with which we may purchase all? Man must expend his very self before God can dwell in him.

[408]Such is the principle [b] by which you must determine the character of men at birth. But it is not enough to learn of signs that hold sway in the decans of other signs, and which signs are thus installed in each; you must be sure to notice individual degrees which are themselves either numbed with ice or scorched by fire and some which, though free from either of these defects, are nevertheless sterile, being marred by excessive or insufficient moisture.[c] For all the signs that rise possess a mixture of influences and a variety of texture. Nothing is uniform. Look at the stretches of land and sea, and rivers that course past banks of varying scenery: failings abound everywhere, and imperfection is found next to excellence. Thus barren soil appears in the midst of rich fields and, with hardly any indication of discontinuity, abruptly breaks the natural order; and what but lately was a haven of the sea now becomes a yawning maelstrom, and the charm of the waters we admired vanishes in a moment; the same river flows now over rocks and now through plains and, as it creates or looks for a path, runs forward or doubles back on its tracks. So too in the sky do the degrees of the signs display diversity: just as sign is at variance with sign, so it is at variance with itself, and but a minute shift causes it to withhold its influences and salutary effects: whatever is brought forth in these degrees is born to a life of frustration or perishes or experiences benefits

hae mihi signandae proprio sunt carmine partes.
sed quis tot numeros totiens sub lege referre,
tot partes iterare queat, tot dicere summas,
perque paris causas faciem mutare loquendi ?
dum canimus verum, non aspera ponere, ut illis
incidimus, sic verba piget ; sed gratia derit,
in vanumque labor cedit quem despicit auris.
sed mihi per carmen fatalia iura ferenti
et sacros caeli motus ad iussa loquendum est,
nec fingenda datur, tantum monstranda figura.
ostendisse deum nimis est : dabit ipse sibimet
pondera. nec fas est verbis splendescere mundum :
rebus erit maior. nec parva est gratia nostri
oris, si tantum poterit signare canenda.
accipe damnandae quae sint per sidera partes.
Lanigeri pars quarta nocet nec sexta salubris ;
septima par illi ac decima est decimaeque secunda
quaeque duas duplicant summas septemque novemqu
unaque viginti numeris pars addita laedit
et quinta et duram consummans septima partem.
Tauri nona mala est, similis cui tertia pars est
post decimam nec non decimae pars septima iuncta ;
bisque undena notans et bis duodena nocentes

433 per patris
440 suspendere
445 pars . . . decumae [est]
446 duplicat
449 quo
451 nocens

[a] If it succeeds in only signifying in hexameters what, had the technical language not been intractable, it should have adorned with all the flowers of poetry. The reading *cavenda*, "the places to beware of," which at first sight seems certain, is refuted by Bühler 485. Moreover, infants about to be born can hardly "beware of" the injurious degrees.

[b] Aries : 4, 6, 7, 10, 12, 14, 18, 21, 25, 27.

[c] According to Housman the words *decimae secunda*

spoilt by much that is disagreeable. These degrees I must now specify in fitting verse. Yet who could so oft express in metre's laws so many numerals, repeat so many degrees, put into words so many totals, and vary the style of utterance in treating of the same themes? *In giving an accurate account we need not feel that writing down inelegant* phrases, just as we chance on them, calls for apology; but this entails a lack of charm, and effort rejected by the ear is effort vainly spent. Yet, as I seek to expound in verse the laws of destiny and the sacred motions of the skies, my words must conform to what is bidden: I am not permitted to fashion, but only to describe, the pattern. To show the deity is more than enough: he himself will establish his authority. Nor is it right to glorify heaven with words: in the reality it will prove even greater. Still the charm of my muse is not contemptible, if it can only signify what ideally should be sung.[a] Learn then of the degrees, sign by sign, which merit condemnation.

[444]The fourth degree of the Ram [b] inflicts harm, and the sixth is also unwholesome; the seventh is as bad as the sixth, and so are the tenth, that second to the tenth,[c] and those which double the numbers seven and nine; added to the count of a score one degree is damaging, so too a fifth and a seventh, the last completing the adverse portion of the sign.

[449]The ninth degree of the Bull [d] is malign, and the third after the tenth is like it, and again the seventh attached to the tenth; the degree which signifies twice eleven and that which signifies twice twelve are

should signify "eleventh," but in this context of numbers "twelfth" seems the more natural interpretation.

[d] Taurus: 9, 13, 17, 22, 24, 26, 28, 30.

quaeque decem trisque ingeminat fraudatque duobus
triginta numeros et tu, tricesima summa, es.

Pestifera in Geminis pars prima et tertia signi,
septima non melior, ter quintae noxia par est,
unaque bis denis brevior nocet unaque maior,
et similis noxae veniet vicesima quinta
cumque duae subeunt vel cum se quattuor addunt.

Nec Cancri prima immunis nec tertia pars est
nec sexta ; octava est similis, decimaque peracta
prima rapit, nec ter quintae clementior usus ;
septima post decimam luctum et vicesima portat
et quinta accedens et septima nonaque summa.

Tu quoque contactu primo, Nemeaee, timendus,
et quarta sub parte premis ; bis quinta salubri
terque caret caelo, vicesima et altera laedit ;
e tribus appositis vitiat totidemque secutis
ultima, nec prima melior tricesima pars est.

Erigones nec pars prima est nec sexta nec una
ad decimam nec quarta nec octava utilis umquam ;
proxima viginti numeris et quarta timenda est,
et quae ter decimam claudit sors ultima partem.

Et quinta in Chelis et septima inutilis astri,
tertia et undecimae decimaeque est septima iuncta

452 fraudata
454 signis
455 quina et
460 octavae [est] decumaeque
467 victum est
473 aestu
474 septima . . . et tertia

---

[a] Gemini : 1, 3, 7, 15, 19, 21, 25, 27, 29.
[b] Cancer : 1, 3, 6, 8, 11, 15, 17, 20, 25, 27, 29.
[c] Leo : 1, 4, 10, 15, 22, 25, 28, 30.
[d] Virgo : 1, 6, 11, 14, 18, 21, 24, 30.

harmful, as well as that which doubles three and ten, that which robs of two the count of thirty, and you, O number thirty.

[454]Baneful in the Twins [a] are the first and third degrees of the sign, the seventh is no better, and the hurt of the treble-fifth is just as great; injurious are the degrees numbering one less than and one more than twice ten, whilst the coming of the twenty-fifth will prove equally harmful, both of itself and when two or four are added to it.

[459]Neither first nor third nor sixth degree of the Crab [b] is free from ill; the eighth resembles these; after the completion of a decade the first carries off its prey, and the behaviour of the treble-fifth is not more merciful; the seventeenth brings grief, so does the twentieth, and also the fifth, seventh, and ninth degrees following.

[464]With you, too, Lion of Nemea, [c] first contact is to be dreaded, and fearful is the onset of your fourth degree; double-fifth and treble-fifth preclude a salutary sky, and noxious is the twenty-second; of the following three the last brings infection, and so does the last of a further three, whilst the thirtieth is no improvement on the first.

[469]Of the Maid [d] neither the first nor the sixth degree is ever beneficial, nor of the second decade the first, fourth, and eighth; beyond a count of twenty the first and fourth are full of terror as well as the last portion which concludes the sign in the thirtieth degree.

[473]In the Balance [e] the fifth confers no benefit, nor does the sign's seventh; so, too, the third past the eleventh, the seventh past the tenth, as also past a

[e] Libra: 5, 7, 14, 17, 24, 27, 29, 30.

quartaque bis denis actis et septima et ambae
quae numerum claudunt nona et tricesima partes.
Scorpios in prima reus est, cui tertia par est
et sexta et decima et quae ter tibi quinta notatur,
undecimam geminans et quae vicesima quinta est
octavoque manet numero nonumque capessit.
Si te fata sinant, quartam ne selige partem
Centauri; fuge et octavam; sex bisve peractis
octo, bis aut denis, metuendus dicitur aer,
cumque iterum duodena refert aut terna decemque
aut septena quater, vel cum ter dena figurat.
Nec pars optanda est Capricorni septima; nona
consentit decimamque sequens quam tertia signat
et tribus aut una quae te, vicesima, fraudat
quaeve auget quinto numero vel septima fertur.
Pars est prima nocens fundentis semper Aquari,
damnanda et decimae succedens prima peractae
tertiaque et quinta et numero quae condita nono est
et post viginti prima et vicesima quinta
cumque illa quartam accumulans vicesima nona.
Tertia per geminos et quinta et septima Pisces,
undecima et decimae metuenda est septima iuncta;
et quinta in quinos numeros revocata duasque
accipiens ultra summas metuenda feretur.
Hae partes sterilem ducunt et frigore et igni
aera vel sicco vel quod superaverit umor,
si rapidus Mavors ignes iaculatur in illum

478 [tibi] 2. 232 parsque marina nitens
482 bis sexque 490 est

---

[a] Scorpio: 1, 3, 6, 10, 15, 22, 25, 28, 29.
[b] Sagittarius: 4, 8, 12, 16, 20, 24, 26, 28, 30.
[c] That is, writes "XXX."
[d] Capricorn: 7, 9, 13, 17, 19, 25, 27.
[e] Aquarius: 1, 11, 13, 15, 19, 21, 25, 29.
[f] Pisces: 3, 5, 7, 11, 17, 25, 27.

score the fourth, the seventh, and the pair which close the tale, the ninth past a score and the thirtieth degree.

477The Scorpion [a] is arraigned in his first degree, matching which are the third and sixth and tenth and that which you note as fifth thrice over, the double of the eleventh, and of the twenties the fifth, that which resides at number eight, and the occupant of nine.

481Were Fortune to give you leave, choose not the Centaur's [b] fourth degree ; shun, too, the eighth ; double six or eight or ten, the atmosphere is termed fraught with dread ; and so it is when he twice reckons twelve or ten and three, when he reckons four times seven, and when he thrice draws the sign of ten. [c]

486The seventh of Capricorn [d] is not a degree to be desired ; at one with it are the ninth, that indicated as third after the tenth, and those which rob you, O twentieth, of three or one or increase you by five or seven.

2. 232The first degree of the ever-pouring Waterman [e] spells trouble, and meriting condemnation are the first, third, fifth, and ninth after the completion of a decade, the first after a score, the twenty-fifth and, with increase of four, the twenty-ninth.

494In the sign of the two Fishes [f] third, fifth, and seventh are formidable degrees, as well as the eleventh, and the seventh joined to the tenth ; that which multiplies five times five and that which receives a further two will also be found fraught with dread.

498These degrees are allotted an atmosphere made sterile by reason of cold and fire or because of drought or superabundant moisture, be it scorching Mars who

Saturnusve suam glaciem *Phoebeve propinquis*
quem trahit a terris rorem Phoebusve calores.

Nec te perceptis signorum cura relinquat
partibus : in tempus quaedam mutantur, et ortu
accipiunt proprias vires ultraque remittunt.

Namque, ubi se summis Aries extollet ab undis
et cervice prior flexa quam cornibus ibit,
non contenta suo generabit pectora censu
et dabit in praedas animos solvetque pudorem :
tantum audere iuvat. sic ipse in cornua fertur
ut ruat aut vincat. non illos sedibus isdem
mollia per placidas delectant otia curas,
sed iuvat ignotas semper transire per urbes
scrutarique novum pelagus totius et esse
orbis in hospitio. testis tibi Laniger ipse,
cum vitreum findens auravit vellere pontum
orbatumque sua Phrixum per fata sorore
Phasidos ad ripas et Colchida tergore vexit.

At, quos prima creant nascentis sidera Tauri,
feminei incedunt. nec longe causa petenda est,
si modo per causas naturam quaerere fas est :
aversus venit in caelum divesque puellis,
Pleiadum parvo referens glomeramine sidus.
accedunt et ruris opes, propriaque iuvencum
dote per inversos exornat vomere campos.

501 [ve] sumet
505 extollit
510 aut] et . . . ullus
519 iuceat
522 glomerabile

---

[a] Saturn is icy as being the planet farthest away from Earth ; Phoebe (the Moon) is nearest to it. [b] Helle.

launches his flames upon it or Saturn his native ice *or Phoebe the dews she gathers from neighbouring Earth* [a] or Phoebus his heat.

502 But having noted the injurious degrees of the signs you must not relax your attention : there are some degrees which undergo a temporary change of character, acquiring at their ascension special powers which they thereafter relinquish.

505 Thus, when the Ram emerges above the surface of the waves and the curve of his neck appears before his horns, he will give birth to hearts that are never content with what is theirs ; he will engender minds bent on plunder and will banish all sense of shame : such is their desire for venture. Even thus does the ram himself rush forth with lowered horns, resolved to win or die. Not for them the gentle ease of a fixed abode with none but peaceful cares ; it is ever their delight to travel through unknown cities, to explore uncharted seas, and enjoy the whole world's hospitality. The Ram himself gives you evidence of this : once furrowing a trail through the glassy sea, he tinged it with the gold of his fleece, when on his back he carried Phrixus, bereft of his sister [b] by fate's decree, and brought him to the banks of the Phasis and to Colchis.

518 But they who are given life by the rising of the Bull's foremost stars walk with an effeminate gait. The cause is not far to seek, at least if one may seek in causes an explanation of nature : it rises into the sky hind-before with a bevy of maidens, for it brings with it the stars of the Pleiades massed in a tiny cluster. The Bull is also attended by the wealth of the countryside and furnishes its young with its own special endowment amid fields upturned by the plough.

Sed, Geminos aequa cum profert unda tegitque
parte, dabit studia et doctas producet ad artes.
nec triste ingenium sed dulci tincta lepore
corda creat, vocisque bonis citharaeque sonantis
instruit, et dotes cantus cum pectore iungit.
At, niger obscura Cancer cum nube feretur,
qua velut exustus Phoebeis ignibus ignis
deficit et multa fuscat caligine sidus,
lumina deficient partus, geminamque creatis
mortem fata dabunt: se quisque et vivit et effert.
Sicui per summas avidus produxerit undas
ora Leo et scandat malis hiscentibus orbem,
ille patri natisque reus, quas ceperit ipse,
non legabit opes, censumque immerget in ipso.
tanta fames animumque cibi tam dira cupido
corripit, ut capiat semet nec compleat umquam,
inque epulas funus revocet pretiumque sepulcri.
Erigone surgens, quae rexit saecula prisca
iustitia rursusque eadem labentia fugit,
alta per imperium tribuit fastigia summum,
rectoremque dabit legum iurisque sacrati
sancta pudicitia divorum templa colentem.
Sed, cum autumnales coeperunt surgere Chelae,
felix aequato genitus sub pondere Librae.
iudex examen sistet vitaeque necisque

529 saltus
531 exutus
533 arctus
535 siqui
549 examen] extranea

---

[a] There being no single degree at the middle of the sign, the influences here described are to be referred to the 15th and 16th.

[b] The cluster Praesaepe, which, like all clusters in astrology, threatens blindness (*cf.* 2. 259 f.).

[c] His resources.

[d] Erigone, properly the daughter of Icarius (*cf.* 2. 31 f.), is

525When ocean displays and conceals equal portions of the Twins,[a] it will bestow zeal for study and direct men to learned arts. It creates no gloomy disposition, but hearts imbued with a pleasant charm, and furnishes them with blessings of voice and tuneful lyre, combining with wit a dowry of melody.

530But when that part of the Crab rises which is dimmed by a sombre cloud,[b] where his own fire fails, as though burnt out by the Sun's, and darkens the signs with impenetrable fog, the sight of those born then will fail, and fate will condemn them to death twice over : each one buries himself while still alive.

535The man to whom the ravenous Lion has displayed its countenance through the topmost waves as it scales with jaws agape the arc of heaven, that man, heedless of father and of sons alike, will not pass on his inheritance but will swallow his patrimony in his body. Such devouring hunger and such a dreadful passion for food take hold of his spirit that he consumes his very self [c] without ever sating it and devotes to his table even his funeral expenses and the price of a tomb.

542At her rising Erigone, who reigned [d] with justice over a bygone age and fled when it fell into sinful ways, bestows high eminence by bestowing supreme power ; she will produce a man to direct the laws of the state and the sacred code, one who will tend with reverence the hallowed temples of the gods.

547When autumn's Claws begin to rise, blessed is he that is born under the equilibrium of the Balance. As judge he will set up scales weighted with life and

here confused with Astraea (*cf.* Aratus, *Phaen.* 98 ff.), who is often (though not elsewhere by Manilius) identified with Virgo.

imponetque iugum terris legesque rogabit.
illum urbes et regna trement nutuque regentur
unius et caeli post terras iura manebunt.
 Scorpios extremae cum tollet lumina caudae,
siquis erit stellis tum suffragantibus ortus,
urbibus augebit terras iunctisque iuvencis
moenia succinctus curvo describet aratro,
aut sternet positas urbes inque arva reducet
oppida et in domibus maturas reddet aristas.
tanta erit et virtus et cum virtute potestas.
 Nec non Arcitenens, prima cum veste resurgit,
pectora clara dabit bello, magnisque triumphis
conspicuum patrias victorem ducet ad arces,
altaque nunc statuet nunc idem moenia vertet.
sed nimium indulgens rebus Fortuna secundis
invidet in facie saevitque asperrima fronti.
horrendus bello Trebiam Cannasque lacumque
ante fugam tali pensabat imagine victor.
 Ultimus in caudae Capricornus acumine summo
militiam in ponto dictat puppisque colendae
dura ministeria et tenui discrimine mortis.
 Quod si quem sanctumque velis castumque probum
hic tibi nascetur cum primus Aquarius exit.
 Ne velit et primos animus procedere Pisces,

570 vitae discrimen inertis 573 neve sit

---

[a] These verses allude to Augustus, at whose birth (Sept. 22, 63 B.C.) the first part of Libra occupied the Horoscope. Whilst line 552 is indecisive in determining whether Augustus was alive or dead at the time of the composition of Book 4, other evidence settles the matter: 764 ff. require Tiberius to be the reigning emperor.

[b] Hannibal, who lost an eye (Livy 22. 2. 11: Sagittarius is one-eyed, *cf*. 2. 260). As at 39 the poet wrongly dates the battle of Lake Trasimene after Cannae.

death ; he will impose the weight of his authority upon the world and make laws. Cities and kingdoms will tremble before him and be ruled by his command alone, whilst after his sojourn on earth jurisdiction in the sky will await him.[a]

553When the Scorpion uplifts the stars which shine at the end of its tail, the man then born with the blessing of the planets will enrich the world with cities and, with robes hitched up and driving a team of oxen, will trace the circuit of the walls with curved plough ; else he will level the cities which have been erected and turn towns back into fields, and produce ripe corn where houses stood. Such will be his worth and such the power which is joined thereto.

560As for the Archer, when the foremost portion of his cloak rises, he will give birth to hearts renowned in war and will conduct the conqueror, celebrating great triumphs in the sight of all, to his country's citadels. Such a one will build high walls one moment and pull them down the next. But if Fortune favours them too generously with success, the mark of her envy is to be seen on their faces, for she works cruel havoc upon their features. So was it that a dread warrior [b] paid for his victories at the Trebia, Cannae, and the Lake, even before the hour of his retreat, with such disfigurement.

568The last part of Capricorn, which consists of the sting at the end of its tail, prescribes for its children service upon the seas and the handling of ships, a hardy calling and one which is ever close to death.

571But if you would have a man that is pious, pure, and good, you will find him born when the first portion of the Waterman rises above the horizon.

573Lest your mind yearn also for the first portion of

garrulitas odiosa datur linguaeque venenum
verba maligna novas mutantis semper ad aures
criminaque ad populum populi ferre ore bilingui.
nulla fides inerit natis, sed summa libido
ardentem medios animum iubet ire per ignes.
scilicet in piscem sese Cytherea novavit,
cum Babyloniacas summersa profugit in undas
anguipedem alatos umeros Typhona ferentem,
inseruitque suos squamosis Piscibus ignes.
nec solus fuerit geminis sub Piscibus ortus :
frater erit dulcisve soror, materve duorum.

Nunc age diversis dominantia sidera terris
percipe. sed summa est rerum referenda figura.
quattuor in partes caeli discribitur orbis,
nascentem lapsumque diem mediosque calores
teque, Helice. totidem venti de partibus isdem
erumpunt secumque gerunt per inania bella.
asper ab axe ruit Boreas, fugit Eurus ab ortu,
Auster amat medium solem Zephyrusque profectum.
hos inter binae mediis e partibus aurae
exspirant similis mutato nomine flatus.
ipsa natat tellus pelagi lustrata corona
cingentis medium liquidis amplexibus orbem,
inque sinus pontum recipit, qui vespere ab atro
admissus dextra Numidas Libyamque calentem

[574] moventum [579] notavit
[576] crimina per [588] ipsumque

[a] Venus : see Introduction p. xxvi.

[b] The poet's conception of the world is schematically represented in the frontispiece : see Introduction pp. xx and lxxxix.

[c] That is, east, west, south, and north respectively.

the Fishes to come forth, be told that their gift is hateful loquacity, a poisonous tongue which ever passes on slanderous talk to fresh ears, and an eagerness to carry to the people on treacherous lips the people's indiscretions. Trustworthiness will not be found in this sign's progeny ; instead a consuming desire urges their fevered minds to go through fire to attain their ends. Certain it is that the goddess of Cythera[a] changed herself into a fish when she plunged into the waters of Babylon to escape from snake-footed Typhon of the winged shoulders; and she has implanted in the scaly Fishes the fire of her own passions. No birth under the two Fishes will be marked by singleness ; a brother will be born as well, or a darling sister, or else the mother of twins.

585 Now you must learn the constellations which bear dominion over the several parts of the earth. But a general sketch of the world must first be given.[b] The round heavens are divided into four parts, which accord with the day's rising, its setting, the noontide heat, and you, Great Bear.[c] From these four quarters as many winds sally forth, and battle against each other in the empty skies. Biting Boreas rushes from the pole whilst Eurus comes flying from the dawn[d] : Auster delights in the midday sun and Zephyrus in the sun's departure. From each of the intervals between them two breezes emit their blasts, like in kind but differing in name. The land itself is afloat, encircled by the crown of ocean which clasps the round world within it in an embrace of water. The land also receives the sea to its bosom, which,[e] admitted from the dusky west, washes on the right

[d] For the inaccuracy of the use of these names, see Introduction p. xc. [e] The Mediterranean Sea.

alluit et magnae quondam Carthaginis arces
litoraque in Syrtes revocat sinuata vadosas
rursusque ad Nilum derectis fluctibus exit.
laeva freti caedunt Hispanas aequora gentes
teque in vicinis haerentem, Gallia, terris
Italiaeque urbes dextram sinuantis in undam
usque canes ad, Scylla, tuos avidamque Charybdin.
hac ubi se primum porta mare fudit, aperto
enatat Ionio laxasque vagatur in undas
et, prius ut, laeva se fundens circuit omnem
Italiam, Hadriaco mutatum nomina ponto,
Eridanique bibit fluctus vetat aequore bellum
Illyricum Epirumque lavat claramque Corinthum
et Peloponnesi patulas circumvolat oras ;
rursus et in laevum refluit vastoque recessu
Thessaliae fines et Achaica praeterit arva.
hinc penitus iuvenisque fretum mersaeque puellae
truditur invitum, faucesque Propontis aperto
Euxino iniungit ponto Maeotis et undis,
quae tergo coniuncta manet fontemque ministrat.
inde ubi in angustas revocatus navita fauces
Hellespontiacis iterum se fluctibus effert,
Icarium Aegaeumque secat laevaque nitentis
miratur populos Asiae totidemque tropaea
quot loca et innumeras gentes Taurumque minantem
fluctibus et Cilicum populos Syriamque perustam

600 revocans
606 [aperto]
608 in laevas effundens
609 hadriam comutatus nomine
610 secat
615 hic pontus
616 propontidos apto
618 pontemque
619 r.] iterum se
620 i. se] revocatus
623 aurumque

[a] The Straits of Messina. [b] The River Po. [c] Phthiotis.

against Numidia and torrid Libya and the towers of once mighty Carthage; it pulls back its shores so that they meander into the shoals of the Syrtes, and emerges again with billows that proceed straight to the Nile. On the left the waves of the sea beat against the peoples of Spain, against you, Gaul, adjacent in lands nearby, and against the cities of Italy, which curves round towards the right portion of the sea as far as your dogs, Scylla, and the greedy maw of Charybdis.[a] As soon as the sea pours forth from this gateway, it floats out into the open Ionian where it spreads over a wide expanse, and turning to the left as before, its name now changed to the Adriatic sea, it completes the circuit of Italy and drinks the waters of the Eridanus[b]; with its flood it restrains Illyria from war, bathes Epirus and renowned Corinth, and hastens round the wide bays of the Peloponnese; yet again it flows back to the left, and in a vast recession passes by the confines of Thessaly and the fields of Achaea.[c] Beyond this point the channel of the youth and the maid who was drowned[d] thrusts inward a reluctant passage; then the Propontis connects its strait to the spacious Euxine sea and the waves of the Maeotis, which, joined to the rear of the Euxine, provides it with a source. When the sailor recalled therefrom to the narrow strait emerges again from the billows of the Hellespont, he cleaves the Icarian and Aegean seas and marvels at the peoples of sleek Asia[e] on the left, in every place a trophy, the innumerable nations of the barbarians, the heights of Taurus menacing the waves, the peoples of Cilicia, parched Syria, and the

[d] Hellespont; Phrixus; Helle.
[e] The Greeks of Asia Minor are meant.

ingentique sinu fugientis aequora terras,
donec in Aegyptum redeunt curvata per undas
litora Niliacis iterum morientia ripis.
haec medium terris circumdat linea pontum
atque his undarum tractum constringit harenis.
mille iacent mediae diffusa per aequora terrae.
Sardiniam in Libyco signant vestigia plantae,
Trinacria Italia tantum praecisa recessit,
adversa Euboicos miratur Graecia montes,
Aegaeis Crete civem sortita Tonantem
Aegyptique Cypros pulsatur fluctibus amnis.
has praeter terras, celebrat quas maxima fama,
totque minore solo tamen emergentia ponto
litora, inaequalis Cycladas Delonque Rhodonque
Aulidaque et Tenedon vicinaque Corsica terris
litora Sardiniae primumque intrantis in orbem
Oceani victricem Ebusum et Balearica rura,
innumeri surgunt scopuli montesque per altum.

Nec tantum ex una pontus sibi parte reclusit
faucibus abruptis orbem ; nam litora plura
impulit oceano Phorcys, sed montibus altis
est vetitus totam ne vinceret aequore terram.
namque inter borean ortumque aestate nitentem
in longum angusto penetrabilis aequore fluctus
pervenit et patulis tum demum funditur arvis
Caspiaque Euxini similis facit aequora ponti.

634 et genetrix
635 aegyptia . . . omnis
636 minora
637 et aequalis
644 pocius

[a] Phoenicia and adjacent lands.
[b] Jupiter, born and brought up in Crete (*cf.*, for example, Virgil, *Aen.* 3. 104 f.).

lands which form a great gulf in their flight from the sea.[a] At last the shore curves through the waves and returns to Egypt, coming to an end again with the banks of the Nile. This is the coastline which encompasses the midland ocean, these are the shores which confine the expanse of its waves. A thousand islands lie in the middle of these far-spread waters. Sardinia has the shape of a footprint in the Libyan sea, whereas the triangle of Sicily is distant from Italy only so far as serves to cut it off; Greece is awed as it faces the mountains of Euboea, whilst Crete, who counts the Thunderer [b] her citizen, is battered by the waves of the Aegean sea, as Cyprus is by those of Egypt's river. *Besides these, which are the most important islands*, and all those strands, though of smaller size, which rise from the sea, the unequal Cyclades, Delos, Rhodes, Aulis, Tenedos, Corsica, whose shore lies near the land of Sardinia, Ebusus, which triumphs over [c] Ocean at its first entrance into the circle of our lands, and the Balearic fields—besides these, rocks and peaks past counting spring up over the deep.

642Nor in one place alone has the sea burst through a gap and opened a way for itself into the land; for the Sea-god has confronted many a shore with the thrust of his waves, but has been prevented by high mountains from overwhelming the whole earth with his flood. For, penetrating by a narrow channel between the north and the shining summer orient,[d] the ocean makes a long arm inland, then at last spreads over low-lying fields and, in semblance like the

[c] By not being submerged.
[d] *I.e.* where the Sun rises in summer, Caecias: thus the compass-point indicated is Aquilo (*cf.* frontispiece).

altera sub medium solem duo bella perinde
intulit Oceanus terris. nam Persica fluctus
arva tenet, titulum pelagi praedatus ab isdem
quae rigat ipse locis, latoque infunditur orbe.
nec procul in mollis Arabas terramque ferentem
delicias variaeque novos radicis odores
leniter affundit gemmantia litora pontus,
et terrae mare nomen habet. media illa duobus.

* * * * *

quondam Carthago regnum sortita sub armis,
ignibus Alpinas cum contudit Hannibal arces,
fecit et aeternum Trebiam Cannasque sepulcris
obruit et Libyam Latias infudit in urbes.
huc varias pestes diversaque monstra ferarum
congessit bellis natura infesta futuris.
horrendos angues habitataque membra veneno
et mortis pastu viventia, crimina terrae,
et vastos elephantas habet, saevosque leones
in poenas fecunda suas parit horrida tellus
et portentosos cercopum ludit in ortus
ac sterili peior siccas infestat harenas,
donec ad Aegypti ponat sua iura colonos.
inde Asiae populi divesque per omnia tellus:
auratique fluunt amnes gemmisque relucet
pontus, odoratae spirant medicamina silvae:

652 tenent 661 italas 665 partu
660 aeterna 663 concessit 667 bella

---

[a] Manilius falls into a common error (*e.g.* Strabo 2. 5. 18) in making the Caspian or Hyrcanian Sea an arm of Ocean, though Herodotus (1. 202. 4) had recognized it as a lake.

[b] Persian Gulf. [c] Arabian Gulf (*i.e.* Red Sea).

[d] By heating the rock with bonfires of wood and then pouring on acid to crack it (Livy 21. 37).

Euxine, forms the Caspian sea.[a] In like manner, towards the midday sun, Ocean has launched two other attacks against the land; for its tide occupies Persian plains, having stolen from the very region it waters a name for its gulf, and floods in to form a wide circle[b]; and at no great distance, into the land of the unmanly Arabs, which produces delicacies and the exotic perfumes of many a plant, the sea gently pours shores rich with pearls, and the waters take their name from the country which lies between the two gulfs.[c]

Earth's mainland is divided into three continents: Libya, Asia, and Europe. Over the first [658] Carthage once obtained sovereignty by force of arms, what time Hannibal blasted with fire the Alpine peaks,[d] immortalized the Trebia, covered Cannae with graves, and poured Libya into the cities of Latium. Hither has nature assembled many a kind of plague and all manner of monstrous beasts to be the foe of future invaders. Libya shelters horrible serpents and, a reproach to the land, creatures which make a home for poisons and live by feeding on death, and gigantic elephants; and, so fruitful to its own hurt, this grim country breeds ferocious lions and jests in giving birth to hideous monkeys. Worse than were it sterile, it makes its desert sands a bane until it lays down its authority before the inhabitants of Egypt.[e] Thereafter come the peoples of Asia, and a land that is rich in everything: rivers flow with gold and the sea sparkles with gems; aromatic woods breathe forth

[e] This verse, like 4. 627, shows that the Nile forms the boundary between Libya and Asia, as the Tanais forms the boundary between Europe and Asia (*cf.* Pliny, *N.H.* 3. 3).

India notitia maior, Parthique vel orbis
alter, et in caelum surgentis moenia Tauri
totque illum circa diverso nomine gentes
ad Tanain Scythicis dirimentem fluctibus orbes
Maeotisque lacus Euxinique aspera ponti.
[aequora et extremum Propontidos Hellespontum]
hanc Asiae metam posuit natura potentis.
quod superest Europa tenet, quae prima natantem
fluctibus excepitque Iovem taurumque resolvit,
ponere passa suos ignes, onerique iugavit.
ille puellari donavit nomine litus
et monumenta sui titulo sacravit amoris.
maxima terra viris et fecundissima doctis
artibus : in regnum florentes oris Athenae ;
Sparta manu, Thebae divis, et rege vel uno
princeps Pella domus, Troiani gratia belli ;
Thessalia Epirosque potens vicinaque ripa
Illyris, et Thrace Martem sortita colonum,
et stupefacta suos inter Germania partus ;
Gallia per census, Hispania maxima bellis ;
Italia in summa, quam rerum maxima Roma
imposuit terris caeloque adiungitur ipsa.
Hos erit in fines orbis pontusque vocandus,

676 diviso
677 tantam scythicas
683 pondere suo signi . . . iuravit
687 urbibus
689 illa
690 ripis

[a] The River Don (*cf.* previous note).

[b] [679], interpolated to procure a noun for *aspera* : ". . . sea and the Hellespont which closes the Propontis."

[c] Alluding to the story of Jupiter and Europa.

[d] The girl Europa.

[e] Bacchus, son of Jupiter and Semele, daughter of Cadmus ; and Hercules, son of Jupiter and Alcmena, wife of Amphitryon.

balm. Thence on to India, too vast to be known, and Parthia, verily another world, and the ramparts of the Taurus that rises to the sky, and all the many races of different name that dwell around it, reaching as far as the Scythian Tanais,[a] which separates two continents with its waters, the meres of Maeotis, and the inclement Euxine.[b] This is the limit nature has set to the might of Asia. The rest of the world belongs to Europe: it first received Jupiter as he swam the waves [c] and gave the bull release, suffering it to set down its desire [d] and uniting it with its burden. The god bestowed the girl's name on the shore, which by that title he consecrated as a memorial of his love. It is the continent most renowned for heroes and most productive of learned arts. There is Athens, distinguished by its sovereignty over eloquence; Sparta, pre-eminent for its feats of arms; Thebes, for the gods [e] it bore; and Pella, for but a single king [f] of its royal house, return for help it gave in the Trojan war [g]; Thessaly, Epirus, and the contiguous Illyrian littoral, powerful lands all three; Thrace, who counts Mars a citizen,[h] and Germany, who stands struck with wonder at the stature of her sons around her; Gaul unrivalled for her wealth, Spain for her bellicosity; and finally Italy, which Rome, capital of the world, has made mistress of the earth, herself made one with heaven.[i]

[696]These then are the boundaries which land and

[f] Alexander.

[g] A contingent of the Paeones (*cf.* 1. 770).

[h] Mars (Ares) was specially associated with Thrace (*cf.* Homer, *Il.* 13. 301).

[i] The Romans deified their city (DEA ROMA in inscriptions), and Augustus permitted the dedication of a temple to himself and Urbs Roma (Tacitus, *Ann.* 4. 37).

quem deus in partes et singula dividit astra
ac sua cuique dedit tutelae regna per orbem
et proprias gentes atque urbes addidit altas,
in quibus assererent praestantis sidera vires.
ac, velut humana est signis discripta figura,
et, quamquam communis eat tutela per omne
corpus, et in proprium divisis artubus exit
(namque Aries capiti, Taurus cervicibus haeret,
bracchia sub Geminis censentur, pectora Cancro,
te scapulae, Nemeaee, vocant teque ilia, Virgo,
Libra colit clunes et Scorpios inguine regnat,
et femina Arcitenens, genua et Capricornus amavit,
cruraque defendit Iuvenis, vestigia Pisces),
sic alias aliud terras sibi vindicat astrum.

Idcirco in varias leges variasque figuras
dispositum genus est hominum, proprioque colore
formantur gentes, sociataque iura per artus
materiamque parem privato foedere signant.
flava per ingentis surgit Germania partus,
Gallia vicino minus est infecta rubore,
asperior solidos Hispania contrahit artus.
Martia Romanis urbis pater induit ora
Gradivumque Venus miscens bene temperat artus,
perque coloratas subtilis Graecia gentes
gymnasium praefert vultu fortisque palaestras,

719 genus

---

[a] *Cf.* 2. 456-465.
[b] Romulus, son of Mars and Ilia.

sea are to be summoned to observe, for the creator has divided the world into portions, distributing it among the individual signs. To each guardian power he has given a special region of the world to rule, bestowing also the peoples and mighty cities proper to them, wherein the signs should claim their predominant influences. And just as the human frame is apportioned among the signs,[a] and the protection they afford, though collectively extending over the whole body, is in addition exercised separately over the limbs allocated among them (the Ram is attached to the head, the Bull to the neck; the arms are reckoned as under the Twins' domain, the breast under the Crab's; the shoulders appeal to you, Nemean, and to you, Maiden, the belly; the Balance attends the loins, and the Scorpion is lord of the groin; the Archer has bestowed his love upon the thighs, Capricorn upon the knees, whilst the Youth is protector of the shanks and the Fishes of the feet), so in like manner do different signs lay claim to different lands.

711For this reason the human race is so arranged that its practices and features vary: nations are fashioned with their own particular complexions; and each stamps with a character of its own the like nature and anatomy of the human body which all share. Germany, towering high with tall offspring, is blond; Gaul is tinged to a less degree with a near-related redness; hardier Spain breeds close-knit, sturdy limbs. The Father [b] of the City endows the Romans with the features of Mars, and Venus joining the War-god fashions them with well-proportioned limbs. Quick-witted Greece proclaims in the tanned faces of its peoples the gymnasium and

et Syriam produnt torti per tempora crines.
Aethiopes maculant orbem tenebrisque figurant
perfusas hominum gentes ; minus India tostos
progenerat ; | tellusque natans Aegyptia Nilo
lenius irriguis infuscat corpora campis
iam propior | mediumque facit moderata tenorem.
Phoebus harenosis Afrorum pulvere terris
exsiccat populos, et Mauretania nomen
oris habet titulumque suo fert ipsa colore.
adde sonos totidem vocum, totidem insere linguas
et mores pro sorte paris ritusque locorum ;
adde genus proprium simili sub semine frugum
et Cererem varia redeuntem messe per urbes
nec paribus siliquas referentem viribus omnis,
nec te, Bacche, pari donantem munere terras
atque alias aliis fundentem collibus uvas,
cinnama nec totis passim nascentia campis ;
diversas pecudum facies propriasque ferarum
et duplici clausos elephantas carcere terrae.
quot partes orbis, totidem sub partibus orbes,
ut certis discripta nitent regionibus astra
perfunduntque suo subiectas aere gentes.

Laniger in medio sortitus sidera mundo,
lance ubi sol aequa pensat noctemque diemque
Cancrum inter gelidumque Caprum per tempora veris,
asserit in vires pontum quem vicerat ipse,

[724] tostis
[728] poenus
[742] et
[743] aequore
[745] [caprum]

---

[a] Mauretania is derived (*cf.* Isidore, *Orig.* 14. 5. 10) from μαῦρος "dark."

[b] Africa and India.

[c] Hellespont : the girl is Helle, her brother Phrixus.

the manly wrestling-schools. Curly hair about the temples betrays the Syrian. The Ethiopians stain the world and depict a race of men steeped in darkness; less sun-burnt are the natives of India; the land of Egypt, flooded by the Nile, darkens bodies more mildly owing to the inundation of its fields: it is a country nearer to us and its moderate climate imparts a medium tone. The Sun-god dries up with dust the tribes of Africans amid their desert lands; the Moors derive their name from their faces,[a] and their identity is proclaimed by the colour of their skins. Add the sounds of as many voices as there are races; include in your reckoning as many languages and customs and practices appropriate to the regions allotted to those races. Add the several characteristic kinds of fruits that grow from a like seed, and also the goddess of food, who comes round with harvests which vary from state to state and produces an unequal yield of every kind of vegetable. Nor do you, Bacchus, endow the lands with a similar bounty, but bring forth different grapes on different hills. Cinnamon does not grow far and wide in every field. Consider the different breeds of cattle, the peculiar species of wild beasts, and the elephant, restricted to two confines on earth.[b] For all that they are reckoned parts, there are as many worlds as there are parts of the world, even as the signs shine upon the special regions to which they have been allocated and imbue with their climate the peoples that lie beneath.

[744]The Ram, whose stars are allotted place in the middle of the firmament, *where with even balance the Sun levels night and day* in springtime halfway between the Crab and chilly Capricorn, claims for his influence the sea[c] which he overcame himself, when after the

virgine delapsa cum fratrem ad litora vexit
et minui deflevit onus dorsumque levari.
illum etiam venerata colit vicina Propontis
et Syriae gentes et laxo Persis amictu
vestibus ipsa suis haerens Nilusque tumescens
in Cancrum et tellus Aegypti iussa natare.
Taurus habet Scythiae montes Asiamque potentem
et mollis Arabas, silvarum ditia regna.
Euxinus Scythicos pontus sinuatus in arcus
sub Geminis te, Phoebe, colit ; vos Thracia, fratres,
ultimus et sola vos tranans colit Indica Ganges.
ardent Aethiopes Cancro, cui plurimus ignis :
hoc color ipse docet. Phrygia, Nemeaee, potiris
Idaeae matris famulus regnoque feroci
Cappadocum Armeniaeque iugis ; Bithynia dives
te colit et Macetum tellus, quae vicerat orbem.
Virgine sub casta felix terraque marique
est Rhodos, hospitium recturi principis orbem,
tumque domus vere Solis, cui tota sacrata est,
cum caperet lumen magni sub Caesare mundi ;
Ioniae quoque sunt urbes et Dorica rura,
Arcades antiqui celebrataque Caria fama.
quod potius colat Italiam, si seligat, astrum
quam quod cuncta regit, quod rerum pondera novit,

756 bost hrachia fratris
757 solidos ganges et
transcolit india cancer
760 regnique ferocis
765 tuque
769 seligis

[a] The summer solstice, when the Sun reaches Cancer.

[b] *Cf.* Strabo 2. 5. 22.

[c] The Sun being the dominant deity of Gemini (*cf.* 2. 439 f.).

[d] Cybele.

[e] Tiberius, who withdrew from public life to the island of Rhodes in 6 B.C., returning in A.D. 2.

[f] Ionia doubtless includes Attica in addition to the Ionian islands and the coast of Asia Minor.

girl had slipped off he bore her brother to the shore and wept over the reduction of his burden and the relief to his back. He is also reverently worshipped by neighbouring Propontis, by the Syrian people, by loose-robed Persia, a nation hampered by its raiment, by the Nile, whose waters swell to meet the Crab,[a] and by Egypt's land, then bidden to be flooded. The Bull holds the mountains of Scythia, powerful Asia, and the effeminate Arabs, whose realm is rich in woods. The Euxine sea, the shores of which curve in the shape of a Scythian bow,[b] worships you, Phoebus, in the person of the Twins [c]; you, brothers, does Thrace worship and farthest Ganges which waters the fields of India. The Ethiopians burn beneath the Crab, whose heat is fiercest, and this the colour of their skins makes plain. You, Lion of Nemea, attendant of the Idaean mother,[d] hold sway over Phrygia, the warlike realm of the Cappadocians, and the mountain-ridges of Armenia; wealthy Bithynia worships you, and so does the land of Macedon, which once conquered the world. Beneath the chaste Maid Rhodes prospers on land and sea, the erstwhile abode of him [e] who was to rule the world as emperor: the whole island is consecrated to the Sun, and Rhodes was in very truth its house at the time when it received into its care the light of the mighty universe in the person of Caesar. Beneath the Maid's dominion are likewise the cities of Ionia,[f] the Dorian fields, the ancient Arcadians,[g] and Caria, renowned in story.[h] What sign would better have the care of Italy, if Italy could choose, than that which controls all, knows the

[g] The Arcadians, who took a pride in their antiquity, were called προσέληνοι, "older than the Moon" (Aristotle, *Frag.* 591).

[h] *Cf.* Herodotus 1. 171.

designat summas et iniquum separat aequo,
tempora quo pendent, coeunt quo noxque diesque ?
Hesperiam sua Libra tenet, qua condita Roma
orbis et imperium retinet discrimina rerum,
lancibus et positas gentes tollitque premitque,
qua genitus Caesar melius nunc condidit urbem
et propriis frenat pendentem nutibus orbem.
inferius victae sidus Carthaginis arces
et Libyam Aegyptique latus donataque rura
Cyrenes lacrimis radicis Scorpios acris
eligit, Italiaeque tamen respectat ad undas
Sardiniamque tenet fusasque per aequora terras.
Cnosia Centauro tellus circumdata ponto
paret, et in geminum Minois filius astrum
ipse venit geminus. celeris hinc Creta sagittas
asserit intentosque imitatur sideris arcus.
insula Trinacriae fluitantem ad iura sororem
subsequitur Triviae sub eodem condita signo,
proximaque Italiae tenui divisa profundo
ora paris sequitur leges nec sidere rupta est.
tu, Capricorne, regis quidquid sub sole cadente
est positum gelidamque Helicen quod tangit ab illo,
Hispanas gentes et quot fert Gallia dives ;

[776] melius] -que meus
[780] tyrrhenos
radiat . . . arces
[781] eruit
[788] Triviae] crentens
[789] italia et
[792] expositum
[793] quod

[a] According to Tarutius Firmanus (Cicero, *De Div.* 2. 98) the Moon was in Libra at the foundation of Rome.

[b] Tiberius was born (Suetonius, *Tib.* 5) Nov. 15, 42 B.C., when the Moon was in Libra (at Augustus's birth the Moon was in Capricorn, *cf.* 2. 507 ff., Suetonius, *Aug.* 94).

[c] Hammonia.

[d] The juice of silphium (see, for example, the commentators

weights of things, marks totals, and separates the unequal from the equal, the sign in which the seasons are balanced and the hours of night and day match each other? Italy belongs to the Balance, her rightful sign: beneath it Rome and her sovereignty of the world were founded,[a] Rome, which controls the issue of events, exalting and depressing nations placed in the scales: beneath this sign was born the emperor,[b] who has now effected a better foundation of the city and governs a world which hangs on his command alone. Scorpio, the sign beneath, chooses the towers of vanquished Carthage, Libya, the flank of Egypt,[c] and the territory of Cyrene, gifted with the tears of a pungent root[d]; yet its backward glance reaches Italian waters, and takes in Sardinia and the islands scattered over the sea. The sea-girt land of Cnossus is subject to the Centaur, and so under a two-formed sign comes the son of Minos, two-formed himself.[e] Hence is it that Crete claims the swift arrow and imitates the drawn bow of the constellation. Situate under the same sign, Sicily follows her sister island[f] in being afloat to the authority of Diana[g]; and the adjacent shore of Italy,[h] separated by a deep but narrow channel, obeys the same laws and is not sundered from the influences of the sign. You, Capricorn, rule all that lies beneath the setting sun and all that stretches thence to touch the frozen north, together with the peoples of Spain and of

on Catullus 7. 4 and the edition of Apicius by Flower and Rosenbaum [*The Roman Cookery Book*] 29).

[e] The Minotaur, half-man, half-bull, supposititious son of Minos, the offspring of his wife Pasiphae's union with a bull.

[f] Crete.

[g] *Cf.* 2. 444.

[h] Housman refers this to Magna Graecia and not just Bruttium.

teque feris dignam tantum, Germania, matrem
asserit ambiguum sidus terraeque marisque
aestibus assiduis pontum terrasque sequentem.
sed Iuvenis nudos formatus mollior artus
Aegyptum ad tepidam Tyriasque recedit in arces
et Cilicum gentes vicinaque Caribus arva.
Piscibus Euphrates datus est, ubi ab his ope sumpta
cum fugeret Typhona Venus subsedit in undis,
et Tigris et rubri radiantia litora ponti.
magna iacet tellus magnis circumdata ripis
Parthis et a Parthis domitae per saecula gentes,
Bactraque et Aethiopes, Babylon et Susa Ninosque,
nominaque innumeris vix complectenda figuris.

Sic divisa manet tellus per sidera cuncta,
e quibus in proprias partes sunt iura trahenda ;
namque eadem, quae sunt signis, commercia servant,
utque illa inter se coeunt odioque repugnant,
nunc adversa polo, nunc et coniuncta trigono,
quaeque alia in varios affectus causa gubernat,
sic terrae terris respondent, urbibus urbes,
litora litoribus, regnis contraria regna ;
sic erit et sedes fugienda petendaque cuique,

796 tenentem
797 nudo . . . artu
798 ad t.] alepidam [in arces]
799 vicina et aquarius
800 [ab] pisces uruptor
803 [a]
804 [et] aetherios apimosque
813 orbibus orbes

---

[a] The defeat of Varus (*cf.* 1. 899 ff.) still rankled in Roman minds.

[b] On the uncertain coast of the North Sea *cf.* Pliny, *N.H.* 16. 2, Lucan 1. 409 ff.

[c] Than Capricorn's.

[d] To Phoenicia.

[e] Lycia and Pamphylia.

wealthy Gaul; and you, Germany, fit only to breed wild beasts,[a] are claimed by a sign uncertain whether it belongs to land or sea, since you follow now sea, now land, as the tides continuously ebb and flow.[b] But the Youth's unclad limbs are of a softer [c] mould, and moving nearer Egypt's warmth [d] he betakes himself to the walls of Tyre, the peoples of Cilicia, and the plains which border on Caria.[e] On the Fishes was bestowed the Euphrates, when in flight from Typhon Venus accepted their aid and hid beneath its waters,[f] and likewise the Tigris and the glistening shores of the Red Ocean.[g] Surrounded by these vast coasts is the vast land of Parthia, and the nations vanquished over the ages by the Parthians, the Bactrians and Ethiopians,[h] Babylon and Susa and Nineveh, and places where names could scarce adequately be conveyed by countless turns of speech.[i]

807Thus is the world for ever distributed among the twelve signs, and from the signs themselves must the laws prevailing among them be applied to the areas they govern; for these areas maintain between themselves the same relationships as exist between the signs; and just as the signs unite with each other or clash in enmity, now confronting one another across the sky and now linked in triangular federation, or as some other principle directs them to their various feelings, even so is land joined with land, city with city, and shores are at war with shores, realms with realms.[j] So must every man shun or seek a place to

[f] The story has been alluded to at 579 ff.: the aid given by the Fishes was their shape (*cf.* 2. 33). [g] Indian Ocean.
[h] The Asiatic Ethiopians (*cf.* Herodotus 3. 94. 1; 7. 70. 1).
[i] Because the native names are so exotic.
[j] Manifestly modelled on Virgil, *Aen.* 4. 628 f.

sic speranda fides, sic et metuenda pericla,
ut genus in terram caelo descendit ab alto.

Percipe nunc etiam quae sint ecliptica Graio
nomine, quod certos quasi delassata per annos
non numquam cessant sterili torpentia motu.
scilicet immenso nihil est aequale sub aevo
perpetuosque tenet flores unumque tenorem,
mutantur sed cuncta die variantque per annos ;
et fecunda suis absistunt frugibus arva
continuosque negant partus effeta creando,
rursus quae fuerant steriles ad semina terrae
post nova sufficiunt nullo mandante tributa.
concutitur tellus validis compagibus haerens
subducitque solum pedibus ; natat orbis in ipso
et vomit Oceanus pontum sitiensque resorbet
nec sese ipse capit. sic quondam merserat urbes,
humani generis cum solus constitit heres
Deucalion scopuloque orbem possedit in uno.
nec non, cum patrias Phaethon temptavit habenas,
arserunt gentes timuitque incendia caelum
fugeruntque novas ardentia sidera flammas
atque uno metuit condi natura sepulcro.
in tantum longo mutantur tempore cuncta
atque iterum in semet redeunt. sic tempore certo
signa quoque amittunt vires sumuntque receptas.

823 diu
824 subsistunt
830 venit . . . resolvit
837 timent *M*: timuit *GL*

[a] For example, says Housman, a man born in Italy (Libra) will do better to settle in Cilicia (Aquarius, linked in triangular federation with Libra) than in Syria (Aries, which confronts Libra across the sky) : in the former he may expect loyalty, in the latter, peril.

[b] From ἐκλείπειν, "to fail."

live in, so hope for loyalty or be forewarned of peril, according to the character which has come down to earth from high heaven.[a]

818Mark now, too, the signs which are, to use a Greek term, ecliptic,[b] because over certain periods of years, as though worn out with toil, they sometimes lag with unavailing movement and fail. Assuredly, nothing remains constant through the vastness of eternity, keeping its prime for ever and holding a single course, but all things change with the passage of time and vary over the years. Fertile fields withhold their crops and refuse to supply a continuous yield, exhausted by production ; again, lands which had given no return for seed sown furnish unexpected tribute afterwards, and that unbidden. The earth is shaken to its foundations for all the strength of the framework which holds it together, and the ground gives way beneath our feet [c] ; the land is afloat upon its very self,[d] as Ocean vomits forth its seas and gulps them back in thirst, and cannot contain the whole of itself. So on a time it flooded every city, when Deucalion was left sole heir of the human race, and his occupation of a single rock made him master of the world. Moreover, when Phaethon [e] made trial of his father's reins, nations were burnt up and heaven feared to be set on fire ; the blazing stars fled before fresh flames,[f] and nature was afraid of interment in a single grave. All things undergo such changes over long periods and then return to their normal states again. Even thus at a certain time the signs too lose their powers and on recovering them exert them

[c] A reference to earthquakes.
[d] *I.e.* has its coastal areas inundated.
[e] *Cf.* 1. 736 ff. [f] The flames of the Sun.

causa patet, quod, Luna quibus defecit in astris
orba sui fratris noctisque immersa tenebris,
cum medius Phoebi radios intercipit orbis
nec trahit assuetum quo fulget Delia lumen,
haec quoque signa suo pariter cum sidere languent
incurvata simul solitoque exempta vigori
et velut elatam Phoeben in funere lugent.
ipsa docet titulo se causa: ecliptica signa
dixere antiqui. pariter sed bina laborant,
nec vicina loco sed quae contraria fulgent,
sicut Luna suo tum tantum deficit orbe
cum Phoebum adversis currentem non videt astris.
nec tamen aequali languescunt tempore cuncta,
sed modo in affectus totus producitur annus,
nunc brevius lassata manent, nunc longius astra
exceduntque suo Phoebeia tempora casu.
atque, ubi perfectum est spatium quod cuique dicatur
impleruntque suos certa statione labores
bina per adversum caelum fulgentia signa,
tum vice bina labant ipsis haerentia casus,
quae prius in terras veniunt terrasque relinquunt,
sidereo non ut pugnet contrarius orbi
sed, qua mundus agit cursus, inclinet et ipse,
amissasque negant vires, nec munera tanta
nec similis reddunt noxas. locus omnia vertit.

Sed quid tam tenui prodest ratione nitentem
scrutari mundum, si mens sua cuique repugnat

843 medios
844 adcetum
848 titulos causae
860 vicina casus] signis
864 negat
865 reddit
866 iam

---

[a] The Sun's. [b] The Moon.

[c] For example, when Aries and Libra have finished suffering from an eclipse, its ill-effects pass to Pisces and Virgo.

anew. The cause is plain : whatever signs have seen the Moon eclipsed, reft of her brother and plunged into the darkness of night, when the Earth comes between to cut off Phoebus' [a] rays and Delia [b] draws not the light wherewith she is wont to shine, those signs too become as stricken as the planet occupying them, bowed the while in grief and deprived of their customary powers, and they mourn Phoebe [b] as though she had been carried out for burial. The cause is revealed by its name, for the ancients called these signs ecliptic. However, signs suffer together in pairs, not of those situated side by side but of those which shine at each other from opposing quarters, since the Moon's orb only fails when she does not see Phoebus as he courses in the opposite sign. But the signs are not all enfeebled for an equal duration, but sometimes a whole year is spent in this condition, sometimes they remain exhausted for a shorter period, and sometimes for a longer one, when their affliction outlives the Sun's yearly cycle. Now when the period allotted to each is accomplished, and the pair of signs which shine at each other across heaven have at the appointed limit discharged their tour of toil, then the two adjacent signs which precede them in rising above and setting beneath the Earth's horizon succumb by a shift of affliction, in such a way that this does not battle against the rotation of the starry sphere but advances in the direction of heaven' motion [c] ; and the signs lose their influences and no longer exert them, neither imparting their usual blessings nor inflicting their wonted hurt : all these changes are caused by the proximity of the eclipse.

[866]But what avail is it to search out the secrets of the shining firmament with such subtle reasoning, if a

spemque timor tollit prohibetque a limine caeli ?
"conditur en" inquit "vasto natura recessu
mortalisque fugit visus et pectora nostra,
nec prodesse potest quod fatis cuncta reguntur,
cum fatum nulla possit ratione videri."
quid iuvat in semet sua per convicia ferri
et fraudare bonis, quae nec deus invidet ipse,
quosque dedit natura oculos deponere mentis ?
perspicimus caelum, cur non et munera caeli ?
mens humana potest propria discedere sede
inque ipsos penitus mundi descendere census
seminibusque suis tantam componere molem
et partum caeli sua per nutricia ferre
extremumque sequi pontum terraeque subire
pendentis tractus et toto vivere in orbe.
[quanta et pars superet rationem discere noctis]
iam nusquam natura latet ; pervidimus omnem
et capto potimur mundo nostrumque parentem
pars sua perspicimus genitique accedimus astris.
an dubium est habitare deum sub pectore nostro
in caelumque redire animas caeloque venire,
utque sit ex omni constructus corpore mundus
aeris atque ignis summi terraeque marisque
hospitium menti totum quae infusa gubernet,
sic esse in nobis terrenae corpora sortis
sanguineasque animas animo, qui cuncta gubernat
dispensatque hominem? quid mirum, noscere mundum
si possunt homines, quibus est et mundus in ipsis

869 enim quid
880 portum
882 superest
889 aetheris
890 spiritum et totum rapido quae iussa gubernent

---

[a] *I.e.* man.
[b] About the sky.
[c] [882] (modelled on Germanicus 573) : ". . . and to learn how to calculate the remaining portion of the night." But

man's spirit resists and fear banishes confidence and bars access to the gate of heaven? "See," he objects, "nature is buried in deep concealment and lies beyond our mortal gaze and ken; it cannot profit us that all is governed by fate, since the rule of fate cannot by any means be seen." What boots it to assail oneself with self-reproach, to deprive oneself of benefits ungrudged by God himself, and to renounce that mental vision which nature has bestowed? We perceive the skies, then why not the skies' gifts too? *The mind of man has the power to leave its proper abode* and penetrate to the innermost treasures of the sky; to construct the mighty universe from its component seeds; to transport the offspring of heaven [a] about the places from which it came [b]; to make for Ocean's farthest horizon, descend to the inverted parts of the Earth, and inhabit the whole wide world.[c] Now nature holds no mysteries for us; we have surveyed it in its entirety and are masters of the conquered sky; we perceive our creator, of whom we are part, and rise to the stars, whose children we are. Can one doubt that a divinity dwells within our breasts and that our souls return to the heaven whence they came? [d] Can one doubt that, just as the world, composed of the elements of air and fire on high and earth and water, houses an intelligence which, spread throughout it, directs the whole,[e] so too with us the bodies of our earthly condition and our life-blood house a mind which directs every part and animates the man? Why wonder that men can comprehend heaven, when heaven exists in their very beings and

the poet is representing man as performing *physical* actions beyond the power of his body.

[d] *Cf.* 2. 105-125. [e] *Cf.* 1. 247-251, 2. 60-83.

exemplumque dei quisque est in imagine parva?
an cuiquam genitos, nisi caelo, credere fas est
esse homines? proiecta iacent animalia cuncta
in terra vel mersa vadis, vel in aere pendent,
omnibus una quies venter*que venusque voluptas,*
mole valens sola corpus censumque per artus,
et, quia consilium non est, et lingua remissa.
unus in inspectus rerum viresque loquendi
ingeniumque capax variasque educitur artes
hic partus, qui cuncta regit: secessit in urbes,
edomuit terram ad fruges, animalia cepit
imposuitque viam ponto, stetit unus in arcem
erectus capitis victorque ad sidera mittit
sidereos oculos propiusque aspectat Olympum
inquiritque Iovem; nec sola fronte deorum
contentus manet, et caelum scrutatur in alvo
cognatumque sequens corpus se quaerit in astris.
huic in tanta fidem petimus, quam saepe volucres
accipiunt trepidaeque suo sub pectore fibrae.
an minus est sacris rationem ducere signis
quam pecudum mortes aviumque attendere cantus?
atque ideo faciem caeli non invidet orbi
ipse deus vultusque suos corpusque recludit
volvendo semper seque ipsum inculcat et offert,
ut bene cognosci possit doceatque videntis,
qualis eat, cogatque suas attendere leges.
ipse vocat nostros animos ad sidera mundus
nec patitur, quia non condit, sua iura latere.
quis putet esse nefas nosci, quod cernere fas est?

901 [in]
903 orbes
905 arce
916 reducit
918 videndis

---

[a] Astronomy.
[b] Astrology.

each one is in a smaller likeness the image of God himself? Are we to believe that man is born of aught but heaven? All the other animals lie prostrate on the earth or submerged in water, or else hover in the air; all alike have only sleep and food *and sex for their delights; the strength of an animal is measured only by its size* and its value by its limbs, and since it has no intelligence it lacks speech, too. The breed of man, who rules all things, is alone reared equal to the inquiry into nature, the power of speech, breadth of understanding, the acquisition of various skills: he has left the open air for city-life, tamed the land to yield him its fruits, made the beasts his slaves, and laid a pathway on the sea; he alone stands with the citadel of his head raised high and, triumphantly directing to the stars his star-like eyes, looks ever more closely at Olympus and inquires into the nature of Jove himself; nor does he rest content with the outward appearance of the gods,[a] but probes into heaven's depths [b] and, in his quest of a being akin to his own, seeks himself among the stars. I ask for heaven a faith as great as that so oft accorded birds and entrails that quiver beneath their native breast. Is it then a meaner thing to derive reason from the sacred stars than to heed sacrifice of beast and cry of bird? God grudges not the earth the sight of heaven but reveals his face and form by ceaseless revolution, offering, nay impressing, himself upon us to the end that he can be truly known, can teach his nature to those who have eyes to see, and can compel them to mark his laws. Of itself the firmament summons our minds to the stars, and in not concealing its ordinances shows that it would have them known. Who then would deem it wrong to understand what it is right

nec contemne tuas quasi parvo in pectore vires :
quod valet, immensum est. sic auri pondera parvi
exsuperant pretio numerosos aeris acervos ;
sic adamas, punctum lapidis, pretiosior auro est ;
parvula sic totum pervisit pupula caelum,
quoque vident oculi minimum est, cum maxima cerna
sic animi sedes tenui sub corde locata
per totum angusto regnat de limite corpus.
materiae ne quaere modum, sed perspice vires,
quas ratio, non pondus, habet : ratio omnia vincit.
ne dubites homini divinos credere visus,
iam facit ipse deos mittitque ad sidera numen,
maius et Augusto crescet sub principe caelum.

[923] corpore

---

[a] Julius Caesar was deified in 42 B.C., Augustus in A.D. 14.

for us to see? Scorn not your powers as if proportionate to the smallness of the mind: its power has no bounds. Thus a small amount of gold exceeds in value countless heaps of brass; thus the diamond, a stone no bigger than a dot, is more precious than gold; thus the tiny pupil of the eye takes in the whole of heaven, and eyes owe their vision to that which is so very small, whilst what they behold is so very large; thus the seat of the mind, though set within the puny heart, exercises from its constricted abode dominion over the whole body. Seek not to measure the material, but consider rather the power which reason has and mere substance not: reason is what triumphs over all. Be not slow to credit man with vision of the divine, for man himself is now creating gods and raising godhead to the stars,[a] and beneath the dominion of Augustus will heaven grow mightier yet.[b]

[b] The last verse does not by itself reveal whether Augustus is alive or dead, but that the latter is the case, with Tiberius on the throne, clearly emerges from 764 ff.

BOOK FIVE

THE *last book of the Astronomica is the most artistic, and best illustrates the poet's ingenious rhetoric. The versification of arithmetic or geometry or bare astrological doctrine is here avoided; and for the most part the book consists of witty and colourful portraits of the different types of person produced by the influences of the extra-zodiacal constellations. After several hundred lines Manilius, perhaps taking a lead from Virgil in the last book of the Georgics, abandons the proportions of his scheme to narrate the story of Perseus and Andromeda; comparison with Ovid's treatment of the myth, however, redounds to our poet's disadvantage and reveals how much the insistence on verbal point has deprived his composition of the power to engage our feelings. Towards the end of the work some common ancestor of our manuscripts has evidently suffered the loss of several pages, a loss which leaves in some doubt whether or no Manilius dealt, as he had promised, with planetary influences. The last topic to be handled is the magnitudes of the stars, not a matter of astrological significance, but one which enables the poet to conclude with an imaginative description of the republic of the sky.*

LIBER QUINTUS

Hic alius finisset iter signisque relatis
quis adversa meant stellarum numina quinque
quadriiugis et Phoebus equis et Delia bigis
non ultra struxisset opus, caeloque rediret
ac per descensum medios percurreret ignes
Saturni, Iovis et Martis Solisque, sub illis
post Venerem et Maia natum te, Luna, vagantem.
me properare etiam mundus iubet omnia circum
sidera vectatum toto decurrere caelo,
cum semel aetherios ausus conscendere currus
summum contigerim sua per fastigia culmen.
hinc vocat Orion, magni pars maxima caeli,
et ratis heroum, quae nunc quoque navigat astris,
Fluminaque errantis late sinuantia flexus
et biferum Cetos squamis atque ore tremendo
Hesperidumque vigil custos et divitis auri
et Canis in totum portans incendia mundum
araque divorum, cui votum solvit Olympus ;
illinc per geminas Anguis qui labitur Arctos
Heniochusque memor currus plaustrique Bootes

[8] viam [10] iussus

---

[a] *I.e.* the Sun and the Moon. [b] Mercury.
[c] Among the southern constellations. [d] Argo.
[e] Not subsequently mentioned. See 1. 439, note.
[f] Cetus has the scaly tail (and rear) of a fish, the jaws (and front) of a land animal.

BOOK 5

Here would another have ended his journey in the skies: having dealt with the signs against which move the deities of five planets together with Phoebus' team of four and Diana's pair,[a] he would have carried his work no further. Here would another leave the celestial sphere and make his descent through the intervening fires of Saturn, Jupiter, Mars, and the Sun, and you, O Moon, beneath them, whose wandering path follows Venus and Maia's son.[b] But me the heavens entice to hasten on a tour of all the stars in the sky as well, since but once have I ventured to mount an aerial car and reached the summit of the dome along its upward slopes. On this side [c] Orion beckons to me, in the mighty sky the mightiest constellation; the heroes' ship [d] that amid the stars sails even now; the Rivers [e] whose winding coils meander far and wide; the Whale which with its scales and dread jaws has part of two creatures [f]; the wakeful warder [g] of the Hesperides and the golden treasure; the Dog who brings fire upon the entire universe; and the altar of the gods, at which Olympus pays its vows. On yonder side [h] the Dragon beckons which glides between the two Bears; the Charioteer still minding

[g] Hydra, not subsequently mentioned.
[h] Among the northern constellations.

atque Ariadnaeae caelestia dona coronae,
victor et invisae Perseus cum falce Medusae
Andromedanque necans genitor cum coniuge Cepheus,
quaque volat stellatus Equus celerique Sagittae
Delphinus certans et Iuppiter alite tectus,
ceteraque in toto passim labentia caelo.
quae mihi per proprias vires sunt cuncta canenda,
quid valeant ortu, quid cum merguntur in undas,
et quota de bis sex astris pars quaeque reducat.

Vir gregis et ponti victor, cui parte relicta
nomen onusque dedit nec pelle immunis ab ipsa,
Colchidos et magicas artes qui visere Iolcon
Medeae iussit movitque venena per orbem,
nunc quoque vicinam puppi, ceu naviget, Argo
a dextri lateris ducit regione per astra.
sed tum prima suos puppis consurgit in ignes,
quattuor in partes cum Corniger extulit ora.
illa quisquis erit terris oriente creatus,
rector erit puppis clavoque immobilis haerens
mutabit pelago terras ventisque sequetur
fortunam totumque volet tranare profundum
classibus atque alios menses altumque videre
Phasin et in cautes Tiphyn superare ruentem.
tolle sitos ortus hominum sub sidere tali,

30 f.: *see after* 709
[34] colchon
[35] medeam
[36] puppim celi
[38] cum
[40] illis
[44] aliumque
[45] trementem
[46] istos

[a] Subsequently mentioned only in connection with Andromeda, not on his own account.

[b] Manilius does not in fact deal with the influences of the extra-zodiacal constellations at their setting and in several instances fails to specify the zodiacal degree of their rising.

his car and Bootes his wain ; the heavenly gift of Ariadne's crown ; Perseus,[a] slayer of the abominable Medusa, blade yet in hand ; with his wife, Cepheus sacrificing his daughter Andromeda ; the region of the sky where fly the Horse of stars, the Dolphin seeking to outstrip the swift Arrow, and Jupiter in swan's disguise ; together with the other stars that glide at large throughout the heavens. I must tell of the powers peculiar to all these constellations, their influences both when rising and when they sink into the waves, and which degree of the zodiac brings each of them back above the horizon.[b]

[32]Lord of the flock and conqueror of the sea, to which, a horn lost and robbed even of its fleece, it gave its burden [c] and a name, the Ram, which bade the magic arts of Colchian Medea journey to Iolcos and spread her poisons throughout the world, even now draws Argo by the poop to its side, as though still on the seas, through the stars on its right.[d] But the foremost part of the poop emerges to show its fires only when the Ram has brought four degrees of his countenance above the horizon. Whoever is born on earth at the rising of Argo will be the captain of a ship ; holding fast to his helm, he will forsake dry land for the sea and pursue his fortune with the winds ; he will crave to traverse the entire ocean with his fleet, visit foreign climes and the deep Phasis, and better the speed of Tiphys towards the rocks.[e] Take away the births of men situate beneath

For a tabulation of the poet's doctrine of the *paranatellonta*, see tables 5 and 6, pp. xciv f.

[c] Helle.

[d] An error : Argo is a southern constellation and rises on the left of the zodiac, but never contemporaneously with Aries.

[e] The Symplegades.

sustuleris bellum Troiae classemque solutam
sanguine et appulsam terris ; non invehet undis
Persida nec pelagus Xerxes facietque tegetque ;
versa Syracusis Salamis non merget Athenas,
Punica nec toto fluitabunt aequore rostra
Actiacosque sinus inter suspensus utrimque
orbis et in ponto caeli fortuna natabit.
his ducibus caeco ducuntur in aequore classes
et coit ipsa sibi tellus totusque per usus
diversos rerum ventis arcessitur orbis.

Sed decima lateris surgens de parte sinistri
maximus Orion magnumque amplexus Olympum,
quo fulgente super terras caelumque trahente
ementita diem nigras nox contrahit alas,
sollertis animos, velocia corpora finget
atque agilem officio mentem curasque per omnis
indelassato properantia corda vigore.
instar erit populi totaque habitabit in urbe
limina pervolitans unumque per omnia verbum
mane salutandi portans communis amicus.

Sed, cum se terris Aries ter quinque peractis
partibus extollit, primum iuga tollit ab undis
Heniochus clivoque rotas convellit ab imo,
qua gelidus Boreas aquilonibus instat acutis.

[50] vera *GL* : ut ra *M*

---

[a] Of Iphigenia and Protesilaus.

[b] By cutting a canal through Athos.

[c] By the vastness of his fleet (*cf.* 3. 20) ; some interpreters see an allusion to the bridging of the Hellespont, but Housman considers this unlikely.

[d] Another mistake (repeated by Firmicus, *Math.* 8. 6. 2) : Orion cannot rise with Aries. Moreover enthusiasm for hunting, chiefest of the influences of Orion, is conspicuously absent from this paragraph : the mobility here predicted

this constellation, and you will take away the Trojan war and the fleet which both set sail and made landfall with bloodshed [a]; then will no Xerxes launch Persia on the main or open up a new sea [b] and cover over the old [c]; no reversal of Salamis at Syracuse will overwhelm Athens, nor will Punic prows ride every sea, the world within Actium's bays hang in the balance between opposing forces, and heaven's destiny float at the mercy of the waves. Men born under Argo are the guides who guide our ships over the trackless deep; it is through them that land meets land and the whole world's wares are summoned with the winds to supply men's divers needs.

[57]Now on the Ram's left flank and together with its tenth degree rises Orion [d]; mightiest of constellations he girdles with his course the mighty skies: when Orion shines over the horizon drawing heaven in his train, night feigns the brightness of day and folds its dusky wings. Orion will fashion alert minds and agile bodies, souls prompt to respond to duty's call, and hearts which press on with unflagging energy in spite of every trial. A son of Orion's will be worth a multitude and will seem to dwell in every quarter of the city; flying from door to door with the one word [e] of morning greeting, he will enjoy the friendship of all.

[67]But when the Ram raises himself above the earth with thrice five degrees completed, then straightway the Charioteer lifts his team from ocean and wrests his wheels up from the downward slope of the horizon where icy Boreas lashes us with his bitter blasts. He

rather reflects the Ram's swift courses (*cf.* Aratus, *Phaen.* 225 ff.). Orion's endowment is to be sought in that of the Belt (175 ff.). [e] *Ave! Cf.* Martial 1. 55. 6; 4. 78. 4.

ille dabit proprium studium caeloque retentas
quas prius in terris agitator amaverat artes :
stare levi curru moderantem quattuor ora
spumigeris frenata lupis et flectere equorum
praevalidas vires ac torto stringere gyro ; 7
at, cum laxato fugerunt cardine claustra,
exagitare feros pronumque antire volantis
vixque rotis levibus summum contingere campum
vincentem pedibus ventos, vel prima tenentem
agmina in obliquum cursus agitare malignos 8
obstantemque mora totum praecludere circum,
vel medium turbae nunc dextros ire per orbes
fidentem campo, nunc meta currere acuta
spemque sub extremo dubiam suspendere casu.
nec non alterno desultor sidere dorso 8
quadrupedum et stabilis poterit defigere plantas,
pervolitans et equos ludet per terga volantum ;
aut solo vectatus equo nunc arma movebit,
nunc leget in longo per cursum praemia circo.
quidquid de tali studio formatur habebit. 9
hinc mihi Salmoneus (qui caelum imitatus in orbe,
pontibus impositis missisque per aera quadrigis
expressisse sonum mundi sibi visus et ipsum
admovisse Iovem terris, dum fulmina fingit

[87] perquolabit [et] [94] de fulmine
[89] licet

[a] That is, the constellation is identified with Erichthonius.

will impart his own enthusiasms and the skills, still retained in heaven, which as driver of a chariot he once took pleasure in on earth.[a] The Charioteer will enable his son to stand in a light chariot and hold in check the four mouths curbed with foam-flecked bits, guide their powerful strength, and keep close to the curve round which they wheel. Again, when the bolts have been drawn and the horses have escaped from the starting-pens, he will urge on the spirited steeds and, leaning forward, he will seem to precede them in their swift career; hardly touching the surface of the track with his light wheels, he will outstrip the winds with his coursers' feet. Holding first place in the contest he will drive to the side in a baulking course and, his obstruction delaying his rivals, deny them the whole breadth of the circus-track; or if he is placed mid-way in the press, he will now swing to a course on the outside, trusting in the open, now sharply on the inside graze the turning-post, and will leave the result in doubt to the very last moment. As a trick-rider too he will be able to settle himself now on one, now on a second horse, and plant his feet firmly upon them: flying from horse to horse he will perform tricks on the backs of animals in flight themselves; or mounted on a single horse he will now engage in exercise of arms, now whilst still riding pick up gifts scattered along the length of the circus. He will possess virtuosity in all that is connected with such pursuits. Of this constellation, I think, Salmoneus may be held to have been born: imitating heaven on earth, he imagined that by setting his team of four on a bridge of bronze and driving it across he had expressed the crash of the heavens and had brought to earth Jove's very self;

sensit, et immissos ignes super ipse secutus 9
morte Iovem didicit) generatus possit haberi.
hoc genitum credas de sidere Bellerophonten
imposuisse viam mundo per signa volantem,
cui caelum campus fuerat, terraeque fretumque
sub pedibus, non ulla tulit vestigia cursus. 10
his erit Heniochi surgens tibi forma notanda.

Cumque decem partes Aries duplicaverit ortus,
incipient Haedi tremulum producere mentum
hirtaque tum demum terris promittere terga
qua dexter Boreas spirat. ne crede severae 10
frontis opus fingi, strictosque hinc ora Catones
abruptumque pari Torquatum et Horatia facta.
maius onus signo est, Haedis nec tanta petulcis
conveniunt : levibus gaudent lascivaque signant
pectora ; et in lusus facilis agilemque vigorem 11
desudant ; vario ducunt in amore iuventam ;
in vulnus numquam virtus sed saepe libido
impellit, turpisque emitur vel morte voluptas ;
et minimum cecidisse malum est, quia crimine vincunt.
nec non et cultus pecorum nascentibus addunt 11
pastoremque suum generant, cui fistula collo
haereat et voces alterna per oscula ducat.

Sed, cum bis denas augebit septima partes
Lanigeri, surgent Hyades. quo tempore natis

95 immensos
101 conanda
106 signi
incoda catonis
107 patri
110 facilis] agiles
114 victum

---

[a] Some sources identify the Charioteer with Bellerophon (Schol. Aratus, *Phaen.* 161).
[b] *Cf.* 1. 365.
[c] Men like Cato the Censor.
[d] Who killed his son (Livy 8. 7. 14 ff.).
[e] Who killed his sister (Livy 1. 26. 3).

however, while counterfeiting thunderbolts he was struck by real ones and, falling after the fires he had flung himself, discovered in death that Jove existed. You may well believe that under this constellation was born Bellerophon,[a] who flew amid the stars and laid a road on heaven : the sky was the field over which he sped, whilst land and sea lay far beneath his feet, and his path was unmarked by footprints. By examples such as these are you to mark the rising figure of the Charioteer.

[102]When at his rising the Ram has completed twice ten degrees, the Kids [b] will begin to display their quivering chins, promise to mankind of their shaggy backs at a later hour, on the right horizon where the north wind blows. Think not that hence is fashioned a product of severe mien, that hence are born stern-faced Catos,[c] an inflexible Torquatus,[d] and men to repeat the deed of Horatius.[e] Such a charge would be too much for the sign, nor does such greatness befit the frisking Kids, who rejoice in frivolity and stamp their young with a wanton breast. These abandon themselves to playful sport and nimble activity, and spend their youth in fickle loves. Though honour never inspires them to shed their blood, lust often drives them to do so, and their base desires cost even their lives : to perish thus is their least disgrace, since their triumph is a triumph of vice. The Kids also give to those born under them the custody of flocks, and beget a shepherd of their own kind, one to hang a pipe round his neck and draw from its different stops melodious strains.

[118]The Hyades will rise when twenty degrees of the Ram are augmented by a seventh.[f] Those born

[f] A gross error : the group is situated in the sign of the Bull.

nulla quies placet, in nullo sunt otia fructu, 12
sed populum turbamque petunt rerumque tumultus.
seditio clamorque iuvat, Gracchosque tenentis
rostra volunt Montemque Sacrum rarosque Quirites ;
pacis bella probant curaeque alimenta ministrant.
immundosque greges agitant per sordida rura ; 12
et fidum Laertiadae genuere syboten.
hos generant Hyades mores surgentibus astris.

Ultima Lanigeri cum pars excluditur orbi,
quae totum ostendit terris atque eruit undis,
Olenie servans praegressos tollitur Haedos 13
egelido stellata polo, qua dextera pars est,
officio magni mater Iovis. illa Tonanti
fida alimenta dedit pectusque implevit hiantis
lacte suo, dedit et dignas ad fulmina vires.
hinc trepidae mentes tremebundaque corda creantur 13
suspensa ad strepitus levibusque obnoxia causis.
his etiam ingenita est visendi ignota cupido,
ut nova per montes quaerunt arbusta capellae,
semper et ulterius pascentes tendere gaudent.

Taurus, in aversos praeceps cum tollitur ortus, 14
sexta parte sui certantis luce sorores
Pleiadas ducit. quibus aspirantibus almam
in lucem eduntur Bacchi Venerisque sequaces

128 orbis
131 quae
133 fundamenta
135 fidae
136 suspensas trepidus
137 ingeniest
143 educunt

[a] The Hyades are a stormy star-group (*cf.* Pliny, *N.H.* 18. 247).

[b] *Cf.* Livy 3. 52. 5-7.

[c] Eumaeus : see Introduction p. xxiv.

[d] Capella (1. 366) : Olenian either as being on the left arm (ὠλένη) of the Charioteer, or as the daughter of Olenus

at this time take no pleasure in tranquillity and set no store by a life of inaction; rather they yearn for crowds and mobs and civil disorders.[a] Sedition and uproar delight them; they long for the Gracchi to harangue from the platform, for a secession to the Sacred Mount, leaving but a handful of citizens at Rome [b]; they welcome fights which break the peace, and provide sustenance for fears. They herd their foul droves over the untilled countryside, for this constellation also begot Ulysses' trusty swineherd.[c] Such are the qualities engendered by the Hyades at the rising of their stars.

128When the zodiac releases the Ram's last degree, which thus displays the whole of him to the earth, plucking him from Ocean's waves, there rises the Olenian goat,[d] keeping watch over the Kids which stray ahead, enstarred on the right in the cold north sky for her services as foster-mother of mighty Jove. She gave the Thunderer sound nourishment, satisfying with her own milk the infant's hungry body, and giving him therewith sufficient strength to wield his bolts. Of the Goat are born anxious minds and trembling hearts which start at every noise and are apt to flutter at the slightest cause. Inborn in them, too, is a longing to explore the unknown, even as goats seek fresh shrubs on mountain slopes and rejoice, as they browse, to move ever further afield.

140As he emerges in his backwards rising with head hanging down, the Bull brings forth in his sixth degree the Pleiades, sisters who vie with each other's radiance. Beneath their influence devotees of Bacchus and Venus are born into the kindly light,

(Schol. Aratus, *Phaen.* 164), or from Olenus in Achaea, its residence (Strabo 8. 7. 5).

perque dapes mensasque super petulantia corda
et sale mordaci dulcis quaerentia risus.
illis cura sui cultus frontisque decorae
semper erit: tortos in fluctum ponere crines
aut vinclis revocare comas et vertice denso
fingere et appositis caput emutare capillis
pumicibusque cavis horrentia membra polire
atque odisse virum teretisque optare lacertos.
femineae vestes, nec in usum tegmina plantis
sed speciem, fictique placent ad mollia gressus.
naturae pudet, atque habitat sub pectore caeca
ambitio, et morbum virtutis nomine iactant.
semper amare parum est: cupient et amare videri.

Iam vero Geminis fraterna ferentibus astra
in caelum summoque natantibus aequore ponti
septima pars Leporem tollit. quo sidere natis
vix alas natura negat volucrisque meatus:
tantus erit per membra vigor referentia ventos.
ille prius victor stadio quam missus abibit;
ille cito motu rigidos eludere caestus,
nunc exire levis missas nunc mittere palmas,
ille pilam celeri fugientem reddere planta
et pedibus pensare manus et ludere fulcro
mobilibusque citos ictus glomerare lacertis,
ille potens turba perfundere membra pilarum
per totumque vagas corpus disponere palmas,
ut teneat tantos orbes sibique ipse reludat

[151] sterilisque
[152] usum] -sunt
[154] caeco

and people whose insouciance runs free at feasts and banquets and who strive to provoke sweet mirth with biting wit. They will always take pains over personal adornment and an elegant appearance: they will set their locks in waves of curls or confine their tresses with bands, building them into a thick topknot, and they will transform the appearance of the head by adding hair to it; they will smooth their hairy limbs with the porous pumice, loathing their manhood and craving for sleekness of arm. They adopt feminine dress, footwear donned not for wear but for show, and an affected effeminate gait. They are ashamed of their sex; in their hearts dwells a senseless passion for display, and they boast of their malady, which they call a virtue. To give their love is never enough: they will also want their love to be seen.

[157]Now when the Twins lift their fraternal stars into the sky and float on the surface of the sea, their seventh degree brings to view the Hare. To those born under this constellation nature all but gives wings and flight through the air—such will be the vigour of limbs which reflect the swiftness of the winds. One man will come off winner in the footrace before even receiving the signal to start; another by his quick movement can evade the hard boxing-glove and now lightly avoid, now land a blow; another can with a deft kick keep in the air a flying ball, exchanging hands for feet and employing in play the body's support, and execute with nimble arms a volley of rapid strokes; yet another can shower his limbs with a host of balls and create hands to spring up all over his body with the result that, without dropping any of the number, he plays against himself

et velut edoctos iubeat volitare per ipsum.
invigilat curis, somnos industria vincit,
otia per varios exercet dulcia lusus.

Nunc Cancro vicina canam, cui parte sinistra
consurgunt Iugulae. quibus aspirantibus orti
te, Meleagre, colunt flammis absentibus ustum
reddentemque tuae per mortem munera matri,
cuius et ante necem paulatim vita sepulta est,
atque Atalantaeos conatum ferre labores,
et Calydonea bellantem rupe puellam
vincentemque viros et quam potuisse videre
virgine maius erat sternentem vulnere primo.
quaque erat Actaeon silvis mirandus, et ante
quam canibus nova praeda fuit, ducuntur et ipsi,
retibus et claudunt campos, formidine montes.
mendacisque parant foveas laqueosque tenacis
currentisque feras pedicarum compede nectunt

172 somnis curas
176 habentibus
183 quamque . . . mutandus

[a] On a sarcophagus of the imperial period a juggler is represented (Daremberg-Saglio 4. 479, fig. 5668) as playing with seven balls, and Housman notes that Manilius may have watched performances of Publius Aelius Secundus, a *pilarius* who is described (*CIL* 6. 8997) as *omnium eminentissimus*.

[b] According to Aelian, *N.A.* 2. 12, the hare sleeps with its eyes open.

[c] The belt of Orion seems to be meant, the three conspicuous stars δεζ Ori, which Aratus, *Phaen.* 587 ff., makes rise with Cancer. The word ought to signify collar-bone, and so (but in the singular) Varro, *De Ling. Lat.* 7. 50, interprets it: there is, however, no star or group of stars between Orion's shoulders (α and γ) which can plausibly be so designated. From the hunting influences here bestowed it is certain that the poet has Orion in mind, and probable that he felt obliged to use another name for the constellation after his egregious error in 58.

and causes the balls to fly about his person as though in answer to his command.[a] Such a man devotes wakeful nights to his concerns, for his energy banishes sleepiness,[b] whilst he spends happy workfree hours in games of divers kinds.

[174]Next shall I sing of the neighbours of the Crab. On its left rise the stars of the Belt.[c] Those born beneath its influence are devoted to you, Meleager, who, consumed by absent flames, at death returned to your mother[d] the gift she gave, whilst your life, even before you died, experienced a lingering funeral-pyre; and to him[e] who undertook to endure the toils brought on him by his wooing of Atalanta; and to Atalanta herself, who joined in the hunt at rocky Calydon and surpassed the male warriors, for by inflicting the first wound she laid low a monster beyond a maiden's strength to look upon. The activity in which Actaeon excited the wonder of the woods, even before he became the novel quarry of his hounds, attracts them, too, and they enclose the plains with nets, the hills with scare-feathers.[f] They prepare treacherous pitfalls and tenacious traps, and capture beasts on the run with the shackles

[d] Althaea. At his birth the Fates ordained for Meleager the length of life of a log burning on the hearth; his mother promptly removed and extinguished it, thus presenting him with the gift of life. This gift he was to return when Althaea, angered at the news that he had killed her brothers, threw on the fire the log she had carefully preserved, causing the absent Meleager to burn slowly on a funeral-pyre, as it were, even before he died. *Cf.* Ovid, *Met.* 8. 451 ff.

[e] Milanion.

[f] Strings with coloured feathers attached, which were shaken and advanced to scare and corner the hunted animal (Seneca, *Dial.* 4. 11. 5).

aut canibus ferrove necant praedasque reportant.
sunt quibus in ponto studium est cepisse ferarum
diversas facies et caeco mersa profundo
sternere litoreis monstrorum corpora harenis
horrendumque fretis in bella lacessere pontum
et colare vagos inductis retibus amnes
ac per nulla sequi dubias vestigia praedas,
luxuriae quia terra parum, fastidit et orbem
venter, et ipse gulam Nereus ex aequore pascit.

At Procyon oriens, cum iam vicesima Cancro
septimaque ex undis pars sese emergit in astra,
venatus non ille quidem verum arma creatis
venandi tribuit. catulos nutrire sagacis
et genus a proavis, mores numerare per urbes,
retiaque et valida venabula cuspide fixa
lentaque correctis formare hastilia nodis,
et quaecumque solet venandi poscere cura
in proprios fabricare dabit venalia quaestus.

Cum vero in vastos surget Nemeaeus hiatus,
exoritur candens latratque Canicula flammas
et rabit igne suo geminatque incendia solis.
qua subdente facem terris radiosque vomente
divinat cineres orbis fatumque supremum
sortitur, languetque suis Neptunus in undis,
et viridis nemori sanguis decedit et herbis.
cuncta peregrinos orbes animalia quaerunt
atque eget alterius mundus ; natura suismet

[194] dubitat
[195] fastidiet
[203] contextis
[204] quicumque *M*: quod- *GL*
[207] candens] -que canis
[208] rapit
[209] movente *G*: -em *LM*
[210] dimicat in

---

[a] So Aratus, *Phaen.* 595, whereas Nigidius (Servius, *Georg.* 1. 218) made Canicula a *paranatellon* of Cancer.

[b] The *ecpyrosis* of the Stoics, who held that the universe

of gins or kill them using hound or lance and bring home their catch. Others delight to capture at sea wild creatures of various forms and lay out on the sandy shores the bodies of monsters which had been hidden in the dark depths ; they delight to provoke to war the sea with its menacing swell and to strain river-currents by lowering nets, in their search for an elusive prey which leaves no tracks. And this because earth makes too little provision for luxury, because the appetite disdains dry land, and because Nereus himself feeds gluttony from the sea.

[197]Procyon rises at the moment when Cancer's twenty-seventh degree ascends from the waves to the stars. He bestows upon those born under him not hunting, but its weapons. To rear keen-scented whelps and to tell their class by their pedigree, their qualities by their place of origin ; to produce nets, and hunting-spears tipped with strong points, and pliant shafts with knots smoothed out ; and to manufacture and sell at a profit whatever the art of hunting is likely to require : these are the gifts Procyon will bestow.

[206]But when the lion of Nemea lifts into view his enormous gaping jaws, the brilliant constellation of the Dog appears [a] : it barks forth flame, raves with its fire, and doubles the burning heat of the sun. When it puts its torch to the earth and discharges its rays, the earth foresees its conflagration and tastes its ultimate fate [b] : Neptune lies motionless in the midst of his waters, and the green blood is drained from leaves and grass. All living things seek alien climes, and the world looks for another world to repair

would ultimately be engulfed in a cosmic conflagration and all things would return to the condition of primeval fire.

aegrotat morbis nimios obsessa per aestus
inque rogo vivit : tantus per sidera fervor
funditur atque uno censentur lumine cuncta.
haec ubi se ponto per primas extulit oras,
nascentem quam nec pelagi restinxerit unda,
effrenos animos violentaque pectora finget
irarumque dabit fluctus odiumque metumque
totius vulgi. praecurrunt verba loquentis,
ante os est animus nec magnis concita causis
corda micant et lingua rabit latratque loquendo,
morsibus et crebris dentes in voce relinquit.
ardescit vino vitium, viresque ministrat
Bacchus et in flammam saevas exsuscitat iras.
nec silvas rupesque timent vastosque leones
aut spumantis apri dentes atque arma ferarum,
effunduntque suas concesso in corpore flammas.
nec talis mirere artes sub sidere tali :
cernis ut ipsum etiam sidus venetur in astris ;
praegressum quaerit Leporem comprendere cursu.
Ultima pars magni cum tollitur orta Leonis,
Crater auratis surgit caelatus ab astris.
inde trahit quicumque genus moresque, sequetur
irriguos ruris campos amnesque lacusque,
et te, Bacche, tuas nubentem iunget ad ulmos,
disponetve iugis imitatus fronde choreas,
robore vel proprio fidentem in bracchia ducet

[217] ceu sunt in flumine [226] vino] vitio

[a] The impetuosity of the speaker causes him to utter words before he has time to adapt them to his grammar or logic. Here our author perhaps echoes [Longinus], *De Subl.* 19. 1 τὰ λεγόμενα . . . φθάνοντα καὶ αὐτὸν τὸν λέγοντα "words . . . running ahead of the speaker himself" (see Bühler 475 f.).

to ; beset by temperatures too great to bear, nature is afflicted with a sickness of its own making, alive, but on a funeral-pyre : such is the heat diffused among the constellations, and the whole sky consists of the light of a single star. When the Dogstar rises over the rim of the sea, which at its birth not even the flood of Ocean can quench, it will fashion unbridled spirits and impetuous hearts ; it will bestow on its sons billows of anger, and draw upon them the hatred and fear of the whole populace. Words run ahead of the speakers [a] : the mind is too fast for the mouth. Their hearts start throbbing at the slightest cause, and when speech comes their tongues rave and bark, and constant gnashing imparts the sound of teeth to their utterance. Their failings are intensified by wine, for Bacchus gives them strength and fans their savage wrath to flame. No fear have they of woods or mountains, or monstrous lions, the tusks of the foaming boar, or the weapons which nature has given wild beasts : they vent their burning fury upon all legitimate prey. But wonder not at these tendencies under such a constellation : you see how even the constellation itself hunts among the stars, for in its course it seeks to catch the Hare in front.

[234]When the last degree of the mighty Lion appears at its rising, the Bowl comes into view, chased with the gilt of its stars. Whoever derives hence his birth and character will be attracted by the well-watered meadows of the countryside, the rivers, and the lakes. He will join your vines, Bacchus, in wedlock to your elms ; or he will arrange them on props, so that the fronds resemble the figures in a dance ; or, allowing your vine to rely on its own strength, he will lead it to spread out its branches as arms, and

teque tibi credet semperque, ut matre resectum,
abiunget thalamis, segetemque interseret uvis,
quaeque alia innumeri cultus est forma per orbem
pro regione colet. nec parce vina recepta
hauriet, emeritis et fructibus ipse fruetur
gaudebitque mero mergetque in pocula mentem.
nec solum terrae spem credet in annua vota :
annonae quoque vectigal mercesque sequetur
praecipue quas umor alit nec deserit unda.
talis effinget Crater umoris amator.

Iam subit Erigone. quae cum tibi quinque feretur
partibus ereptis ponto, tollentur ab undis
clara Ariadnaeae quondam monumenta coronae
et mollis tribuent artes. hinc dona puellae
namque nitent, illinc oriens est ipsa puella.
ille colet nitidis gemmantem floribus hortum
caeruleumque oleis viridemve in gramine collem.
pallentis violas et purpureos hyacinthos
liliaque et Tyrias imitata papavera luces
vernantisque rosae rubicundo sanguine florem
conseret et veris depinget prata figuris.
aut varios nectet flores sertisque locabit
effingetque suum sidus similisque *coronas*
Cnosiacae faciet ; calamosque in mutua pressos
incoquet atque Arabum Syriis mulcebit odores

241 -que ut] qui
242 adiungit
245 emiseris
251 tibi] ter
260 foliis
264 silvis

---

[a] Semele, daughter of Cadmus : see note on 2. 17.
[b] *I.e.* become a tax-collector.
[c] Papyrus, for example, or sponges.
[d] *I.e.* after Leo.
[e] In her lifetime.
[f] The maiden's gifts are the constellation Corona Borealis, which rises with the sign of the zodiac that is a maiden's self, namely Virgo.

entrusting you to yourself will for ever protect you from the bridal bed, seeing how you were cut from your mother.[a] He will sow corn among the grapes, and will adopt any other of the countless forms of cultivation that exist throughout the world, as the conditions of the district require. He will drink without stint the wine he has produced and enjoy in person the well-earned fruits of his labours ; neat wine will incite him to jollity, when he will drown all seriousness in his cups. Nor only on the soil will he stake his hopes for paying his yearly vows : he will also go in pursuit of the grain-tax,[b] and of those wares [c] especially which are nourished by moisture or associated with water. Such are the men to be fashioned by the Bowl, lover of all that is wet.

[251]Next [d] to rise is Erigone. When you behold her ascending with five degrees wrested from the sea, there will emerge from the waves the bright memorial of what was once [e] Ariadne's crown ; and gentle will be the skills herefrom bestowed, for on this side shine a maiden's gifts, on that there rises a maiden's self.[f] The child of the Crown will cultivate a garden budding with bright flowers and slopes grey with olive or green with grass. He will plant pale violets, purple hyacinths, lilies, poppies which vie with bright Tyrian dyes, and the rose which blooms with the redness of blood, and will stipple meadows with designs of natural colour. Or he will entwine different flowers and arrange them in garlands ; he will wreathe the constellation under which he was born, and like *Ariadne's crown will be the crowns he fashions* ; *and stems* he will squeeze together, and distil mixtures therefrom, and will flavour Arabian with Syrian scents and produce unguents which give

et medios unguenta dabit referentia flatus, 2
ut sit adulterio sucorum gratia maior.
munditiae cordi cultusque artesque decorae
et lenocinium vitae praesensque voluptas.
Virginis hoc anni poscunt floresque Coronae.

At, cum per decimam consurgens horrida partem 2
Spica feret prae se vallantis corpus aristas,
arvorum ingenerat studium rurisque colendi
seminaque in faenus sulcatis credere terris
usuramque sequi maiorem sorte receptis
frugibus innumeris atque horrea quaerere messi 2
(quod solum decuit mortalis nosse metallum :
nulla fames, non ulla forent ieiunia terris ;
dives erat census saturatis gentibus *olim*
argenti venis aurique latentibus orbi)
et, si forte labor vires tardaverit, artes
quis sine nulla Ceres, non ullus seminis usus, 2
subdere fracturo silici frumenta superque
ducere pendentis orbes et mergere farra
ac torrere focis hominumque alimenta parare
atque unum genus in multas variare figuras.
et, quia dispositis habitatur spica per artem 2
frugibus, ac structo similis componitur ordo,
seminibusque suis cellas atque horrea praebet,
sculpentem faciet sanctis laquearia templis
condentemque novum caelum per tecta Tonantis.
haec fuerat quondam divis concessa figura, 2

[267] [cordi] . . . decori [286] destructos
[281] facturos liti

---

[a] The bright star Spica cannot be said to rise together with any part of Virgo, being in fact part of Virgo itself, located according to Ptolemy in its 27th degree.

off a mingled fragrance, that the charm of the perfume be enhanced by the blending. His heart is set upon elegance, fashion, and the art of adornment, upon gracious living and the pleasure of the hour. Such is the endowment prescribed by the years of the Maid and the flowers of the Crown.

270Now when the bristly Corn-Ear rises together with the Virgin's tenth degree,[a] bearing before it the spikes which palisade its grain, it engenders a love of fields and of agriculture. Its son is destined to invest seed in the furrowed soil for interest, to obtain usury greater than the principal by reaping a crop past reckoning, so that he will look in vain for barns to hold the crop. This is the only metal it were right for mortals to know: then would there be no hunger or scarcity on earth; rich indeed was the lot of those well-fed folk *when on a time deposits of silver and gold lay hidden* from mankind. Upon him whose strength is perchance enfeebled by toil it bestows skill in occupations without which there would be no bread or any benefit from the grain: putting the corn beneath the rock which will crush it, he will turn over it the upper millstone, moisten the flour and bake it at the hearth; he will prepare men's daily food and mould the selfsame substance into a host of different shapes. And, because the ear of corn acts as a dwelling for grains skilfully arranged, and its structure is designed like a building, since it provides cells and granaries to the seedlings within it, Spica will produce a man who carves panelled ceilings in the sacred temples, creating a second heaven in the Thunderer's abode. Such ornamentation was once reserved for the gods;

nunc iam luxuriae pars est : triclinia templis
concertant, tectique auro iam vescimur auro.
Sed parte octava surgentem cerne Sagittam
Chelarum. dabit haec iaculum torquere lacertis,
et calamum nervis, glaebas et mittere virgis, 2
pendentemque suo volucrem deprendere caelo,
cuspide vel triplici securum figere piscem.
quod potius dederim Teucro sidusve genusve,
teve, Philoctete, cui malim credere parti ?
Hectoris ille faces arcu taedamque fugavit, 3
mittebat saevos ignes quae mille carinis.
hic sortem pharetra Troiae bellique gerebat,
maior et armatis hostis subsederat exul.
quin etiam ille pater tali de sidere cretus
esse potest, qui serpentem super ora cubantem 3
infelix nati somnumque animamque bibentem
sustinuit misso petere ac prosternere telo.
ars erat esse patrem ; vicit natura periclum
et pariter iuvenem somnoque ac morte levavit
tunc iterum natum et fato per somnia raptum. 3
At, cum secretis improvidus Haedus in antris
erranti similis fratrum vestigia quaerit
postque gregem longo producitur intervallo,

294 et
300 taedam] tela *L*: -um *M*: -o *G*
301 saevos] -que suos quae] et
302 ortam pharetram bellumque
311 astris

[a] Those of the Greeks at Troy. See Homer, *Il*. 8. 266-334 ; 15. 436-485.

[b] It was decreed that Troy could not be taken without the arrows of Hercules : these were held by Philoctetes, who, afflicted with a noisome wound in the foot, had been aban-

today it is one of our extravagances: dining-rooms rival temples in splendour, and under a roof of gold we now take our meals off gold.

293 Mark now the Arrow rising in the eighth degree of the Balance. It will bestow the skill of hurling the javelin with the arm, of shooting the arrow from the string and missiles from rods, and of hitting a bird on the wing in the sky that is its home or piercing with three-pronged spear the fish that deemed itself so safe. What constellation or nativity should I rather have given Teucer? To what degree should I prefer to assign Philoctetes? His bow enabled Teucer to repel the flaming torches of Hector which threatened to pour fell fire upon a thousand ships.[a] Carrying in his quiver the fate of Troy and the Trojan War, Philoctetes,[b] who tarried in exile, proved a foe more potent than an armoured host. Under this constellation indeed may well have been born that luckless parent [c] who caught sight of a serpent couched upon his son's face and sapping the life-blood of the sleeping child, but nerved himself to let fly a shaft at it and succeeded in killing the reptile. Fatherhood supplied his skill; a natural instinct overcame the danger and delivered the boy from sleep and death alike, given then a second life and snatched whilst dreaming from the grave.

311 But when the heedless Kid,[d] like one astray in secluded dells, looks for his brethren's tracks and rises at a distance far behind the flock, he fashions an

doned by the Greeks in Lemnos; subsequently healed and brought to Troy, he slew many of the Trojans, including Paris. Unlike most of the Greeks Philoctetes, as an archer, wore no armour.

[c] Alcon (*cf.* Valerius Flaccus 1. 398 ff., Servius, *Buc.* 5. 11).

[d] See Introduction pp. xciv f.

sollertis animos agitataque pectora in usus
effingit varios nec deficientia curis
nec contenta domo. populi sunt illa ministra
perque magistratus et publica iura feruntur.
non illo coram digitos quaesiverit hasta,
defueritque bonis sector, poenamque lucretur
noxius et patriam fraudarit debitor aeris.
cognitor est urbis. nec non lascivit amores
in varios ponitque forum suadente Lyaeo,
mobilis in saltus et scaenae mollior arte.

Nunc surgente Lyra testudinis enatat undis
forma per heredem tantum post fata sonantis,
qua quondam somnumque fretis Oeagrius Orpheus
et sensus scopulis et silvis addidit aures
et Diti lacrimas et morti denique finem.
hinc venient vocis dotes chordaeque sonantis
garrulaque in modulos diversa tibia forma
et quodcumque manu loquitur flatuque movetur.
ille dabit cantus inter convivia dulcis
mulcebitque sono Bacchum noctemque tenebit.
quin etiam curas inter secreta movebit
carmina furtivo modulatus murmure vocem,
solus et ipse suas semper cantabit ad aures,
sic dictante Lyra, cum pars vicesima sexta
Chelarum surget, quae cornua ducet ad astra.

Quid regione Nepae vix partes octo trahentis

318 digito quae iuverit
322 suadetque
326 ferens
329 horeaeque *M*: bor- *GL*
337 hic distante

---

[a] Literally simply "the spear": a spear was fixed in the ground at public auctions (originally as a sign of booty won in war), and the word *hasta* came to denote the auction, *sub hasta vendere* being used much as we might say "bring under the hammer." [b] Mercury (Eratosthenes, *Catast.* 24).

adroit mind and a spirit which occupies itself with business of every kind, which falters not under cares, and which is never satisfied with home. Such men are the servants of the state, passing through the magistracies and judicial offices. When he is there, the auctioneer's spear [a] will not look in vain for a lifted finger or confiscated goods lack a bidder ; when he is there, no criminal will go scot-free or any debtor to the treasury defraud the state. He is the city's attorney. In addition, he indulges in a host of love affairs, and at the Wine-god's behest drops public business, when he reveals himself as agile at dancing and more supple than performers on the stage.

[324]Next, with the rising of the Lyre, there floats forth from Ocean the shape of the tortoise-shell, which under the fingers of its heir [b] gave forth sound only after death ; once with it did Orpheus, Oeagrus' son, impart sleep to waves, feeling to rocks, hearing to trees, tears to Pluto, and finally a limit to death. Hence will come endowments of song and tuneful strings, hence pipes of different shapes which prattle melodiously, and whatever is moved to utterance by touch of hand or force of breath. The child of the Lyre will sing beguiling songs at the banquet, his voice adding mellowness to the wine and holding the night in thrall. Indeed, even when harassed by cares, he will rehearse some secret strain, tuning his voice to a stealthy hum ; and, left to himself, he will ever burst into song which can charm no ears but his own. Such are the ordinances of the Lyre, which at the rising of Libra's twenty-sixth degree will direct its prongs to the stars.

[339]At the side of Scorpio risen hardly eight degrees

Ara ferens turis stellis imitantibus ignem, 3
in qua devoti quondam cecidere Gigantes,
nec prius armavit violento fulmine dextram
Iuppiter, ante deos quam constitit ipse sacerdos ?
quos potius fingent ortus quam templa colentis
atque auctoratos in tertia iura ministros, 3
divorumque sacra venerantis numina voce,
paene deos et qui possint ventura videre ?

Quattuor appositis Centaurus partibus effert
sidera et ex ipso mores nascentibus addit.
aut stimulis agitabit onus mixtasque iugabit 3
semine quadrupedes aut curru celsior ibit
aut onerabit equos armis aut ducet in arma.
ille tenet medicas artes ad membra ferarum
et non auditos mutarum tollere morbos.
hoc est artis opus, non exspectare gementis 3
et sibi non aegrum iamdudum credere corpus.

Hunc subit Arcitenens, cuius pars quinta nitentem
Arcturum ostendit ponto. quo tempore natis
Fortuna ipsa suos audet committere census,
regalis ut opes et sancta aeraria servent 3
regnantes sub rege suo rerumque ministri,

350 agitavit 356 aegros

---

[a] Housman identifies the third order as temple-slaves or freedmen, comparing Digest 33. 1. 20. 1 *sacerdoti et hierophylaco et libertis qui in illo templo erunt* : "... priest and sacristan [these being Manilius's "temple dignitaries"] and temple-freedmen."

[b] *I.e. onus* signifies ὄνους.

[c] Playing on the derivation of Centaur from κεντεῖν, "to goad."

[d] Literally "he will load horses with arms (that is his own, he being armed and mounted) or will drive them into arms (that is yoked to war-chariots)."

what of the Altar bearing incense-flame of which its stars are the image? On this in ages past the Giants were vowed to destruction before they fell, for Jupiter armed not his hand with the powerful thunderbolt until he had stood as priest before the gods. Whom rather than temple dignitaries will such risings shape and those enrolled in the third order of service,[a] those who worship in sacred song the divinity of the gods and those who, all but gods themselves, have power to read the future?

[348]After a further four degrees the Centaur rears his stars and from his own nature assigns qualities to his progeny. Such a one will either urge on asses [b] with the goad [c] and yoke together quadrupeds of mixed stock or will ride aloft in a chariot; else he will saddle horses with a fighter or drive them into the fight.[d] Another knows how to apply the arts of healing to the limbs of animals [e] and to relieve the dumb creatures of the disorders they cannot describe for his hearing. His is indeed a calling of skill, not to wait for the cries of pain, but recognize betimes a sick body not yet conscious of its sickness.

[357]Him the Archer follows, whose fifth degree shows bright Arcturus [f] to those upon the sea. To folk born in this hour Fortune herself makes bold to entrust her treasures, so that the wealth of monarchs and temple finances will be in their keeping [g]: they will be kings under kings, and ministers of state, and

[e] These endowments reflect the identification of Centaurus as Chiron (Eratosthenes, *Catast.* 40).

[f] Strictly speaking Arcturus is a star, but the name is occasionally used for the whole constellation of Bootes or Arctophylax, the Bearward.

[g] Custodianship is a suitable endowment for the Bearward.

tutelamque gerant populi, domibusve regendis
praepositi curas alieno limine claudant.

Arcitenens cum se totum produxerit undis,
ter decima sub parte feri formantibus astris 36
plumeus in caelum nitidis Olor evolat alis.
quo surgente trahens lucem matremque relinquens
ipse quoque aerios populos caeloque dicatum
alituum genus in studium censusque vocabit.
mille fluent artes : aut bellum indicere mundo 37
et medios inter volucrem prensare meatus,
aut nidis damnare suis, ramove sedentem
pascentemve super surgentia ducere lina.
atque haec in luxum. iam ventri longius itur
quam modo militiae : Numidarum pascimur oris 37
Phasidos et lucis ; arcessitur inde macellum
unde aurata novo devecta est aequore pellis.
quin etiam linguas hominum sensusque docebit
aerias volucres novaque in commercia ducet
verbaque praecipiet naturae lege negata. 38
ipse deum Cycnus condit vocemque sub illo
non totus volucer, secumque immurmurat intus.
nec te praetereant clausas qui culmine summo
pascere aves Veneris gaudent et reddere caelo
aut certis revocare notis, totamve per urbem 38
qui gestant caveis volucres ad iussa paratas,

362 f. *after* 373
369 -que genus [in]
377 effecta *LM* : convecta *G*
384 caecos

be charged with the guardianship of the people; or, as the stewards of grand houses, they will confine their business to the care of another's home.

[364]When the Archer has fully emerged from the waves, the Swan ascends into heaven with this creature's thirtieth degree, its down and glittering wings figured by stars. Accordingly he who at its rising leaves his mother's womb and beholds the light of day shall make the denizens of the air and the race of birds that is dedicated to heaven the source of his pleasure and profit. From this constellation shall flow a thousand human skills: its child will declare war on heaven and catch a bird in mid-flight, or he will rob it of its nestlings, or draw nets up and over a bird whilst it is perched on a branch or feeds on the ground. And the object of these skills is to satisfy our high living. Today we go farther afield for the stomach than we used to go for war: we are fed from the shores of Numidia and the groves of Phasis; our markets are stocked from the land whence over a new-discovered sea was carried off the Golden Fleece. Nay more, such a man will impart to the birds of the air the language of men and what words mean; he will introduce them to a new kind of intercourse, teaching them the speech denied them by nature's law. In its own person the Swan hides a god [a] and the voice belonging to it; it is more than a bird and mutters to itself within. Fail not to mark the men who delight to feed the birds of Venus in pens on a rooftop, releasing them to their native skies or recalling them by special signs; or those who carry in cages throughout the city birds taught to obey words of command, men whose total wealth consists

[a] As being the disguise of Jupiter; *cf.* 1. 339 f., 5. 25.

quorum omnis parvo consistit passere census.
has erit et similis tribuens Olor aureus artes.
  Anguitenens magno circumdatus orbe draconis,
cum venit in regione tuae, Capricorne, figurae,
non inimica facit serpentum membra creatis.
accipient sinibusque suis peploque fluenti
osculaque horrendis iungent impune venenis.
  At, cum se patrio producens aequore Piscis
in caelumque ferens alienis finibus ibit,
quisquis erit tali capiens sub tempore vitam,
litoribus ripisve suos circumferet annos,
pendentem et caeco captabit in aequore piscem,
et perlucentis cupiens prensare lapillos
verticibus mediis oculos immittet avaros
cumque suis domibus concha valloque latentis
protrahet immersus. nihil est audere relictum :
quaestus naufragio petitur corpusque profundo
immissum pariter quam praeda exquiritur ipsa.
nec semper tanti merces est parva laboris :
censibus aequantur conchae, lapidumque nitore
vix quisquam est locuples. oneratur terra profundo.
tali sorte suas artes per litora tractat,
aut emit externos pretio mutatque labores
institor aequoreae varia sub imagine mercis.
  Cumque Fidis magno succedunt sidera mundo

[394] producet in
[531] cuperet
[532] emittet
[404] rapidumque notori

---

[a] For such performing birds, see Pliny, *N.H.* 10. 116.

[b] Implying that a diver's life was usually an unenviable one (Seneca, *Dial.* 4. 12. 4).

[c] *I.e.* all the rich men have beggared themselves to buy pearls.

[d] On the identity of *Fides* see Introduction p. xcvi. The synonymous *fidicula*, occasionally used to designate the

of a little sparrow.[a] These and like skills will be the gift of the golden Swan.

[389]When Ophiuchus, encircled by the serpent's great coils, rises beside the figure of Capricorn, he renders the forms of snakes innocuous to those born under him. They will receive snakes into the folds of their flowing robes, and will exchange kisses with these poisonous monsters and suffer no harm.

[394]But when the Southern Fish rises into the heavens, leaving its native waters for a foreign element, whoever at this hour takes hold of life will spend his years about sea-shore and river-bank: he will capture fish as they swim poised in the hidden depths; he will cast his greedy eyes into the midst of the waters, craving to gather pellucid stones and, immersed himself, will bring them forth together with the homes of protective shell wherein they lurk. No peril is left for man to brave: profit is sought by means of shipwreck, and the diver who has plunged into the depths becomes, like the booty, the object of recovery. And not always small is the gain to be derived from this dangerous labour[b]: pearls are worth fortunes, and because of these splendid stones there is scarcely a rich man left.[c] Dwellers on land are burdened with the treasures of the sea. A man born to such a lot plies his skill along the shore; or he purchases at a fixed wage another's labour and sells for a profit what it has brought him, a pedlar in the many different forms of sea products.

[409]When the constellation of the Lute[d] rises into the mighty heavens, there shall be born a man

constellation of the Lyre, also signifies an instrument of torture; and it is clearly this significance which furnishes the poet with his character-sketch.

quaesitor scelerum veniet vindexque reorum, 41
qui commissa suis rimabitur argumentis
in lucemque trahet tacita latitantia fraude.
hinc etiam immitis tortor poenaeque minister
et quisquis verove favet culpamve perodit
proditur atque alto qui iurgia pectore tollat. 41

Caeruleus ponto cum se Delphinus in astra
erigit et squamam stellis imitantibus exit,
ambiguus terrae partus pelagique creatur.
nam, velut ipse citis perlabitur aequora pinnis
nunc summum scindens pelagus nunc alta profundi 42
et sinibus vires sumit fluctumque figurat,
sic, venit ex illo quisquis, volitabit in undis.
nunc alterna ferens in lentos bracchia tractus
conspicuus franget spumanti limite pontum
et plausa resonabit aqua, nunc aequore mersas
diducet palmas furtiva biremis in ipso, 42
nunc in aquas rectus veniet passuque natabit
et vada mentitus reddet super aequora campum ;
aut immota ferens in tergus membra latusque
non onerabit aquas summisque accumbet in undis
pendebitque super, totus sine remige velum. 43
illis in ponto iucundum est quaerere pontum,
corporaque immergunt undis ipsumque sub antris
Nerea et aequoreas conantur visere Nymphas,
exportantque maris praedas et rapta profundo
naufragia atque imas avidi scrutantur harenas. 43

414 vero[ve] 426 passumque
418 pelagoque 430 totum . . . votum
425 furtivo remis 432 qui mergunt

---

[a] Housman, less probably : "and champion the unjustly accused." But see Bühler 485.

[b] Of litigants.

to investigate wrong-doing and punish the guilty [a]: he will get to the bottom of crimes by sifting the evidence for them and bring to light all that lies hidden under the silence of deceit. Hence, too, are begotten the merciless torturer, the dispenser of penalties, whoever insists on the truth and abominates evil, and the man to remove dissensions from the depths of the heart.[b]

[416]When the sea-dark Dolphin ascends from Ocean to the heavens and emerges with its scales figured by stars, birth is given to children who will be equally at home on land and in the sea. For just as the dolphin is propelled by its swift fins through the waters, now cleaving the surface, now the depths below, and derives momentum from its undulating course, wherein it reproduces the curl of waves, so whoever is born of it will speed through the sea. Now lifting one arm after the other to make slow sweeps *he will catch the eye as he drives a furrow of foam through the sea* and will sound afar as he thrashes the waters; now like a hidden two-oared vessel he will draw apart his arms beneath the water; now he will enter the waves upright and swim by walking and, pretending to touch the shallows with his feet, will seem to make a field of the surface of the sea; else, keeping his limbs motionless and lying on his back or side, he will be no burden to the waters but will recline upon them and float, the whole of him forming a sail-boat not needing oarage. Other men take pleasure in looking for the sea in the sea itself: they dive beneath the waves and try to visit Nereus and the sea nymphs in their caves; they bring forth the spoils of the sea and the booty that wrecks have lost to it, and eagerly search the sandy bottom.

par ex diverso studium sociatur utrumque
in genus atque uno digestum semine surgit.
adnumeres etiam illa licet cognata per artem
corpora, quae valido saliunt excussa petauro
alternosque cient motus, elatus et ante
nunc iacet atque huius casu suspenditur ille,
membrave per flammas orbesque emissa flagrantis,
quae delphina suo per inane imitantia motu
molliter ut liquidis per humum ponuntur in undis
et viduata volant pinnis et in aere ludunt.
at, si deficient artes, remanebit in illis
materies tamen apta ; dabit natura vigorem
atque alacris cursus campoque volantia membra.

Sed regione means Cepheus umentis Aquari
non dabit in lusum mores. facit ora severae
frontis is ac vultus componit pondere mentis.
pascentur curis veterumque exempla revolvent
semper et antiqui laudabunt verba Catonis.
componet teneros etiam qui nutriat annos
et dominum dominus praetextae lege sequatur
quodque agat id credat, stupefactus imagine iuris,
tutorisve supercilium patruive rigorem.
quin etiam tragico praestabunt verba coturno,
cuius erit, quamquam in chartis, stilus ipse cruentus,
nec minus hae scelerum facie rerumque tumultu

440 delatus et ille 451 [is]
444 [quae] delphinamque 460 haec

[a] On the Farnese globe Cepheus is depicted in the garb of a tragic actor.
[b] Than the audience at a performance.

From their different sides swimmers and divers share an equal enthusiasm for both pursuits, for their enthusiasm, though displayed in different ways, springs from a single source. With them you may also reckon men of cognate skill who leap in the air, thrown up from the powerful spring-board, and execute a see-saw movement, he who was first lifted on high now finding himself on the ground and by his descent raising the other aloft; or hurl their limbs through fire and flaming hoops, imitating the dolphin's movement in their flight through space, and land as gently on the ground as they would in the watery waves: they fly though they have no wings and sport amid the air. Even if the Dolphin's sons lack these skills, they will yet possess a physique suited to them; nature will endow them with strength of body, briskness of movement, and limbs which fly over the plain.

[449]But Cepheus, rising beside the dripping Waterman, will not engender dispositions inclined to sport. He fashions faces marked by a stern demeanour, and moulds a countenance whereon is depicted gravity of mind. Such men will live on worry and will incessantly recall the traditions of a bygone age and commend old Cato's maxims. Cepheus will also create a man to bring up boys of tender age: he will lord it over his lord by virtue of the law which governs a minor and, bemused by this semblance of power, will mistake for reality the role of arrogant guardian or stern uncle which he plays. Offspring of Cepheus will also furnish words for the buskin of tragedy,[a] whose pen, if only on paper, is drenched in blood; and the paper no less [b] will revel in the spectacle of crime and catastrophe in human affairs. They will

gaudebunt. vix una trium memorare sepulcra
ructantemque patrem natos solemque reversum
et caecum sine nube diem, Thebana iuvabit
dicere bella uteri mixtumque in fratre parentem,
quin et Medeae natos fratremque patremque,
hinc vestes flammas illinc pro munere missas
aeriamque fugam natosque ex ignibus annos.
mille alias rerum species in carmina ducent;
forsitan ipse etiam Cepheus referetur in actus.
et, siquis studio scribendi mitior ibit,
comica componet laetis spectacula ludis,
ardentis iuvenes raptasque in amore puellas
elusosque senes agilisque per omnia servos,
quis in cuncta suam produxit saecula vitam
doctior urbe sua linguae sub flore Menander,
qui vitae ostendit vitam chartisque sacravit.
et, si tanta operum vires commenta negarint,
externis tamen aptus erit, nunc voce poetis
nunc tacito gestu referensque affectibus ora,
et sua dicendo faciet, | scaenisque togatos

461 vix una trium] atri luxum
463 nube] sole
465 quin et] queret
467 nectos *GL*: notos *M*
468 aliae
469 actis
479 referetque
482b togatus

---

[a] Thyestes unwittingly ate his three sons, whom, their extremities cut off, his brother Atreus served up to him as a meal (*cf.* Apollodorus, *Epit.* 2. 13): the burial (incomplete because the sons were not completely eaten) took place in the father's stomach (Cicero, *De Off.* 1. 97, who perhaps quotes the *Atreus* of Accius).

[b] Between Eteocles and Polynices (*cf.* Aeschylus, *Septem contra Thebas*).

[c] Oedipus (*cf.* Sophocles, *Oedipus Tyrannus*).

[d] Referring (out of chronological order) to her murder of her children; to her father Aeetes, whose pursuit she delayed

delight to tell of scarce one burial accorded three[a]: the father belching forth the flesh of his sons, the sun fled in horror, and the darkness of a cloudless day; they will delight to narrate the Theban war between a mother's issue,[b] and one[c] who was both father and brother to his children; the story of Medea's sons, her brother and her father, the gift which was first robe and then consuming flame, the escape by air, and youth reborn from fire.[d] A thousand other scenes from the past will they include in their plays; perhaps Cepheus himself will also be brought upon the stage.[e] If anyone is born with the urge to write in lighter vein, he will compose for presentation at the merry games scenes of comedy about the loves of headstrong youths and abducted maidens, hoodwinked old men, and slaves of infinite resource. In such plays Menander made his own day live for all generations: a man whose eloquence surpassed that of his native Athens (and that when its language attained its richest bloom), he held up a mirror to life and enshrined the image in his works. Should his powers not rise to such masterpieces, the child of Cepheus will yet be fitted to perform those of others: he will interpret the poet's words, now by his voice, now by silent gesture and expression, and the lines he declaims he will make his own. On the stage he will take the part of Romans or the mighty heroes of

by tearing her brother Absyrtus to pieces and throwing his mangled limbs in the way; to the poisonous robe sent as a wedding gift to Glauce (or Creusa), daughter of Creon, king of Corinth, which killed both the bride and her father; to the dragon-chariot sent by the Sun-god for Medea's escape; and to Aeson, Jason's father, whose youth Medea magically restored (see Ovid, *Met.* 7. 162 ff.). *Cf.* Euripides, *Medea.*

[e] *Cf.* Euripides, *Andromeda.*

aut magnos heroas aget, | solusque per omnis
ibit personas et turbam reddet in uno ;
omnis fortunae vultum per membra reducet,
aequabitque choros gestu cogetque videre
praesentem Troiam Priamumque ante ora cadentem.

Nunc Aquilae sidus referam, quae parte sinistra
rorantis iuvenis, quem terris sustulit ipsa,
fertur et extentis praedam circumvolat alis.
fulmina missa refert et caelo militat ales
bis sextamque notat partem fluvialis Aquari.
illius in terris orientis tempore natus
ad spolia et partas surget vel caede rapinas
nec pacem bello, civem discernet ab hoste,
cumque hominum derit strages, dabit ille ferarum.
ipse sibi lex est, et qua fert cumque voluntas
praecipitant vires ; laus est contemnere cuncta.
et, si forte bonis accesserit impetus ausis,
improbitas fiet virtus, et condere bella
et magnis patriam poterit ditare triumphis.
et, quia non tractat volucris sed suggerit arma
immissosque refert ignes et fulmina reddit,
regis erit magnive ducis per bella minister
ingentisque suis praestabit viribus usus.

494 pace ac *LM* : pacem a *G*   499 poterunt ornare
493 dederit   502 magister

---

[a] An error : Aquila, a northern constellation, rises on Aquarius's right.

[b] The poet here identifies Aquarius as Ganymede.

[c] Another error (shared with Hyginus, *Poet. Astr.* 2. 16) :

myth ; he will assume every role himself, one after another, and in his single person represent a crowd ; he will draw over his limbs the aspect of fortune's every vicissitude and his gestures will match the songs of the chorus ; he will convince you that you see Troy's actual fall and Priam expiring before your very eyes.

486 Now I shall tell of the constellation of the Eagle : it rises on the left [a] of the youth who pours, whom once it carried off from earth,[b] and with wings outspread it hovers above its prey.[c] This bird brings back the thunderbolts which Jupiter has flung and fights in the service of heaven : its appearance marks the twelfth degree of the river-pouring Waterman. He that is born on earth in the hour of its rising will grow up bent on spoil and plunder won even with bloodshed ; he will draw no line between peace and war, between citizen and foe, and when he is short of men to kill, he will engage in butchery of beast. He is a law unto himself, and rushes violently wherever his fancy takes him ; in his eyes to show contempt for everything merits praise. Yet, should perchance his aggressiveness be enlisted in a righteous cause, depravity will turn into virtue, and he will succeed in bringing wars to a conclusion and enriching his country with glorious triumphs. And, since the Eagle does not wield, but supplies weapons, seeing that it brings back and restores to Jupiter the fires and bolts he has hurled, in time of war such a man will be the aide of a king or of some mighty general, and his strength will render them important service.

Aquila hovers above Capricorn and Sagittarius rather than Aquarius.

At, cum Cassiope bis denis partibus actis
aequorei iuvenis dextra de parte resurgit,
artifices auri faciet, qui mille figuris
vertere opus possint caraeque acquirere dotem
materiae et lapidum vivos miscere colores.
hinc Augusta nitent sacratis munera templis,
aurea Phoebeis certantia lumina flammis
gemmarumque umbra radiantes lucibus ignes.
hinc Pompeia manent veteris monumenta triumphi
et Mithridateos vultus induta tropaea,
non exstincta die semperque recentia flammis.
hinc lenocinium formae cultusque repertus
corporis atque auro quaesita est gratia frontis
perque caput ducti lapides per colla manusque
et pedibus niveis fulserunt aurea vincla.
quid potius matrona velit tractare creatos
quam factum revocare suos quod possit ad usus?
ac, ne materies tali sub munere desit,
quaerere sub terris aurum furtoque latentem
naturam eruere omnem orbemque invertere praedae
imperat et glaebas inter deprendere gazam
invitamque novo tandem producere caelo.
ille etiam fulvas avidus numerabit harenas
perfundetque novo stillantia litora ponto

[507] carnique 514: *see after* 542
[512] ub *M*: vili *GL* [515] lues

[a] Elsewhere Manilius employs the conventional ancient spelling of the queen's name: *Cassiepia.* The variant *Cassiope* is exclusive to this passage and points to a special source, possibly Ovid, *Met.* 4. 738.

[b] Augustus claimed (*Res Gestae* 20) to have restored eighty-two temples, and is said (Suetonius, *Aug.* 30. 2) to have presented to the temple of Capitoline Jupiter in a single donation sixteen thousand pounds of gold and fifty million sesterces' worth of jewels.

[504]When after the appearance of twenty degrees of the watery youth Cassiope [a] rises on his right, she will produce goldsmiths who can turn their work into a thousand different shapes, endow the precious substance with yet greater value, and add thereto the vivid hue of jewels. From Cassiope come the gifts of Augustus which gleam in the temples he consecrated,[b] where the blaze of gold rivals the sun's brightness and the fires of gems flash forth light out of shadow. From Cassiope come the memorials of Pompey's triumph of old [c] and the trophies which bear the features of Mithridates [d]: they remain [e] to this very day, spoils undimmed by the passage of time, their sparkle as fresh as ever. From Cassiope come the enhancement of beauty and devices for adorning the body: from gold has been sought the means to give grace to the appearance; precious stones have been spread over head, neck, and hands; and golden chains have shone on snow-white feet. What products would a grand lady like Cassiope prefer her sons to handle rather than those she could turn to her own employments? And that material for such employment should not be lacking, she bids men look for gold beneath the ground, uproot all which nature stealthily conceals, and turn earth upside down in search of gain; she bids them detect the treasure in lumps of ore and finally, for all its reluctance, expose it to a sky it has never seen. The son of Cassiope will also count greedily the yellow sands, and drench a dripping beach with a new

[c] His third triumph, over Mithridates, whose collection of jewels is mentioned by Pliny, *N.H.* 37. 11 ff.

[d] Images of Mithridates are mentioned by Appian, *Bell. Mithr.* 116 f.

[e] In the temple of Capitoline Jupiter.

parvaque ramentis faciet momenta minutis
Pactolive leget census spumantis in aurum ; 530
aut coquet argenti glaebas venamque latentem 53
eruet et silicem rivo saliente liquabit ;
aut facti mercator erit per utrumque metalli, 53
alterum et alterius semper mutabit ad usus.
talia Cassiope nascentum pectora finget.

Andromedae sequitur sidus, quae Piscibus ortis
bis sex in partes caelo venit aurea dextro.
hanc quondam poenae dirorum culpa parentum 54
prodidit, infestus totis cum finibus omnis
incubuit pontus, fluitavit naufraga tellus,
et quod erat regnum pelagus fuit. una malorum 51
proposita est merces, vesano dedere ponto 54
Andromedan, teneros ut belua manderet artus.
hic hymenaeus erat, solataque publica damna 54
privatis lacrimans ornatur victima poenae
induiturque sinus non haec ad vota paratos,
virginis et vivae rapitur sine funere funus.
at, simul infesti ventum est ad litora ponti,
mollia per duras panduntur bracchia cautes ; 55
astrinxere pedes scopulis, iniectaque vincla,
et cruce virginea moritura puella pependit.
servatur tamen in poena vultusque pudorque ;
supplicia ipsa decent ; nivea cervice reclinis
molliter ipsa suae custos est visa figurae. 55

530 protulit ut legeret
531 f.: *see after* 398
533 ad *M*: et *GL*
542 timuit naufragia
545 solaque
546 poena
555 visa] ipsa

[a] See Introduction p. xxvii, under 27 CEPHEUS.

flood; he will make small weights to measure the tiny grains, or else will collect the wealth of gold-foaming Pactolus ; or he will smelt lumps of silver, separating the hidden metal and causing the mineral to flow forth in a running stream ; otherwise he will become a trader of the metals produced by these two craftsmen, ever ready to change coinage of the one metal into wares of the other. Such are the inclinations which Cassiope will fashion in those born under her.

[538]There follows the constellation of Andromeda, whose golden light appears in the rightward sky when the Fishes have risen to twelve degrees. Once on a time the sin of cruel parents [a] caused her to be given up for sacrifice, when a hostile sea in all its strength burst upon every shore, the land was shipwrecked in the flood, and what had been a king's domain was now an ocean. From those ills but one price of redemption was proposed, surrender of Andromeda to the raging main for a monster to devour her tender limbs. This was her bridal ; relieving the people's hurt by submitting to her own, she is tearfully adorned as victim for the avenging beast and dons attire prepared for no such troth as this ; and the corpseless funeral of the living maiden is hurried on its way. Then as soon as the procession reaches the shore of the tumultuous sea, her soft arms are stretched out on the hard rocks ; they bound her feet to crags and cast chains upon her ; and there to die on her virgin cross the maiden hung. Even in the hour of sacrifice she yet preserves a modest demeanour : her very sufferings become her, for, gently inclining her snow-white neck, she seemed to have full charge of her pose. The folds of her robe

defluxere sinus umeris fugitque lacertos
vestis et effusi scapulis haesere capilli.
te circum alcyones pinnis planxere volantes
fleveruntque tuos miserando carmine casus
et tibi contextas umbram fecere per alas.
ad tua sustinuit fluctus spectacula pontus
assuetasque sibi desit perfundere rupes,
extulit et liquido Nereis ab aequore vultus
et, casus miserata tuos, roravit et undas.
ipsa levi flatu refovens pendentia membra
aura per extremas resonavit flebile rupes.
tandem Gorgonei victorem Persea monstri
felix illa dies redeuntem ad litora duxit.
isque, ubi pendentem vidit de rupe puellam,
deriguit, facie quem non stupefecerat hostis,
vixque manu spolium tenuit, victorque Medusae
victus in Andromeda est. iam cautibus invidet ipsis
felicisque vocat, teneant quae membra, catenas ;
et, postquam poenae causam cognovit ab ipsa,
destinat in thalamos per bellum vadere ponti,
altera si Gorgo veniat, non territus illa.
concitat aerios cursus flentisque parentes
promissu vitae recreat pactusque maritam
ad litus remeat. gravidus iam surgere pontus
coeperat ac longo fugiebant agmine fluctus
impellentis onus monstri. caput eminet undas
scindentis pelagusque vomit, circumsonat aequor
dentibus, inque ipso rapidum mare navigat ore ;
hinc vasti surgunt immensis torquibus orbes
tergaque consumunt pelagus. sonat undique Phorcys

558 ter
576 ira
582 movit

slipped from her shoulders and fell from her arms, and her streaming locks covered her body. You, princess, halcyons in circling flight lamented and with plaintive song bewailed your fate, shading you by linking their spans of wing. To look at you the ocean checked its waves and ceased to break, as was its wont, upon the cliffs, whilst the Nereids raised their countenance above the surface of the sea and, weeping for your plight, moistened the very waves. Even the breeze, refreshing with gentle breath your pinioned limbs, resounded tearfully about the cliff-tops. At length a happy day brought to those shores Perseus returning from his triumph over the monstrous Gorgon. On seeing the girl fastened to the rock, he, whom his foe had failed to petrify with her aspect, froze in his tracks and scarcely kept his grasp of the spoil: the vanquisher of Medusa was vanquished at the sight of Andromeda. Now he envies the very rocks and calls the chains happy to clasp such limbs. On learning from the maiden's lips the cause of her punishment, he resolves to go through war against the sea to win her hand, undaunted though a second Gorgon come against him. He quickly cuts a path through the air and by his promise to save their daughter's life awakens hope in the tearful parents; with the pledge of a bride he hastens back to the shore. Now had a heavy surge begun to rise and long lines of breakers were fleeing before the thrust of the massive monster. As it cleaves the waves, its head emerges and disgorges sea, the waters breaking loudly about its teeth and the swirling sea afloat in its very jaws; behind rise its huge coils like rings of an enormous neckchain, and its back covers the whole sea. Ocean clamours

atque ipsi metuunt montes scopulique ruentem.
infelix virgo, quamvis sub vindice tanto
quae tua tunc fuerat facies! quam fugit in auras
spiritus! ut toto caruerunt sanguine membra,
cum tua fata cavis e rupibus ipsa videres 59
adnantemque tibi poenam pelagusque ferentem
quantula praeda maris! quassis hic subvolat alis
Perseus et semet caelo iaculatur in hostem
Gorgoneo tinctum defigens sanguine ferrum.
illa subit contra versamque a gurgite frontem 59
erigit et tortis innitens orbibus alte
emicat ac toto sublimis corpore fertur.
sed, quantum illa subit, semper, iaculata profundo,
in tantum revolat laxumque per aethera ludit
Perseus et ceti subeuntis verberat ora. 60
nec cedit tamen illa viro, sed saevit in auras
morsibus, et vani crepitant sine vulnere dentes;
efflat et in caelum pelagus mergitque volantem
sanguineis undis pontumque exstillat in astra.
spectabat pugnam pugnandi causa puella, 6
iamque oblita sui metuit pro vindice tali
suspirans animoque magis quam corpore pendet.
tandem confossis subsedit belua membris
plena maris summasque iterum remeavit ad undas
et magnum vasto contexit corpore pontum, 6
tum quoque terribilis nec virginis ore videnda.
perfundit liquido Perseus in marmore corpus,
maior et ex undis ad cautes pervolat altas

592 quantis
593 [semet] caelo pendens
595 versaque

in every quarter, and the very mountains and crags quake at the creature's onset. What terror then, unhappy maiden, was expressed on your countenance, defended though you were by such a champion! How all your breath fled into the air! How all the blood ebbed from your limbs, when from the cleft in the rocks you beheld with your own eyes your fate, the avenging monster swimming towards you and driving the waves before it, how helpless you a victim for the sea! Hereupon with a flutter of winged sandals Perseus flies upwards and from the skies hurls himself at the foe, driving home the weapon stained with the Gorgon's blood. The beast rises to meet him, rears its head, twisting it out of the water, leaps aloft upon its support of winding coils, and towers high in the air with all its bulk. But as much as it rises hurtling up from the deep, always so much does Perseus fly higher and mock the sea-beast through the yielding air, and strike its head as it attacks. Yet not submitting to the hero the monster bites furiously at the breezes, though its teeth snap vainly and inflict no wounds; it spouts forth sea towards heaven, drenches its winged assailant with a blood-stained deluge, and sends in spray the ocean to the stars. The princess watches the duel of which she is the prize and, no longer mindful of herself, sighs with fear for her gallant champion: her feelings more than her body hang in suspense. At last, its frame riddled with stabs, through which the sea fills its body, the beast sinks, returns once more to the surface, and covers the mighty ocean with its massive corpse, still a fearful sight, and not for a maiden's eyes to look on. Having bathed his body in pure water, Perseus, a greater warrior now, flies from the

solvitque haerentem vinclis de rupe puellam
desponsam pugna, nupturam dote mariti. 61
hic dedit Andromedae caelum stellisque sacravit
mercedem tanti belli, quo concidit ipsa
Gorgone non levius monstrum pelagusque levavit.

Quisquis in Andromedae surgentis tempora ponto
nascitur, immitis veniet poenaeque minister 62
carceris et duri custos, quo stante superbe
prostratae iaceant miserorum in limine matres
pernoctesque patres cupiant extrema suorum
oscula et in proprias animam transferre medullas.
carnificisque venit mortem vendentis imago 62
accensosque rogos, cui stricta saepe securi
supplicium vectigal erit, qui denique posset
pendentem e scopulis ipsam spectare puellam,
vinctorum dominus sociusque in parte catenae
interdum, poenis ut noxia corpora servet. 63

Piscibus exortis cum pars vicesima prima
signabit terrae limen, fulgebit et orbi,
aerius nascetur Equus caeloque volabit,
velocisque dabit sub tali tempore partus
omne per officium vigilantia membra ferentis. 63
hic glomerabit equo gyros dorsoque superbus
ardua bella geret rector cum milite mixtus ;

626 cui] et   630 ut] in

---

[a] The following passage is clearly based on Cicero, *Fifth Verrine* 118 ff.

[b] The Horse seems to be here identified as Pegasus, and to have wings as on the Farnese globe.

sea to the lofty crags and releases from the chains which bind her to the rock the girl whose betrothal was sealed by his readiness to fight and who could now become a bride thanks to the bridegroom's dowry of her life. Thus did Perseus win place in heaven for Andromeda and hallow in a constellation the prize of that glorious battle, wherein a monster no less terrible than the Gorgon herself perished and in perishing relieved the sea of a curse.

619The man whose birth coincides with the rising of Andromeda from the sea will prove merciless, a dispenser of punishment, a warder of dungeon dire; he [a] will stand arrogantly by while the mothers of wretched prisoners lie prostrate on his threshold, and the fathers wait all night to catch the last kisses of their sons and receive into their inmost being the dying breath. From the same constellation comes the figure of the executioner, ready to take money for a speedy death and the rites of a funeral pyre; for him execution means profit, and oft will he bare his axe; in short, he is a man who could have looked unmoved on Andromeda herself fettered to the rock. Governor of the imprisoned he occasionally becomes a fellow convict, chained to criminals so as to save them for execution.

631When the twenty-first degree of the rising Fishes illuminates earth's threshold and shines upon the world, the winged Horse [b] will appear and gallop aloft in the heavens. In that hour it will bring forth people endowed with swiftness of movement and limbs alert to perform every task. One man will cause his horse to wheel round in caracoles, and proudly mounted on its back he will wage war from on high, commander and soldier in one. Another

hic stadium fraudare fide poteritque videri
mentitus passus et campum tollere cursu.
nam quis ab extremo citius revolaverit orbe
nuntius extremumve levis penetraverit orbem ?
vilibus ille etiam sanabit vulnera sucis
quadrupedum, et medicas herbas in membra ferarum
noverit, humanos et quae nascentur ad usus.

Nixa genu species et Graio nomine dicta
Engonasin, cui nulla fides sub origine constat,
dextra per extremos attollit lumina Pisces.
hinc fuga nascentum, dolus insidiaeque creantur,
grassatorque venit media metuendus in urbe.
et, si forte aliquas animus consurget in artes,
in praerupta dabit studium, vendetque periclo
ingenium, ac tenuis ausus sine limite gressus
certa per extentos ponet vestigia funes
et caeli meditatus iter vestigia perdet
saepe nova et pendens populum suspendet ab ipso.

Laeva sub extremis consurgunt sidera Ceti
Piscibus Andromedan ponto caeloque sequentis.
hoc trahit in pelagi caedes et vulnera natos
squamigeri gregis, extentis laqueare profundum
retibus et pontum vinclis artare furentis ;

638 fidem
640 quamvis
641 vel lebis *LM* : vel bis *G*
643 herbas] artes
646 guicula vides
655 saepe nova] et peneva
660 armare furentem

[a] On a sloping tightrope.

will possess the ability to rob the racecourse of its true length: such is his speed that he will seem to dissemble the movement of his feet and make the ground vanish before him. Who more swiftly could fly back from the ends of the earth as messenger or with light foot to the earth's ends make his way? He will also heal a horse's wounds with the sap of common plants, and will know the herbs which bring aid to an animal's limbs and those which grow for the use of man.

645 The figure on bended knee and called by the Greek name of Engonasin, about whose origin no certainty prevails, brings forth its stars on the right simultaneously with the last portion of the Fishes. Of this constellation is begotten the desertion, craftiness, and deceit characteristic of its children, and from it comes the thug who terrorizes the heart of the city. If perchance his mind is moved to consider a profession, Engonasin will inspire him with enthusiasm for risky callings, with danger the price for which he will sell his talents: daring narrow steps on a path without thickness he will plant firm feet on a horizontal tightrope; then, as he attempts an upward route to heaven,[a] he will repeatedly slide back from footholds newly gained and, suspended in mid-air, he will keep a multitude in suspense upon himself.

656 On the left, as the last portion of the Fishes rises, appears the constellation of the Whale, pursuing Andromeda in heaven as on the sea. This monster enlists its sons in an onslaught on the deep and a butchery of scaly creatures; theirs will be a passion for ensnaring the deep with nets spread wide and for straitening the sea with bonds; they will

et velut in laxo securas aequore phocas
carceribus claudent raris et compede nectent
incautosque trahent macularum nemine thynnos.
nec cepisse sat est : luctantur corpora nodis
exspectantque novas acies ferroque necantur,
inficiturque suo permixtus sanguine pontus.
tum quoque, cum toto iacuerunt litore praedae,
altera fit caedis caedes : scinduntur in artus,
corpore et ex uno varius discribitur usus.
illa datis melior, sucis pars illa retentis.
hinc sanies pretiosa fluit floremque cruoris
evomit et mixto gustum sale temperat oris ;
illa putris turbae strages confunditur omnis
permiscetque suas alterna in damna figuras
communemque cibis usum sucumque ministrat.
aut, cum caeruleo stetit ipsa simillima ponto
squamigerum nubes turbaque immobilis haeret,
excipitur vasta circum vallata sagena
ingentisque lacus et Bacchi dolia complet
umorisque vomit socias per mutua dotes
et fluit in liquidam tabem resoluta medullas.
quin etiam magnas poterunt celebrare salinas
et pontum coquere et ponti secernere virus,
cum solidum certo distendunt margine campum
appelluntque suo deductum ex aequore fluctum
claudendoque negant abitum : sic suscipit undas
area et epoto per solem umore nitescit.

663 nomine *LM*: numine *G*
669 corpora
677 turbaeque
680 volet socia . . . dote
681 liquidas tabes
683 vires
686 abitum] tum
sic] demum
687 ponto

[a] The best *garum*, called *haemation* (*Geoponica* 20. 46. 6).
[b] *Hallec.* [c] The common *garum*, also called *liquamen*.

confine in spacious prisons seals which deem themselves as safe as in the open sea and shackle them fast in fetters ; the unwary tunny they will draw along in a network of meshes. And their capture is not the end : the fish struggle against their bonds, await new assaults, and suffer death by the knife ; and the sea is dyed, mixed with blood of its own. Furthermore, when the victims lie dead along the shore, a second slaughter is perpetrated on the first ; the fish are torn into pieces, and a single body is divided to serve separate ends. One part is better if its juices are given up, another if they are retained. In the one case a valuable fluid is discharged, which yields the choicest part of the blood : flavoured with salt, it imparts a relish to the palate.[a] In the other case all the pieces of the decaying carcase are blended together and merge their shapes until every distinguishing feature has been lost : they provide food with a condiment of general use.[b] Or when, presenting the very likeness of the dark-hued sea, a shoal of the scaly creatures has come to a stop and cannot move for their numbers, they are surrounded and drawn from the water by a huge drag-net, and fill large tanks and wine-vats ; their common endowment of liquid is exuded upon each other, for their inward parts melt and issue forth as a stream of decomposition.[c] Moreover, such men will be able to fill great salt-pans, to evaporate the sea, and to extract the sea's venom : they prepare a wide expanse of hardened ground and surround it with firm walls, next conduct therein waters channelled from the nearby sea and then deny them exit by closing sluice-gates : so the floor holds in the waves and begins to glisten as the water is drained off by

congeritur siccum pelagus mensisque profundi
canities detonsa maris, spumaeque rigentis
ingentis faciunt tumulos, pelagique venenum,
quo perit usus aquae suco corruptus amaro,
vitali sale permutant redduntque salubre.

At, revoluta polo cum primis vultibus Arctos
ad sua perpetuos revocat vestigia passus
numquam tincta vadis sed semper flexilis orbe,
[aut Cynosura minor cum prima luce resurgit
et pariter vastusve Leo vel Scorpius acer
nocte sub extrema promittunt iura diei]
non inimica ferae tali sub tempore natis
ora ferent, placidasque regent commercia gentes.
ille manu vastos poterit frenare leones
et palpare lupos, pantheris ludere captis,
nec fugiet validas cognati sideris ursas
inque artes hominum perversaque munera ducet;
ille elephanta premet dorso stimulisque movebit
turpiter in tanto cedentem pondere punctis;
ille tigrim rabie solvet pacique domabit,
quaeque alia infestant furiis animalia terras

689 sed nota
691 quod erit
692 salubrem
706 cunctis
708 silvis

the sun. When the sea's dry element has collected, Ocean's white locks are shorn for use at table, and huge mounds are made of the solid foam ; and the poison of the deep, which prevents the use of sea-water, vitiating it with a bitter taste, they commute to life-giving salt and render a source of health.

[693]Now when, after completing a revolution round the pole, the Bear [a] with muzzle foremost replaces her unceasing steps in her former tracks, never immersed in Ocean but ever turning in a circle,[b] to those born at such a time wild creatures will show no hostile face, and in their dealings with animals these men will find them submissive to their rule. Such a one will be able to control huge lions with a gesture, to fondle wolves, and to play with captive panthers ; so far from shunning the powerful bears that are the kin of the constellation, he will train them to human accomplishments and feats foreign to their nature ; he will seat himself on the elephant's back and with a goad will direct the movements of a beast which disgraces its massive weight by yielding to tiny jabs ; he will dispel the fury of the tiger, training it to become a peaceful animal, whilst all the other beasts which molest the earth with their savageness

[a] *I.e.* the Great Bear. Since the circumpolar constellations never disappear below the horizon, they cannot properly be said to rise above it simultaneously with the horoscoping degree ; however, as the poet implies in the text, they are held to be *paranatellonta* when at the bottom of their orbit (*i.e.* nearest the horizon) and on the point of ascending.

[b] [696-698], interpolated (perhaps under the influence of Aratus, *Phaen.* 303 ff.) to bring in mention of other animal constellations : ". . . or when the Cynosure, the Lesser Bear, restarts her climb at the break of day and when likewise the monstrous Lion or fierce Scorpion towards night's close gives the day promise of her rights, . . ."

iunget amicitia secum, catulosque sagacis

* * * * *

has stellis proprias vires et tempora rerum
constituit magni quondam fabricator Olympi.

* * * * *

tertia Pleiadas dotavit forma sorores
femineum rubro vultum suffusa pyropo,
invenitque parem sub te, Cynosura, colorem,
et quos Delphinus iaculatur quattuor ignes
Deltotonque tribus facibus, similique nitentem
luce Aquilam et flexos per lubrica terga dracones.

[30] has] ab

---

[a] Here, if the editor's conjecture is correct (*cf.* Introduction p. cvii), eight pages of A, *i.e.* 176 lines including chapter-headings, have been lost. The end of the chapter on *paranatellonta* may be fairly surmised from Firmicus, *Math.* 8. 17. 6 f. (*cf.* Introduction p. xcvii). That the missing passage chiefly dealt with planetary influences seems probable in view of the misplaced verses 5. 30 f., which, clearly in Manilius's style (*cf.* 4. 292, 408, 498, 696, etc.) and yet belonging nowhere else, purport to conclude the poet's treatment of the subject promised at 2. 965 and 3. 156 ff.

[b] For this restoration of the poet's treatment of stellar magnitudes see Introduction pp. c ff.

he will join in friendship to himself; keen-scented whelps *he will train . . .*[a]

* * * * *

[30]Such are the special powers and times of influence which in ages past the creator of the mighty firmament appointed for the planets.

Finally, that I may bring my account of heaven to a close, let me tell you of the different degrees of luminosity with which the stars shine.[b] *Brilliants of the first order enable you to identify easily the constellations of the Bull, the Lion, and the Virgin; Bootes, the Lyre, and the Charioteer; Orion, the Dogs, the Ship, and the Southern Fish. Twinkling with a lesser radiance, although still passing bright, are stars which mark the Twins, the Balance, the Scorpion, and the Archer; Helice, the Crown, the Swan, the Horse, and Perseus and Andromeda.* [710]The third magnitude has provided the dowry of the Pleiad sisters,[c] spreading over their maiden faces a blush of bronze. It includes as of like degree your brightness, Lesser Bear, and the fires which the Dolphin emits from its four[d] torches and the Triangle from its three; and, as shining with similar brightness, the Eagle[e] and the snakes with their slippery coils.[f] Then out of all the

[c] Actually none of the Pleiades is of the third magnitude (*cf.* Introduction p. ciii and figure 24).

[d] *Cf.* Aratus, *Phaen.* 317 f., who like Manilius omits ε.

[e] It would seem that Manilius forgot to take account of Altair (α Aql), a star of more than second magnitude according to Ptolemy, *Synt.* 7. 5, vol. 2, pp. 72 f.

[f] Draco and the snake of Ophiuchus, and perhaps Hydra too, though its brightest star, Alphard, is of the second magnitude.

tum quartum sextumque genus discernitur omni
e numero, summamque gradus qui iungit utramque.
maxima pars numero censu concluditur imo,
quae neque per cunctas noctes neque tempore in omni
resplendet vasto caeli summota profundo,
sed, cum clara suos avertit Delia cursus
cumque vagae stellae terris sua lumina condunt,
mersit et ardentis Orion aureus ignes
signaque transgressus permutat tempora Phoebus,
effulget tenebris et nocte accenditur atra.
tum conferta licet caeli fulgentia templa
cernere seminibus minimis totumque micare
stipatum stellis mundum nec cedere summa
floribus aut siccae curvum per litus harenae,
sed, quot eant semper nascentes aequore fluctus,
quot delapsa cadant foliorum milia silvis,
amplius hoc ignes numero volitare per orbem.
utque per ingentis populus discribitur urbes,
principiumque patres retinent et proximum equester
ordo locum, populumque equiti populoque subire
vulgus iners videas et iam sine nomine turbam,
sic etiam magno quaedam res publica mundo est
quam natura facit, quae caelo condidit urbem.
sunt stellae procerum similes, sunt proxima primis
sidera, suntque gradus atque omnia iusta priorum :
maximus est populus summo qui culmine fertur ;

716 quintumque . . . omnem
717 summaque
iungitur angue
718 per minimos
720 respondent alto
722 vaga est illae
724 mutat per
727 [minimis]
[728] = 1. 142
729 spatium

remaining stars one distinguishes a fourth class and a sixth, and the degree which lies between these two magnitudes. The most numerous portion is contained in the lowest order: sunk in the vast depths of heaven they shine neither every night nor in every season; but when the bright Moon diverts her course below the horizon, when the planets hide their light under the earth, when golden Orion has steeped his resplendent fires, and when, after its passage through the signs, the Sun renews the seasons of the year, then do these stars glitter in the darkness and their kindled flames pierce the blackness of night. Then may one see heaven's shining temples teeming with minute points of light and the whole firmament sparkle with dense array of stars: their abundance yields not to the flowers of the field or the grains of sand upon Ocean's winding shore; but, as many the waves which ride in endless procession on the sea, as many the myriad leaves which fall and flutter down in the woods, more numerous even than these are the fires which circle the heavens. And as in great cities the inhabitants are divided into classes, whereof the senate enjoys primacy and the equestrian order importance next to this, and one may see the knights followed by the commons, the commons by the idle proletariat, and finally the innominate throng, so too in the mighty heavens there exists a commonwealth wrought by nature, which has founded a city in the sky. There are luminaries of princely rank and stars which come close to this highest eminence; there are all the grades and privileges of superior orders. But outnumbering all these is the populace which revolves about

[738] respondere [741] victa

cui si pro numero vires natura dedisset,
ipse suas aether flammas sufferre nequiret,
totus et accenso mundus flagraret Olympo. 74

[a] Referring to the Milky Way.

heaven's dome [a]: had nature given it powers consonant with its legions, the very empyrean would be helpless before its fires, and the whole universe would become embroiled in the flames of a blazing sky.

APPENDIX

READINGS NOT GIVEN IN HOUSMAN'S EDITIO MAIOR

Sources will be found by reference to the Select Bibliography, pp. cxiii ff.; readings first published in Housman's editio minor are indicated by "*ed. min.*," those first in this volume by "*ed.*"

BOOK ONE

69 amisso . . . renato *Breiter*
69a *Gain* [1]
72 poterat *Scaliger*
175 ab occasu *Bentley*
227 hesperiis *Goold* [2]
285 solidus *Bentley*
352 crispans *ed. min.*
355 resupina *ed. min.*
430 nec di mortiferum *Goold* [2]
471 turba *Bentley*
517 ferre *Goold* [2]
suppl. before 566: *ed. min.*
726 suturam *Flores*
727 densa *Goold* [2]
795 fasces *Bentley*
813 etenim *Bentley*
829 Phoebusque *Goold* [2]
834 ruptis *Bailey*

BOOK TWO

105a *supplement*: *ed.*
312 [a] *ed.*
594ab *supplement*: *ed.*
621 quae . . . prorumpere *Bailey*
683-686 *transposed*: *Jacob*

BOOK TWO (cont.)

826 terris *ed. min.*
852 ortum *Bailey*

BOOK THREE

106 f. *punctuated*: *Bailey*
268-270 *transposed*: *ed.*
269 quorum *ed.*
419 quater et *Bailey*
527 inconstans *ed.*

BOOK FOUR

42 exitium *Breiter*
generis *ed. min.*
53 te iam *Bailey*
76 pausas *Garrod* [2]
268a *supplement*: *ed.*
424 et *Gain* [2]
467 vitiat *ed.*
473 astri *Gain* [1]
635a *supplement*: *ed.*
876a *supplement*: *ed.*

BOOK FIVE

87 pervolitans et *ed.*
154 caeca *Bentley*
209 vomente *Alton*
515 die *Thomas*
655 saepe *Alton*

INDEX

[*Index vocabulorum omnium*: Fayus; *Index nominum et artis vocabulorum*: Housman, ed.min.]

Entries are as far as possible clued to book and line number of the Latin text. References followed by the letter n direct the reader also to an appended footnote, those followed by another letter (but not f. = "and following") solely to the footnote thus indicated. Asterisks invite the reader to consult, besides the text, the appropriate section of the *Guide to the Poem* given in the Introduction.

ABBREVIATIONS, ancient, 4.210n
Absyrtus, 3.10n; 5.465n
Achaean fields = Phthiotis, 4.614n
Acheron, 1.93
Achilles (1), *Aeacides*, 1.762n; 2.3
Achilles (2), astronomical writer, on the tropic degree, lxxxi
Acron, 1.788k
Actaeon, 5.183
Actium, battle of, 1.914; 5.52
Adriatic sea (*Hadriacus*), 4.609
AEACUS, line of:
Achilles and Ajax, 1.762n
Achilles, 2.3
Aeetes, 3.9n; 5.465n
Aegean sea, 4.621, 634
Aegyptus: *see* Egypt
Aelius Secundus, Publius, 5.171a
Aemilian house, 1.796
Aeneas, favoured by fate, 4.24
Aeschylus, *Septem* 434: 3.16a
Aesculapius, xxvii
Aeson, 3.12c; 5.467d
Aethiopes: *see* Ethiopians
Aetna (1), author of, echoes Manilius, xiv
Aetna (2), mount, 1.854; 2.880
Africa, -ns, 4.728. *See also* Libya
Africus, compass-point: map
Agamemnon, son of Atreus, 1.762
Agrippa, 1.798
Ajax, line of Aeacus, 1.762n
Alcon, story of, 5.304-307n
ALEXANDER THE GREAT:
of Pella, 1.770; 4.688
achievements, 3.22
Pompey a second, 4.53
Alexandria, latitude denoted, 3.271ff.
alliteration, fourfold, 4.212
Alphecca (α CrB), cii; 1.320ff.
Alps, Hannibal crosses, 4.659
Altair (α Aql), passed over, ciii
Althaea, mother of Meleager, 5.177
amantia signa, fig. 9 (xlix); 2.466ff.*
amphibious signs, 2.230-233
Amphitrite, xxvii
ANDROMEDA (*Cepheis*):
and Perseus, xxii, xxviii
listing, xxviii; 1.356
stellar magnitudes, cii
location, 1.350, 436, 616; 5.657
myth, 2.28
episode, 5.540-618
rising influences, 5.619-630
Anguis (1) = Draco, *q.v.*, 1.306; 5.19
Anguis (2) = Hydra, *q.v.*, 1.415
Anguitenens = Ophiuchus, *q.v.*, 5.389
anklets, golden, 5.519

INDEX

antarctic circle, celestial, 1.589-593
Antares (α Sco), cii; 1.268c
Aparctias, compass-point, xc
Apeliotes, compass-point, xc
Apollo, the raven sacred to, xxix
Apollonius Rhodius, 3.9c
apples of the Hesperides, 5.16
Aqua Aquarii: *see* Flumina
AQUARIUS, the Waterman (*Iuvenis, Urna*):
  listing, xxv; 1.272
  stellar magnitudes, civ
  location, 1.441
  characteristics, 2.150-269*
  conjunctions, 2.270-432*
  under Juno, 2.446
  rules shanks, 2.464f.; 4.709
  relationships, 2.466-692*
  dodecatemories, 2.693-748*
  rising and setting times, 3.203-509*
  years given by, 3.577f.
  influences on native, 4.259-272
  decans, 4.354-357
  has power over water, 4.384f.
  injurious degrees, 2.232; 4.490-493
  special degrees, 4.571f.
  lands influenced by, 4.797-799
  paranatellonta, 5.449-537
  is Ganymede, 5.487
aquatic signs, 2.223-225
aqueducts, 4.264f.
AQUILA, the Eagle:
  listing, xxvii; 1.343
  stellar magnitudes, ciii; 5.715
  location, 1.626, 688
  rising influences, 5.486-503
Aquilo, compass-point, xc; 4.646d
ARA, the Altar:
  listing, xxix; 1.421, 431; 5.18
  stellar magnitude, cv
  rising influences, 5.339-347
Arabian gulf (Red sea), 4.654-657
Arabian scents, 5.264
ARABS:
  unmanly, 4.654, 754
  under Taurus, 4.754
Arachne, Minerva's victory over, 4.136
ARATUS:
  complimented, 1.402n
  Phaenomena alluded to, 2.25n
    referred to, 5.58d
    interpolation based on, 5.696-698b
Arcadians, ancient, under Virgo, 4.768
Archer: *see* Sagittarius
Arcitenens: *see* Sagittarius
ARCTI, the Bears:
  =Helice and Cynosura, *qq.vv.*, 5.19
  north, 1.237
  circumpolar signs, 1.275, 283, 314, 524; 3.344, 382
  navigational signs, 1.295-302
  antarctic, 1.443, 451
  location, 1.610, 684
arctic circle, celestial, 1.563A-567
Arctophylax: *see* Bootes
Arctos: *see* Helice
ARCTURUS:
  =α Boo, xxi; 1.318
  =Bootes, *q.v.*, 5.358
ARGO (*Ratis*):
  listing, xxix; 1.412; 5.13
  stellar magnitudes, cii
  location, 1.623, 694
  rising influences, 5.32-56
Ariadne, crown of=Corona, *q.v.*, 5.21
ARIES, the Ram (*Corniger, Laniger*):
  listing, xxiv; 1.263
  stellar magnitudes, civ
  location, 1.615, 674
  characteristics, 2.150-269*
  conjunctions, 2.270-432*
  under Minerva, 2.439
  rules head, 2.456; 4.704
  relationships, 2.466-692*
  dodecatemories, 2.693-748*
  leads zodiac, 2.945
  rising and setting times, 3.203-509*
  years given by, 3.567f.
  as tropic sign, 3.646, 672
  influences on native, 4.124-139
  decans, 4.312-315
  loves wool, 4.380
  injurious degrees, 4.444-448
  special degrees, 4.505-517
  lands influenced by, 4.744-752
  paranatellonta, 5.32-139
Aristotle, on comets, xxxvi; 1.817-866
Armenia, under Leo, 4.761
Arrow: *see* Sagitta
artificial lakes and rivers, 4.264

INDEX

ASIA :
Troy, 1.512
allies of Troy, 1.770
on the map, 4.622, 671, 680
under Taurus, xci ; 4.753
ASPECTS OF LIFE influenced :
bodily accidents, 2.908
brothers, 2.912
business relationships, 3.107
character, 2.831 ; 3.135
children, 2.672, 946 ; 3.132
civil life, 3.105
class, 2.834 ; 3.130
conduct of life, 2.831 ; 3.134
consummations, 2.838
dangers, 3.127
death, 2.952
decision, moment for, 3.154
distinction, 2.814
education, 2.834
faith, good, 2.955
family ties and traditions, 2.671 ; 3.135f.
fathers, fortunes of, 2.934
finances, 2.823 ; 3.123
foreign travel, 3.103
friendships, 3.108, 121
gains, 2.825
health, 2.901 ; 3.139f.
home, 3.98
honours, 2.815
hopes, 2.883
household, 3.99
leisure, 2.840
litigation, 3.111-116
marriage, 2.671, 839, 925 ; 3.120
oratory, 3.112f.
parental worries, 2.947 ; 3.132f.
popularity, 2.816 ; 3.130f.
property, 3.100
religion, 2.840
services, 3.108f.
sickness, 2.902 ; 3.141
social rank, 2.834 ; 3.129
success, 3.146-148
therapy, 3.142
warfare, 3.102
wealth, 2.823 ; 3.123
Astraea, confused with Erigone, xxv ; 4.542f.
ASTROLOGY :
origin, xviif. ; 1.25-66
bibliographical docket, xxxvii
founded by Mercury, 1.30
proved by true predictions, 2.130
ASTRONOMICA :
title modelled on *Georgica*, xif.
date, xii
the poem completed, xiv
manuscripts, cvi-cix
Book 5 the last book, cviiif. ; 5.1
summaries, 3, 81, 161, 221, 299
purple patches, 1.474-531, 758-804 ; 2.749-787 ; 5.540-618
ASTRONOMY :
bibliographical docket, xvi
constellations, xxi ; 1.255-531
early history of, xxif.
Atalanta, 5.179f.
ATHENS :
condemned by Socrates' death, 1.775
plague at, 1.884-891
commands eloquence, 4.687
defeat at Syracuse, 5.50
Menander's city, 5.475
ATHLA (*labores*, *sortes*) :
circle of twelve, 3.43-159*
technical term, 3.162c
Athos, cut through, 3.21n ; 5.49n
Atreus, 3.19c
Atridae = Agamemnon and Menelaus, 1.762
auctioneer's spear (*hasta*), 5.318n
audientia signa, fig. 8 (xlviii) ; 2.466ff.*
Augustus : *see* Caesar (2)
Aulis, island, 4.638
Auriga : *see* Heniochus
Aurora, mother of Memnon, 1.767
Auster, south : map ; 4.592
austri, the south, 1.612 ; 3.256
autumn signs, 2.265-269
ave !, word of greeting, 5.65n
axis of the world described, 1.275-293

BABYLON :
-ian astronomy, lxxxii
waters of = Euphrates, 4.580
under Pisces, 4.804
BACCHUS (*Iacchus*, *Liber*, *Lyaeus*) :
god, 1.417 ; 2.17n ; 4.688n ; 5.241n
vine, 2.20 ; 4.736 ; 5.238
wine, 2.658 ; 3.153 ; 5.143, 227, 322, 333, 679

INDEX

grape, 3.662 ; 4.204
Bactrians, under Pisces, 4.804
Balance : *see* Libra
Balearic islands, 4.640
ball-games, 5.165-167
barren signs, 2.238f.
Bear, Great : *see* Helice
Bellerophon, Heniochus as, 5.97a
Belt of Orion : *see* Jugulae
Belus, xviii ; 1.42a
benefics, xcviii
Bentley, on Manilius, xiif., xv, lxxxvii
Bestia, the Beast (Lupus), xxx
bestial signs, 2.155-157n
bird-catching, 5.370-373
birds, performing, 5.378-380
Bithynia, under Leo, 4.761
blindness, 4.533f.
blood, green, 5.212
Bobbio catalogue lists a copy of Manilius, xiv, cviii
Boethius, confused with Manilius, cviii
Bonincontrius, verses inserted by, 1.38f. ; 2.631
BOOTES (*Arctophylax, Arcturus*):
listing, xxvi ; 1.316
stellar magnitudes, ci
interpolated mention, 1.565
his wagon, 5.20
rising influences, 5.357-363
Boreas, north, xc ; 4.591 ; 5.70, 105
bow : *see* Cretan, Scythian
Bowl : *see* Crater
boxing, 5.163f.
BRUTUS :
Lucius Junius, 1.785
Marcus Junius, 1.908n
Bull : *see* Taurus

Caecias, compass-point, 4.646n
Caecilius Metellus, Lucius, 4.68n
CAESAR (1), GAIUS JULIUS :
adoptive father of Augustus, 1.9, 913
descended from Venus, 1.798f.
mention lost : after 1.801
deified, 1.926 ; 4.934a
murder of, 2.595 ; 4.57-62
CAESAR (2), AUGUSTUS :
emperor when Books 1 and 2 written, xii ; 1.385, 800, 925 ; 2.508f.
apostrophized, *pater patriae*, 1.7
succeeds Julius Caesar, 1.913
birth, 4.552a
deified, 4.934a
gifts of, 5.509n
CAESAR (3), TIBERIUS :
emperor when Book 4 written, xii ; 4.766, 776
domiciled in Rhodes, 4.764n
Callisto : *see* Helice
Calydon, Atalanta at, 5.180
Camillus, Marcus Furius, 1.784n ; 4.86
Campus Martius, exercise in, 3.630
CANCER, the Crab :
listing, xxiv ; 1.266
stellar magnitude, cv
location, 1.568, 622, 672
myth, 2.33
characteristics, 2.150-269*
conjunctions, 2.270-432*
under Mercury, 2.440
rules breast, 2.459f. ; 4.705
relationships, 2.466-692*
dodecatemories, 2.693-748*
rising and setting times, 3.203-509*
years given by, 3.570
as tropic sign, 3.625ff. ; 4.745, 752
influences on native, 4.162-175
decans, 4.323-329
loves trade, 4.381
injurious degrees, 4.459-463
special degrees (Praesaepe), 4.530-534
lands influenced by, 4.758f.
paranatellonta, 5.174-205
CANICULA, the dog (*Canis* Major) :
listing, xxix ; 1.396 ; 5.17
stellar magnitudes, cii
rising and setting, 1.398f.
location, 1.623
rising influences, 5.206-233
Cannae, battle of, 4.37, 566, 660
Canopus (α Car), invisible north of Rhodes, xix (fig. 1) ; 1.216
Capaneus, death of, 3.16a
CAPELLA (α Aur) :
listing, xxviii ; 1.366
stellar magnitude, ci
myth, 2.30
rising influences, 5.128-139
the Olenian (*Olenie*), 5.130n
Caper : *see* Capricornus

INDEX

CAPITOL :
saved by Camillus, 1.784
site of temple, 4.28n
Cappadocians, under Leo, 4.761
CAPRICORNUS, the Sea-goat (*Caper*) :
listing, xxv ; 1.271
stellar magnitudes, civ
location, 1.375, 626, 672
characteristics, 2.150-269*
conjunctions, 2.270-432*
under Vesta, 2.445
rules knees, 2.463f. ; 4.708
relationships, 2.466-692*
Moon in, at Augustus's birth, xii ; 2.509
dodecatemories, 2.693-748*
rising and setting times, 3.203-509*
years given by, 3.576f.
as tropic sign, 3.637, 674 ; 4.745
influences on native, 4.242-258
decans, 4.350-353
has power over fire, 4.384
injurious degrees, 4.486-489
special degrees, 4.568-570
lands influenced by, 4.791-796
paranatellonta, 5.389-448
cardinal points, fig. 12 (lvii) ; 2.788-840*
Caria, under Virgo, 4.768
Carians, neighbours of, 4.799
Carina : *see* Argo
CARTHAGE :
vanquished, 1.792 ; 4.40
Marius in ruins of, 4.48
location, 4.599
once ruled Africa, 4.658
under Scorpio, 4.778
Carthaginians, navigation of, 1.301
Caspian sea, believed to be connected with Ocean, 4.649n
CASSIEPIA, Cassiopeia (*Cassiope*) :
myth, xxii ; 2.28 ; 5.23, 540ff.
listing, xxviii ; 1.354
stellar magnitudes, civ
location, 1.616, 686, 697
upside down, 1.686n
spelling of name, 5.504a
rising influences, 5.504-537
Cassius Longinus, Gaius, 1.908n
Catacthonion, a Greek, 2.46n
Cato the Censor, 5.106n, 453
Cato the Younger, 1.797 ; 4.87
Caudine Forks, treachery of, 1.786n
CENTAURUS (1), the Centaur :
listing, xxix ; 1.418
stellar magnitudes, civ
location, 1.613, 693
rising influences, 5.348-356
is Chiron, 5.353e. *See also* Bestia
Centaurus (2) : *see* Sagittarius
Cepheis = Andromeda, *q.v.*, 1.436
CEPHEUS :
myth, xxii ; 2.29 ; 5.23, 540ff.
listing, xxvii ; 1.354
stellar magnitudes, civ
rising influences, 5.449-485
a character in drama, 5.469
CERES :
wheat, 2.21, 658 ; 3. 629, 664
goddess, rules Virgo, 2.442
crops, 3.152 ; 4.734
bread, 4.251 ; 5.280
CETUS, the Whale (*Pristis*) :
listing, xxx ; 1.433
confused with Bestia, xxxf.
stellar magnitudes, civ
amphibious, 5.15
rising influences, 5.656-692
chariot-racing, tactics in, 5.76-84
Charybdis and Scylla, 4.605
Chelae : *see* Libra
Chiron, xxix ; 5.353e
Choerilus of Iasos, alluded to, 3.22e
Choerilus of Samos, *Persica*, 3.19d
chronocrators, lxxviiif. ; 3.510-559
Chrysippus, on the creation of the universe, xviii ; 1.149-170
CICERO :
read by Manilius, xiii
on the Milky Way, xxxv ; 1.758-761
in heaven, 1.794
Fifth Verrine imitated, 5.621a
CIL 2.4426, Manilius quoted, 4.16
CILICIA :
peoples of, 4.624
under Aquarius, 4.799
Cimbrian slave and Marius, 4.45n
Cincinnatus, Lucius Quinctius, 4.149a
cinnamon, 4.738
circles of the sky, fig. 2 (xxxiii) ; 1.561-630*

INDEX

cisterns, 4.265
city-building, 2.772-783 ; 4.555f.
civil strife, 2.594-600
Claudius, Appius, founder of gens, 1.795
Claws : *see* Libra
Cleopatra, threat to Rome, 1.917
Cloelia, 1.780 ; 4.33n
Cnossian land=Crete, *q.v.*, 4.783
Cnossian memorial of Ariadne, 1.323
Cocles, 1.781 ; 4.32
Colchian, Medea, *q.v.*, 3.9 ; 5.34
Colchis, land, 4.517
colures, 1.609-630
Coma Berenices, xxx
Comedy, New, 5.470-473
COMETS :
  explanations, 1.817-879
  types, 1.835-851
  meteors confused with, 1.827-834 ; 1.847-851
commonwealth of heaven, 5.734-745
compass, Manilius's, 4.587-594
CONSTELLATIONS :
  early schematization, xxif.
  pictorial representation, xxii
  present-day listing, xxiii
  Manilius's listed, xxiv-xxxi
  stellar magnitudes of, c-cv
  zodiacal, 1.256-274
  northern, 1.275-370
  southern, 1.373-455
contrary signs : *see* diametric signs
cookery, sauces used in : *see* *garum, hallec*
Corinth, renowned, 4.611
Corniger=Aries, *q.v.*, 5.39
Corona Australis, xxx
CORONA (Borealis), the Crown :
  listing, xxvif. ; 1.319
  stellar magnitude, cii
  Ariadne's, 5.21
  rising influences, 5.251-269
Corsica, 4.638
Corus, compass-point : map
Corvinus, Marcus Valerius, 1.782n
CORVUS, the Raven :
  listing, xxix ; 1.417
  stellar magnitudes, civ
  sacred to Apollo, 1.417n, 783n
Cossus, wins *spolia opima*, 1.788
Crab : *see* Cancer
CRATER, the Bowl :
  listing, xxix ; 1.418
  stellar magnitude, cv
  rising influences, 5.234-250
Creon, 3.12c; 5.465d
Cretan bow of Sagittarius, 2.241
CRETE :
  on the map, 4.634
  under Sagittarius, 4.783-786
Croesus, on the pyre, 4.64n
Ctesibius, his water-driven model of the heavens, 4.268c
Cupido, the boy-god Love, 4.151
Curiatii, 4.34n
Curius Dentatus, Manius, 1.787n ; 4.148n
Cybele, 2.441n ; 4.760n
Cyclades, of unequal size, 4.637
CYCNUS, the Swan (*Olor*) :
  listing, xxvii ; 1.337
  stellar magnitudes, cii
  location, 1.687
  Jupiter's disguise, 2.31 ; 5.25, 381
  rising influences, 5.364-388
Cyllenian : *see* Mercury (1), (2)
Cynegetica, a Greek, 2.43n
CYNOSURA, Little Bear (Ursa Minor) :
  in Ptolemy's catalogue, xxiif.
  listing, xxvi ; 1.299
  stellar magnitude, ciii ; 5.712
  used in navigation, 1.301f.
  location, 1.628
  nurse of Jupiter, 2.30
  interpolated mention, 5.696
Cyprus, washed by Nile, 4.635
Cyrene, under Scorpio, 4.780
Cytherean=Venus, *q.v.*, 2.33, 439, 922 ; 4.579

Daemonie, Temple 5 : 2.897
Daemonium, Temple 4 : 2.938
Damon and Phintias, [2.586-588]n
Danai, Greek chiefs, 1.765
daylight at Rome, variation of, table 3 (lxxvi) ; 3.443-482*
Dea, Temple 3 : 2.916
decans, fig. 23 (lxxxvi) ; 4.294-407*
Decius Mus, Publius, father, son, and grandson, 1.789n ; 4.86
deformed signs, 2.256-264
DEGREES :
  Eudoxean, 1.563Ab
  injurious, 4.408-501*
  significant zodiacal, 4.502-584*

INDEX

Delia : *see* Moon
Delos, island, 4.637
DELPHINUS, the Dolphin :
  listing, xxvii ; 1.346 ; 5.25
  stellar magnitude, ciii ; 5.713
  rising influences, 5.416-448
DELTOTON, the Triangle (*Trigonum*) :
  listing, xxvii ; 1.353
  stellar magnitude, ciii ; 5.714
  location, 1.615
DEMOCRITUS :
  the atomic theory of, 1.128-131n
  on the Milky Way, 1.755-757*
depravity, turns into virtue, 5.498
depressions, planetary, xcix
Deucalion, 4.833
Deus, Temple 9 : 2.909
diametric signs, fig. 6 (xlv) ; 2.395-432*, 652
DIANA :
  rules Sagittarius, 2.444
  Trivia, 4.788
dining-rooms, splendour of, 5.291f.
Diodorus of Alexandria, on the Milky Way, 1.721-728
Diogenes Apolloniates, on comets, xxxvi ; 1.867-873
Diomedes (*Tydides*), 1.763
DIS :
  portal of, Temple 7 : 2.951
  made to weep by Orpheus, 5.328
diurnal signs, 2.203-222*
divination, primitive methods of, 1.91f. ; 4.911-914
DODECATEMORIES :
  zodiacal, fig. 10 (lii) ; 2.693-737*
  planetary, fig. 11 (liv) ; 2.738-748*
dodecatemory of the Moon, 2.[732-734]*
dodecatropos, fig. 13 (lix) ; 2.856-967*
Dogstar : *see* Sirius
Don, river : *see* Tanais
Dorian fields, under Virgo, 4.767
Dorotheus of Sidon, Manilius's agreements with in allotting lands to zodiac, xcif.
double signs, 2.159-177
DRACO, the Dragon (*Anguis, Serpens*) :
  listing, xxvi ; 1.306
  rising influences (lost after 5.709), xcvif.
  stellar magnitude, civ ; 5.715
  an antarctic, 1.452
  interpolated mention, 1.565
  location, 1.627
  three in the sky, 5.715n

Eagle : *see* Aquila
early man, account of, 1.67-112
EARTH (*Tellus, Terra*) :
  sphericity of, xviii-xx
  theory of four land-surfaces, xx
  suspended in space, 1.173f.
  mother-goddess, 1.421 ; 2.875
  mainland divided into three continents, 4.657Aff.
  once flooded, 4.831-833
  final destruction by fire, 5.210n
earthquakes, 4.828f.
Ebusus (Ibiza), 4.640
ECLIPSES, LUNAR :
  as proof of Earth's sphericity, xixf.
  din raised to avert, 1.227n
ECLIPTIC :
  not at right-angles to the horizon at equator, 3.306d
  definition of : *see note to star-charts*
ecliptic signs, 4.818-865*
EGYPT :
  on the map, 4.626, 635, 670, 798
  complexion of people, 4.726
  under Aries, 4.752
  its flank (Hammonia), 4.779
Egypt's river : *see* Nile
elements, the four, 3.52 ; 4.888f.
ELEPHANTS :
  only in Africa and India, 4.740
  riding on, 5.705
Empedocles, on the origin of the universe, xviii ; 1. 137-144
ENGONASIN, the Kneeler (Hercules) : listing, xxvi ; 1.315
  stellar magnitudes, civ
  rising influences, 5.645-655
Ennius, *Annals* alluded to, 3.25f
eoi, the east, 1.637
Epicureanism, opposed by Manilius, xvii
Epicurus, the atomic theory of, 1.128-131n, 486n
Epirus, 4.611, 690
equator, celestial, 1.575-581

INDEX

equator, observer at, 3.301-322*
equatorial sea : map ; 1.246
equinoctial colure, 1.609-617
equinoxes : *see* precession, tropic degree
Equuleus, xxx
EQUUS, the Horse (Pegasus) :
listing, xxvii ; 1.348
stellar magnitudes, cii
flies in the sky, 5.24, 633
rising influences, 5.631-644
Erechthean, of Erechtheus, legendary king of Athens, 1.884
Erichthonius, Heniochus as, 1.362 ; 5.72n
Eridanus (1)=river Po, 4.610
Eridanus (2), constellation : *see* Flumina
ERIGONE (=Virgo, *q.v.*) :
daughter of Icarius, 2.31f.
confused with Astraea, 4.542f.
Eteocles and Polynices, 3.17n ; 4.83f ; 5.463n
ETHIOPIANS :
black-skinned, 4.723
under Cancer, 4.758
Asiatic Ethiopians, under Pisces, 4.804n
Euboea, mountains of, 4.633
EUDOXUS :
degree-system of, xxxii ; 1.563Ab
tropic degree of, lxxxii
Eumaeus, the faithful swineherd of Ulysses, 5.126n
EUPHRATES, river :
interpolated mention, 1.44
river of Babylon, 4.580
under Pisces, 4.800
Euripides, *Andromeda*, xxii ; 5.469e
Europa and the bull, 2.489-491 ; 4.681-685
Europe, continent, 4.681
Eurus, east, xc ; 4.591
EUXINE SEA :
on the map, 4.617, 649, 678
under Gemini, 4.755
exaltations, planetary, xcix
extra-zodiacal constellations, influences of, 5.32-709*

Fabius Maximus, Quintus, 1.790 ; 4.39
Fabricius Luscinus, Gaius, 1.787

FARNESE GLOBE :
description, xxiii
shows Virgo's wings, xxv ; 2.176b
biped Sagittarius, xxv
Cepheus as actor, xxvii; 5.458a
Equus winged, 5.633b
omits Milky Way, xxxiv
Felix Fortuna, Temple 11: 2.887f.
feminine signs, 2.150-154
fertile signs, 2.236f.
FIDES, the Lute :
a ghost constellation, xxx, xcvi
rising influences, 5.409-415
figura, horoscope, 2.856 ; 3.169
FIRMICUS MATERNUS :
Manilius 5 used by, xiv, xcvi
5.709ff. reconstructed from, xcvii
Fishes : *see* Pisces
fishing, 4.285-289 ; 5.189-194, 297, 658-663
Flood, the legendary, 4.831-833
FLUMINA, the Rivers :
peculiar to Manilius, xxx
listing, xxx, 1.440
stellar magnitudes, civ
rising influences promised, 5.14
forest-fires, 1.856f.
formido, scare-feathers, 5.185n
Fortuna, Temple 10 : 2.927
Fortuna Felix, Temple 11 : 2.887f.
FORTUNE :
goddess, 2.134, 591 ; 4.91, 96, 564 ; 5.359
Lot of, lxiv ; 3.96
diurnal procedure for finding, 3.160-193*
nocturnal procedure, 3.194-202*
fountains, 4.262
friendship, the rarest thing in the world, 2.582

Galaxy : *see* Milky Way
Gallia : *see* Gaul
Ganges river, under Gemini, 4.757
Ganymede, Aquarius as, 5.487
garum, fish-sauce, 5.672a, 681c
GAUL :
on the map, 4.603
unrivalled for her wealth, 4.693
complexion of people, 4.716
under Capricorn, 4.793

INDEX

Gellius, Aulus, on Marsi, xcvii
GEMINI, the Twins:
listing, xxiv; 1.265
stellar magnitudes, cii
location, 1.387, 622, 695
characteristics, 2.150-269*
conjunctions, 2.270-432*
under Apollo, 2.440
rule arms, 2.458f.; 4.705
relationships, 2.466-692*
dodecatemories, 2.693-748*
rising and setting times, 3.203-509*
years given by, 3.569
influences on native, 4.152-161
decans, 4.320-322
love the Muses, 4.381
injurious degrees, 4.454-458
special degrees, 4.525-529
lands influenced by, 4.755-757
paranatellonta, 5.157-173
geometry of the circle, 1.539-560
geography, zodiacal: map; 4.744-817*
Gerbert (Pope Sylvester II), refers to a manuscript of Manilius, cviii
GERMANICUS, *Aratus*:
echoes Manilius, xiv
interpolation based on, 4.882n
GERMANY:
Varian disaster, 1.899
land of swamps, 3.633
stature of her sons, 4.692, 715
under Capricorn, 4.794
uncertain coast of, 4.795f.
Giants, war of, 1.421-432; 2.876-880; 3.5n; 5.341
Glauce, 3.12c; 5.465d
Glory (*Gloria*), occupies MC, 2.812
gluttony, 4.404f.
God, dwells in man, 2.108; 4.886
God (*Deus*), Temple 9: 2.909
Goddess (*Dea*), Temple 3: 2.916
Goethe, on Manilius, xv
gold-prospecting, 5.523-530
Gorgon, Medusa, 1.359; 5.22, 571, 576, 594, 618
Gracchi, rabble-rousers, 5.122
Gradivus=Mars, *q.v.*, 4.719
Great Bear: *see* Helice
GREECE:
vanquished in turn, 1.512
on the map, 4.633
complexion of people, 4.720
GREEK (language):
derivations from, 1.292n; 5.350n
words, Manilius on, 2.694, 829, 888, 897, 909, 917, 937; 3.162; 4.298, 818; 5.645
necessity of using, 3.40f.
the best period of, 5.475
GREEKS:
navigators, 1.298
at Troy, 1.501, 765

Hadriacus pontus = Adriatic 4.609
HAEDI, the Kids:
older than Auriga, xxviii
listing, xxviii; 1.365
stellar magnitude, ci
rising influences, 5.102-117
watched by Capella, 5.130
HAEDUS, the Kid:
a ghost constellation, xxx, xcivf.
rising influences, 5.311-323
haemation, the best *garum*, 5.672a
hair-styles, 5.147-149
hallec, a condiment, 5.675b
Hammonia, under Scorpio, 4.779n
Hannibal, 4.41, 567n, 659
Happy Fortune, Temple 11: 2.887f.
Hare: *see* Lepus
hare, sleeps with its eyes open, 5.172b
haruspicy, 1.92
Hasdrubal, defeat of, 1.791
hasta=auctioneer's spear, 5.318n
HECTOR:
mention lost: after 1.765
killed by Achilles, 2.3
body ransomed, 3.8
reaches Greek ships, 5.300
Heliac shores=Rhodes, 1.217n
HELICE, Great Bear (*Arctos*, *Ursa Major*):
listing, xxvi; 1.296-298
rising influences, xcvif.; 5.693ff.
stellar magnitudes, cii
invisible from antarctic, 1.218
Greeks navigate by, 1.298
confronts Orion, 1.502
above arctic circle, 1.566
location, 1.634
Callisto, daughter of Lycaon, 2.29; 3.359
the north, 4.589, 792

INDEX

Helicon, 1.4 ; 2.50
HELLE :
 sister of Phrixus, xxiv ; 4.516n
 drowned, 4.615n, 747n
 gives name to Hellespont, 5.33n
HELLESPONT :
 bridged by Xerxes, 3.21n ; 5.49c
 on the map, 4.615n, 620
 interpolated mention, 4.679
 named after Helle, 5.33n
HENIOCHUS, the Charioteer (Auriga) :
 listing, xxviii ; 1.362
 stellar magnitudes, ci
 is Erichthonius, 1.362 ; 5.72n
 location, 1.696
 drives chariot in sky, 5.20
 rising influences, 5.67-101
 as Bellerophon, 5.97n. *See also* Haedi, Capella
Heraclitus, on the origin of the universe, xviii ; 1.132-134
Hercules (1), a Theban god, 4.688e
Hercules (2), constellation : *see* Engonasin
Herodotus, on the Caspian sea, 4.649a
HESIOD :
 on the origin of the universe, xviii ; 1.125-127
 his works, 2.12n
 alluded to, 3.5a
Hesperia = Italy, 4.773
Hesperides, apples of the, 5.16
hesperii, the west, 1.[226], 638, 645
Hesperos = Venus, *q.v.*, 1.178
hexagonal signs, fig. 5 (xliv) ; 2.358-384*
hexameter poetry, genres of, xxxvii ; 2.1-48 ; 3.1-26
HIPPARCHUS :
 his star-catalogue, xxii, c
 his discovery of precession of the equinoxes, lxxxii
 moves zodiacal boundaries, lxxxiii
Hispania : *see* Spain
HOMER :
 stars known to, xxii ; xxix
 greatest of bards, works of, 2.1n
 birthplace, 2.7n
 Iliad alluded to, 3.7n
Horatii, triplets, 1.778n
Horatius Cocles, 1.781 ; 4.32
HORATIUS (Publius ?) :
 kills Curiatii, 4.34
 kills sister, 5.107. *See also* Horatii
horizon, definition, 1.648-662
HOROSCOPOS :
 the rising cardinal point, 2.826-830 ; 3.190, 200, 518, 538, 608
 how to determine, 3.203-509*
 -are = to occupy the Hor., 3.296
horsemanship, 5.73-75, 636f.
hour, standard, 3.247-274*
Houses, planetary, xcix
human signs, 2.155-157n
hunting, 5.185-188
HYADES, the Rainers :
 listing, mistranslated as *Suculae* " Piglets," xxiv
 stellar magnitudes, ci
 interpolated mention, 1.371
 rising influences, 5.118-127
HYDRA, the Water-snake (*Anguis* 2) :
 listing, xxix ; 1.415
 stellar magnitude, civ ; 5.715
 location, 1.612
 rising influences promised, 5.16
Hyginus, Manilius errs with, 5.488c

Iacchus = Bacchus, *q.v.*, 1.417
Ibiza (Ebusus), 4.640
Icarian sea, 4.621
Icarius, father of Erigone, xxiv
Idaean mother = Cybele, *q.v.*, 4.760
Iliac = Trojan, 2.1
Iliad : *see* Homer
Illyrian littoral, 4.691
Illyrians, ready to make war, 4.611
inclusive reckoning, 2.654a, 661
INDIA :
 too vast to be known, 4.674
 complexion of people, 4.724
 under Gemini, 4.757
Indian Ocean, under Pisces, 4.806n
influences : *see* aspects of life, native
initiatives, technical term, lxv
injurious degrees, 4.408-501*
insidiantia signa, fig. 9 (xlix) ; 2.466ff.*

intermediate fertility, signs of, 2.240-243
INTERPOLATIONS in the text:
1.38f., 44, 171f., 226, 235, 350f., 357, 371f., 394, 435 (based on Virgil), 594-596, 564-565A, 663-665, 681-683, 707, 766;
2.111-114, 120, 173f., 284-286, 390, 409, 518, 529, 586-588, 631, 732-734, 936, 968-970;
3. 317, 477, 508;
4. 119-121, 679, 882 (based on Germanicus);
5. 696-698, 728
intervalla: *see* quadrants
inverted signs, 2.197-202
Iolcos, 5.34
Ionian cities, under Virgo, 4.767
Ionian sea, 4.607
Iphigenia, alluded to, 5.47a
Iris, rainbow, 1.713
Isis, sistrum of, 1.918
Isocrates, *Panegyric*, 3.22e
ITALY:
on the map, 4.604, 609, 632
mistress of world, 4.694
under Libra, 4.769
Hesperia, 4.773
sea of = Tyrrhenian, 4.781
Magna Graecia, 4.789
Ithacan = Ulysses, 1.764

Jason, his feats, 3.10f.
Jebb, on Manilius, xxi
Jocasta, 3.17n
jugglers, 5.171n
JUGULAE, the Belt:
listing (under Orion), xxviii
so called to hide error, xcvi
stellar magnitude, ci
rising influences, 5.174-196
Julius Caesar: *see* Caesar (1)
JUNO:
origin of the Milky Way, 1.752
rules Aquarius, 2.446
JUPITER (1), god:
Thunderer, 1.104, 800
and Leda, 1.337-340
eagle the bird of, 1.343
catasterizes Heniochus, 1.363
Capella, 1.370
Ara, 1.431
conquers Giants, 1.423; 5.343
father of Sarpedon, 1.767
Capitoline, 1.784; 4.29n; 5.289
infancy of, 2.15; 5.132-134
and Juno, 2.16
and Cynosura, 2.30
rules Leo, 2.441
and Europa, 2.491; 4.681-685
a citizen of Crete, 4.634
ruler of heaven, 4.174, 908
blasts Salmoneus, 5.91-96
JUPITER (2), planet:
abode in Temple 11: 2.890
characteristics, xcviii
listed, 1.807; 5.6
Juvenal, echoes Manilius, xiv
Juvenis: *see* Aquarius

kings of Rome, 1.778

laboris porta, Temples 6, 12: 2.870
LACUNAS in the text after:
1.69, 316, 440, 560, 563, 765, 801;
2.4, 105, 594, 619, 683;
3.216, 417, 497, 549;
4.268, 433, 501, 635, 657, 744, 876, 899;
5.263, 278, 423, 709 (xcvif., cvif.)
Laertiades = Ulysses, *q.v.*, 5.126
land-surfaces of the Earth, xx
Laniger: *see* Aries
Latin language, inadequacy of, 2.897; 3.40-42
LATIUM:
battles in, 4.43
cities of, 4.661
Leander, swims straits, 4.80b
Leda, ravished by Jupiter, 1.340
LEO, the Lion (*Nemeaeus*, *-eeius*):
listing, xxiv; 1.266
stellar magnitudes, ci
myth, 2.32
characteristics, 2.150-269*
conjunctions, 2.270-432*
under Jupiter, 2.441
rules sides, 2.460; 4.706
relationships, 2.466-692*
interpolated mention, 2.529; 5.697
dodecatemories, 2.693-748*
rising and setting times, 3.203-509*
years given by, 3.571
influences on native, 4.176-188
decans, 4.330-333
a hunter, 4.382

injurious degrees, 4.464-468
special degrees, 4.535-541
lands influenced by, 4.759-762
paranatellonta, 5.206-250
LEPUS, the Hare :
listing, xxix ; 1.412
stellar magnitudes, civ
rising influences, 5.157-173
chased by Canicula, 5.233
Leucippus, on the origin of the universe, xviii ; 1.128-131
Liber = Bacchus, *q.v.*, 3.662
Libonotus, compass-point : map
LIBRA, the Balance (*Chelae*) :
listing, xxv ; 1.267
stellar magnitude, cii
interpolated mention, 1.565A ; 2.529
location, 1.611
as tropic sign, 1.674 ; 3.649, 659, 672
characteristics, 2.150-269*
conjunctions, 2.270-432*
under Vulcan, 2.442f.
rules loins, 2.462 ; 4.407
relationships, 2.466-692*
figured as a human, 2.524*
dodecatemories, 2.692-748*
rising and setting times, 3.203-509*
years given by, 5.573f.
influences on native, 4.203-216
decans, 4.338-343
has power over measures, 4.383
injurious degrees, 4.473-476
special degrees, 4.547-552
lands influenced by, 4.769-777
Tiberius's natal sign, 4.776f.
paranatellonta, 5.293-338
LIBYA :
on the map, 4.598
= Carthage, 4.661
monstrous nature of, 4.779
Libyan ruins (= of Carthage), 4.47
Libyan sea, 4.631
LIFESPAN allotted :
by zodiac, fig. 21 (lxxix) ; 3.560-580*
by dodecatropos, fig. 22 (lxxx) ; 3.581-617*
Lion : *see* Leo
liquamen, the common *garum*, 5.681c
LIVIUS (1), TITUS (Livy) :
read by Manilius, xiii
allusions to events in, 1.777ff. ; 4.26ff. ; 5.106f.
Livius (2) Salinator, Marcus, 1.791
[LONGINUS], *De Sublimitate* :
read by Manilius, xiii
referred to, 3.22c
lots : *see* athla, Fortune
Lucan, echoes Manilius, xiv
Lucifer = Venus, *q.v.*, 1.177
LUCRETIUS :
Manilius combats, xvii
Manilius influenced by, 1.884ff.
Luna : *see* Moon
lunar eclipses : *see* eclipses, lunar
Lupus, constellation : *see* Bestia
luxury, protests against, 4.402-405 ; 5.195f., 290-292, 374-377, 404f.
Lyaeus = Bacchus, *q.v.*, 5.322
Lycaon's daughter : *see* Helice
Lychnus, barbaric constellation, xcvii
LYCIA :
Sarpedon, king of, 1.768
under Aquarius, 4.799n
Lycurgus, 1.773
lying signs, 2.253-255
LYRA, the Lyre :
listing, xxvii ; 1.324
stellar magnitudes, ci
inverted, 1.627
rising influences, 5.324-338.
See also Fides

Macedonia, under Leo, 4.762
Macrobius, on the Earth's land-surfaces, xx
Maeotis = Sea of Azov, 4.617, 678
Magna Graecia, under Sagittarius, 4.789n
magnitudes, stellar, 5.710-745*
Magnus (1) : *see* Alexander the Great
Magnus (2) : *see* Pompeius, Gnaeus
Maia, mother of Mercury, 2.943 ; 5.7
malefics, xcviii
MAN :
account of early, 1.67-112
cognate with heaven, 2.105-128 ; 4.886-897
Manetho, *Apotelesmatica*, c
Manger, star-cluster : *see* Praesaepe

INDEX

MANILIUS, MARCUS:
name, xi
datable references in, xii
perhaps of Asiatic descent, xiii
knowledge of Latin literature, xiii
acquainted with [Longinus], xiii
compared with Ovid, xv, 299
vies with Lucretius, xvii
his sources, xviii, lxii-lxv, lxxxi, lxxxix
used star-globe, xxiii
tries to conceal his errors, xxxi, xcvi
curt treatment of planets, xxi, xcviii
versifier of numerals, xxxii, xlif., lii, lxx, lxxxi, lxxxvii
blunders in specifying decans of Pisces, lxxxv; 4.359f.
versifier of diagrams, lxxxix
astronomical errors in Book 5: xciii
confused with Boethius, cviii
his originality, 1.1-6; 2.53-59, 136-140; 3.1-4
MANUSCRIPTS of *Astronomica*:
the tradition, cvi-cix
an ancient edition attested by chapter-headings in, cvi
reconstruction of ancestor of, cvif. *See also* interpolations, lacunas, transpositions
Marcellus, Marcus Claudius, wins *spolia opima*, 1.788n
Marians, Pompey's triumph over, 1.793n
Marius, Gaius, 4.45n
marksmanship, 5.294-310
MARS (1), god (*Gradivus, Mavors*):
rules Scorpio, 2.443
=war, 3.632; 4.402
citizen of Thrace, 4.691n
Romans take after, 4.718
MARS (2), planet (*Mavors*):
abode in Temple 6: lix
characteristics, xcviiif.
listing, 1.807; 5.6
malignity of its flames, 4.500
Marsi, wizards, xcvii
masculine signs, 2.150-154
Mauretania=land of the dark, 4.729
Mavors=Mars, *q.v.*, 2.443; 4.500
Mavortia virgo, Penthesilea, 1.768
MEDEA:
story of, 3.9n; 5.465n
child-killer, 4.82n
poisons the world, 5.35
medicine impotent against fate 4.74-76
Mediterranean sea, outline described, 4.597-629
Medusa=Gorgon, *q.v.*, 5.22
Meleager, story of, 5.176n
Memnon, son of Aurora, 1.767n
Menander, the dramatic work of, 5.470-476
Menelaus, son of Atreus, 1.762
MERCURY (1), god (Cyllenian):
founder of astrology, xvii; 1.30
rules Cancer, 2.440
invents lyre, 5.325n
MERCURY (2), planet (Cyllenian):
characteristics, xcviiif.
listing, 1.808; 5.7
orbit of, 1.871a
abode in Temple 1: 2.943
called Stilbon, 2.944d
meridian, 1.633-647
Messene, 3.14n
Messina, straits of: *see* Scylla
metal-smelting, 5.533f.
Metelli, house of, 1.796
Metellus, Lucius Caecilius, 4.68n
Milanion, 5.179n
MILKY WAY:
locus, origin, xxxivf.; 1.684-761
density, 5.742n
milling, 5.281f.
mind, vast power of the, 2.117-125; 4.877-881, 923-935
mining, 5.523-526
Minos, 4.784
Minotaur, under Sagittarius, 4.784n
Mithridates, Pompey's triumph over, 1.793n; 4.51; 5.510n
monkeys, 4.668
Mons Sacer, 5.123
months, tardy, 2.202n
MOON (*Delia, Luna, Phoebe*):
eclipses as proof of the Earth's sphericity, xixf.; 1.221-234
position at moment of a nativity, lxxxv; 2.726ff.; 3.517,* 590*; 4.122-501*
characteristics, xcviiif.
in space, 1.176, 200
regularity of phases, 1.187, 527

spherical, 1.208
banishes lesser stars, 1.469
rides in chariot, 1.669
listing, 1.808 ; 3.62 ; 5.3, 7
influence on tides, animals, 2.90-104
sister of Phoebus, 2.96
dodecatemory of, 2.[732-734]*
abode in Temple 3 : 2.913
determines Lot of Fortune, 3.186ff.
mention lost : after 4.501
eclipsed, 4.841-847
absence brings out lesser stars, 5.721
Mucius Scaevola, Gaius, 1.779 ; 4.31
Muse, poetry, 2.42
MUSES :
sisters on Helicon, 2.49
Pierides, 2.767 ; 3.3
invoked, 3.3
loved by Gemini, 4.381

NATIVE, TYPES OF :
acrobats, 5.438-445
actors, 5.478-485
agitators, 5.119-124
animal-tamers, xcvii ; 4.234-237
arbitrators, 4.215f.
archers, 5.295-297
astronomers, 4.158-160
athletes, 5.160f.
bakers, 5.283f.
bear-tamers, 5.703f.
boxers, 5.163f.
browsers, 5.137-139
bustlers, 5.61-66
busybodies, 5.314-321
butchers, 4.185
cavalrymen, 5.636f.
charioteers, 5.71-84
city-destroyers, 4.557f.
city-founders, 4.555f.
clothiers, 4.132
comic poets, 5.470-473
composers, 4.154f.
custodians, 5.360-363
divers, 5.431-435
dog-fanciers, 5.200f.
drinkers, 5.244-246
effeminates, 4.519 ; 5.146-155
elephant-riders, 5.705f.
executioners, 5.625-630
exhibitionists, 5.154-156
explorers, 4.512
farmers, 4.140-151 ; 5.272-275
fishermen, 4.285-289 ; 5.658-663
fishmongers, 5.667-681
florists, 5.262f.
fowlers, 5.370-373
furnacemen, 4.250f.
gardeners, 5.256-261
generals, 4.561-565
gladiators, 4.225
gluttons, 4.539-541
goldsmiths, 4.249 ; 5.506-508
gossips, 4.574-576
horse-grooms, 5.642-644
horsemen, 4.231-233 ; 5.636-641
hotheads, 5.220-227
hunters, 4.222-224 ; 5.175-188, 228-230
inquisitors, 5.410-415
investors, 4.167-172
jewellers, 5.508
judges, 4.549
jugglers, 5.168-171
jurists, 4.545
lawyers, 4.209-214
libertines, 5.321-323
lion-tamers, 5.701
merchants, 4.167
military aides, 5.502f.
millers, 5.281f.
mimers, 5.482-485
miners, 4.246f. ; 5.523-526
money-lenders, 4.173f.
musicians, 4.152-155 ; 5.329-331
pearl-divers, 5.398-405
perfumers, 5.263-266
pilots, 4.279
plunderers, 4.508 ; 5.491-496
priests, 5.344
prison-guards, 5.621-624
prophets, 5.347
prospectors, 5.527-530
prowlers, 5.648f.
rowers, 4.284
rulers, 4.550
saints, 4.571f.
salters, 5.682-692
scholars, 4.191-196, 526
seafarers, 4.569f.
seamen, 4.274 ; 5.40-56
shepherds, piping, 5.116f.
shipbuilders, 4.275
silversmiths, 4.249 ; 5.533f.
singers, 5.332-336

INDEX

snake-charmers, xcvii
snake-handlers, 5.391-393
soldiers, 4.220-222
speakers, eloquent, 4.194
stenographers, 4.197-199
students, 4.526
stunt-performers, 5.650-655
swimmers, 5.422-430
swineherds, 5.125f.
tax-collectors, 5.248
temple-guardians, 4.546
torturers, 5.413
tragedians, 5.458-469
treasurers, 5.360-363
tremblers, 5.135
trick-riders (*desultores*), 5.85
tutors, 5.454-457
veterinary surgeons, 5.353-356
wantons, 5.110-114
water-diviners, 4.261
weavers, 4.131
wool-workers, 4.128f.
nativities, infinite variety of, 3.67-70 ; 4.373
Nature : *see* universe
naval warfare, 4.288
navigation halted at setting of Haedi, 1.365n
Nechepso and Petosiris, xviii, lxxxvi
Nemean : *see* Leo
Nepa : *see* Scorpius
NEPTUNE (1), god :
=sea, 2.224 ; 5.211
and Cassiepia, xxvii
and Delphinus, 1.347n
rules Pisces, 2.447
Neptune (2), planet, c
Nereids, 5.540n, 563
Nereus, sea-god, 5.196, 433
Nero, Gaius Claudius, 1.791
Nestor, the Pylian, 1.764n
Nicander, works of, 2.44n
NILE, river :
interpolated mention, 1.44
has seven mouths, 3.273f.
floods Egypt, 3.634 ; 4.726
Pompey killed on shore, 4.50
location, 4.601, 627, 635
under Aries, 4.751
Ninos=Nineveh, under Pisces 4.804
nocturnal signs, 2.203-222*
north pole, observer at, 3.356-384*
Notus, south wind, 1.439
numerals in verse, 4.431-442
Numidia, -ns, 4.598 ; 5.375
nycthemeron, 3.545a
Nymphs, 2.23 ; 5.433

objector, imaginary, 4.387-389, 869-872
observer's fixed circle, 2.788-967
OCEANUS :
(1) god, 4.640, 651, 830
(2) sea, 1.347, 411 ; 4.644
octatropos, lxif.
Oeagrus, father of Orpheus, 5.326
Oedipus, story of, 3.17n ; 5.464n
Oenopides of Chios, on the Milky Way, 1.729-734
Olenie, -ian goat : *see* Capella
Olor : *see* Cycnus
olympiad, five years, 3.596n
Olympus=heaven, *passim* (21, [1] occurrences)
OPHIUCHUS, the Snake-holder (*Anguitenens*) :
protrudes into the zodiac, xxii
listing, xxvii ; 1.331
stellar magnitudes, civ
rising influences, 5.389-393
opposite signs : *see* diametric
order of teaching, proper, 2.755-771
Orestes and Pylades, 2.583
ORION :
known to Homer, xxii
listing, xxviii ; 1.387-393 ; 5.12
stellar magnitudes, ci
mention lost : after 1.440
fronts the Great Bear, 1.502
rising influences (a mistake of the poet's), 5.57-66n
a bright constellation, 5.723. *See also* Jugulae
Ornithogonia, a Greek, 2.43n
Orpheus, 1.325 ; 5.326
orthography of the Latin text, cxi
OVID :
Manilius compared with, 299
influenced by, 5.504a

Pactolus, gold-foaming river, 5.530
Pacuvius, his *Dulorestes*, 2.583-585b
Palamedes, inventor of numbers, 4.206

INDEX

PALLAS (Minerva) :
=olive, 2.21
rules Aries, 2.439 ; 4.135
Pamphylia, under Aquarius, 4.799n
Pan, 2.39
Papirius Cursor, Lucius, 1.786n
Pappus of Alexandria, 4.267c
parallel circles, celestial, 1.563A-593*
paranatellonta, stars or constellations rising simultaneously with particular degrees of the ecliptic, 5.32-709*
Paris, ruins Troy, 4.80c
partes damnandae, 4.408-501*
PARTHIA :
another world, 4.674
under Pisces, 4.803
pater patriae, Augustus, 1.7
Pegasus : *see* Equus
Peleus, of line of Aeacus, 1.762a
Pella, famed for Alexander, 1.770 ; 4.689
Peloponnese, 4.612
Penthesilea (*Mavortia virgo*), 1.768
Pergama=Troy, *q.v.*, 1.501, 765 ; 2.5
PERSEUS :
and Andromeda, xxii, xxviii ; 2.28 ; 5.567ff.
listing, xxviii ; 1.358
stellar magnitudes, cii
interpolated mention, 1.350
location, 1.697
rising influences promised, 5.22
vanquisher of Medusa, 5.567, 571
PERSIA (*Persis*) :
Themistocles conqueror of, 1.776
under Aries, 4.750
Xerxes launches on sea, 5.49
Persian gulf, 4.651-653
Persian Wars, 3.19n ; 5.48n
Petosiris and Nechepso, xviii, lxxxvi
Phaethon, 1.736 ; 4.834
Phasis, river, 4.517 ; 5.45, 376
Philippi, battle of, fought on site of Pharsalus, 1.909n
Philoctetes, story of, 5.299n
Phintias and Damon, [2.586-588]n
Phoebe=Moon, *q.v.*, 2.913 ; 3.197 ; 4.847
PHOEBUS (1), god :
patron of poets, 1.19
raven sacred to, 1.417, 783
rules Gemini, 2.440 ; 4.756
Phoebus (2)=Sun, *q.v.*, *passim* (43 occurrences)
PHOENICIA :
on the map, 4.625n
under Aquarius, 4.798n
Phoenix, compass-point : map
Phorcys, son of Neptune=the ocean, 4.644 ; 5.585
Phrixean sign=Aries, 3.304
Phrixus, 4.516, 615n, 747n
Phrygia, under Leo, xci ; 4.759
Phthiotis (*Achaica arva*), 4.614
Pierides=Muses, *q.v.*, 2.767 ; 3.3
PISCES, the Fishes :
listing, xxvi ; 1.273
Taurus babies in, lxxxiv
stellar magnitudes, civ
myth, 2.33
characteristics, 2.150-269*
conjunctions, 2.270-432*
under Neptune, 2.447
rule feet, 2.465 ; 4.709
relationships, 2.466-692*
dodecatemories, 2.693-748*
rising and setting times, 3.203-509*
years given by, 3.579f.
influences on native, 4.273-291
decans, 4.358-362n
have power over seas, 4.385
injurious degrees, 4.494-497
special degrees, 4.573-584
lands influenced by, 4.800-806
paranatellonta, 5.538-692
PISCIS NOTIUS, the Southern Fish :
listing, xxx ; 1.438
stellar magnitude, cii
rising influences, 5.394-408
PLANETS :
listing, xxxi ; 1.532ff. ; 5.6f.
zodiacal orbits, xxxif.
order of, lxxviii
Manilius shirks dealing with, xcviii
promises to do so, 2.750, 965 ; 3.156-158, 587
seemingly did so, 5.30f., 709a
general astrological doctrine, xcviii-c
newly discovered, c
contrary movement of, 1.15n
Plato, the inspired, 1.774

INDEX

PLEIADES :
an early asterism, xxii
listing, xxiv, fig. 24 (ciii)
stellar magnitude, ciii ; 5.710
interpolated mention, 1.371
a tiny cluster, 4.522
rising influences, 5.140-156
Pliny, refers to a Manilius, xi
Pluto (1), god, 2.951 ; 5.328
Pluto (2), planet, c
Po, river (*Eridanus*), 4.610
Poeni, navigation of, 1.301
Polynices : *see* Eteocles
POMPEIUS (1) Magnus, Gnaeus :
his three triumphs, 1.793n
seas swept clean by, 1.921
apostrophized, 4.50-55
triumph over Mithridates, 5.513
Pompeius (2), Sextus, 1.921n
pontus aequatorius : map ; 1.246
Porphyry, on finding the Dodecatemory of the Moon, liii
porta laboris, Temples 6, 12 : 2.870
Posidonius, Manilius possibly dependent on, xviii, lxxxix
Praesaepe, the Manger, xxii ; 4.530n
precession of the equinoxes, lxxxii-lxxxiv
PRIAM :
father of fifty kings, 2.2
ransoms Hector's body, 3.8
his corpse on the shore, 4.64n
a character in drama, 5.485
PRINCEPS :
Augustus, 1.7 ; (in heaven) 4.935
Tiberius, 4.764
PROCYON (Canis Minor) :
listing, xxix ; 1.412
stellar magnitude, cii
rising influences, 5.197-205
prodigious births, 4.101f.
PROPONTIS :
on the map, 4.616
interpolated mention, 4.679
under Aries, 4.749
proscriptions, referred to, 2.594
prospecting for gold, 5.523-530
Protesilaus, alluded to, 5.48a
PTOLEMY, the astronomer :
his star-catalogue, xxiif., lxxxii, c
creates artificial zodiac, lxxxiii
Tetrabiblos referred to, xcviii

Punic prows, 5.51
Puppis : *see* Argo
Pylades and Orestes, 2.583
Pylian, the = Nestor, 1.764
Pythagoreans, on the Milky Way 1.735-749
Pyxis : *see* Argo

quadrants, of observer's circle, fig. 12 (lvii) ; 2.841-855*
quadrate signs, fig. 4 (xliii) ; 2.287-296*, 652-672, 683-686
Quirinus = Romulus, *q.v.*, 1.801
Quirites, citizens, 5.123

Ram : *see* Aries
Ratis = Argo, *q.v.*, 1.623
Raven : *see* Corvus
Red Ocean : *see* Indian Ocean
Remus, 4.26n
repeated verse, 5.6 = 1.807
reservoirs, 4.264
RHETORICAL EXEMPLA used more than once : *see* Alexander, Camillus, Cocles, Eteocles, Fabius, Giants, Hannibal, Horatius, Medea, Oedipus, Orpheus, Phaethon, Pompeius (1), Salamis, Xerxes
Rhianus, *Messeniaca*, 3.14d
RHODES :
Heliac shores, 1.217
observer at, fig. 17 (lxix) ; 3.247-274*
risings and settings for, table 1 (lxx) ; 3.275-300*
listed with other islands, 4.637
once home of Tiberius, consecrated to Sun, under Virgo, 4.764f.
rising of stars, 1.399b
risings and settings of zodiacal signs, tables 1, 2 (lxx, lxxv) ; 3.275-442*
Rivers : *see* Flumina
Romans, 1.777, 910 ; 3.23 ; 4.67, 718
ROME :
Troy reborn, 1.511f.
walls of, 1.781
saved by Camillus, 1.784
threatened by Cleopatra, 1.917
under Augustus, 1.925
annals of, 3.24
daylight at, 3.443-482*
humble origin, 4.27

Mucius returns to, 4.31
Cocles defends, 4.33
depended on Horatius, 4.36
fights itself, 4.43
taken by Marius, 4.48
capital of world, 4.694
Romulus its father, 4.718
founded under Libra, 4.773n
ruled by Tiberius, 4.776n
ROMULUS :
wins *spolia opima*, 1.788k
Quirinus, 1.801
and Remus, 4.26n
father of Rome, 4.718n
running signs, 2.245f.

Sacred Mount, 5.123
SAGITTA, the Arrow :
listing, xxvii ; 1.342 ; 5.24
stellar magnitudes, cv
rising influences, 5.293-310
SAGITTARIUS, the Archer (*Arcitenens, Centaurus, Sagittifer*) :
listing, xxv ; 1.270
stellar magnitudes, cii
location, 1.691
characteristics, 2.150-269*
conjunctions, 2.270-432*
under Diana, 2.444
rules thighs, 2.463 ; 4.708
relationships, 2.466-692*
dodecatemories, 2.693-748*
rising and setting times, 3.203-509*
years given by, 3.575
influences on native, 4.230-242
decans, 4.347-349
has power over beasts, 4.384
injurious degrees, 4.481-485
special degrees, 4.560-567
lands influenced by, 4.783-790
paranatellonta, 5.357-388
Sagittifer = Sagittarius, *q.v.*, 2.267, 500, 513
Salamis, 5.50
Salmoneus, story of, 5.91-96
salt industry, 5.682-690
Saltus Teutoburgiensis, disaster of, xii ; 1.898-903
SARDINIA :
has shape of footprint, 4.631
close to Corsica, 4.639
under Scorpio, 4.782
Sarpedon, 1.767n
Saturn (1), god, 2.931-935
SATURN (2), planet :
abode in Temple 4 : lviii ; 2.931f.
characteristics, xcviiif.
listing, 1.807 ; 5.6
malignity of its cold, 4.501
sauces, preparation of, 5.667-681
Scaevola, Gaius Mucius, 1.779 ; 4.31
Scales : *see* Libra
Scaliger, on Manilius, xii, xv, lv, lxxxvii
Scipiadae, 1.792n
SCORPIUS, the Scorpion (*Nepa*) :
size, xxii
listing, xxv ; 1.268
stellar magnitudes, cii
location, 1.690
myth, 2.32
characteristics, 2.150-269*
conjunctions, 2.270-432*
under Mars, 2.443
rules groin, 2.462 ; 4.707
relationships, 2.466-692*
dodecatemories, 2.693-748*
rising and setting times, 3.203-509*
years given by, 3.574
influences on native, 4.217-229
decans, 4.344-346
has power over arms, 4.383
injurious degrees, 4.477-480
special degrees, 4.553-559
lands influenced by, 4.778-782
paranatellonta, 5.339-356
interpolated mention, 5.697
Scylla and Charybdis, 4.605
SCYTHIA :
defended by its cold, 3.633
under Taurus, 4.753
Scythian bow, 4.755
Scythian Tanais, 4.677
Sea-goat : *see* Capricornus
seamanship, 4.282-284
secession of the plebs, 5.123
Semele, mother of Bacchus, 2.17c ; 5.241n
Seneca, echoes Manilius, xiv
Septentrio, compass-point, xc
Serpens (1) = Draco, *q.v.*, 1.610
Serpens (2), the Serpent : *see under* Ophiuchus, xxvii
Serranus, Gaius Atilius Regulus (consul 257, 250 B.C.), 4.148
Sertorius, Pompey's triumph over, 1.793n

INDEX

Servius Sulpicius, 4.213
Servius Tullius, 4.66n
setting of stars, 1.399b
SEXTUS EMPIRICUS :
the best introduction to Greek astrology, xxxvii
on planetary doctrine, xcviii
SICILY :
Theocritus a son of, 2.40
Trinacria, 4.632
under Sagittarius, 4.787
sigmatic verse (all words in *-s*), 4.780
SIGNS OF THE ZODIAC :
symbols denoting, xvi
listed, xxivf. ; 1.263-274
characteristics, 2.150-269*
in which Sun is at apogee, 2.201*
conjunctions, 2.270-432*
their guardians, 2.433-452*
allotted parts of the body, 2.453-465* (4.701-709)
relationships, 2.466-692*
their dodecatemories, 2.693-748*
rising and setting times, 3.203-509*
as chronocrators, 3.510-559*
years given by, 3.560-580*
tropic, 3.618-682
influences on natives, 4.122-293
decans, 4.297-407*
injurious degrees, 4.408-501*
special degrees, 4.502-584*
lands influenced by, 4.744-817*
paranatellonta, 5.32-709*
silphium, 4.780n
single signs, 2.158n
Sirius (α CMa), the Dogstar, 1.407f.
sitting signs, 2.249-252
smelting of metals, 5.533f.
snakes, three in the sky, 5.715n
Social War, 4.43n
Socrates, fashioner of Plato, 1.774
Sol : *see* Sun
Solon, the upright, 1.773
solstitial colure, 1.618-630
solstitial degree : *see* tropic degree
Sophocles, *Andromeda*, xxii
Sors Fortunae : *see* Fortune, Lot of
Southern Bears = south pole, 1.443, 451, 590
Southern Fish : *see* Piscis Notius
SPAIN :
on the map, 4.602
unrivalled for wars, 4.693
breeds sturdy limbs, 4.717
under Capricorn, 4.793
Sparta, famed for arms, 4.688
Spica (α Vir), rising influences, 5.270-292
spolia opima, won by Cossus and Marcellus, 1.787n
spring signs, 2.265-269
stades, 3.275n
standing signs, 2.247f.
STARS :
invisible, fig. 1 (xix) ; 1.215-220*
some more distant than others, 1.394a. *See also* constellations
Stilbon = Mercury, abode in Temple 1 : 2.944d
STOIC DOCTRINE :
Manilius an exponent of, xv
on evolution, xvii
on comets, xxxvi
world ruled by divine spirit, 1.247-254; 2.60-135; 3.48-55; 4.886-935
determinism, 4.1-118
Subsolanus, compass-point, xc
Sulpicius Rufus, Servius, 4.213n
summer signs, 2.265-269
summer tropic, celestial, 1.568-574
SUN (*Phoebus*, *Sol*) :
abode in Temple 9 : 2.907
position at moment of a nativity, lxxviii ; 3.514
characteristics, xcviiif.
drives chariot, 1.174, 198
in space, 1.198
spherical, 1.208
listing, 1.807 ; 5.3, 6
Titanian, 1.869
determines Lot of Fortune, 3.188ff.
malignity of its heat, 4.501
parches Africa, 4.728
Tiberius a second, 4.765
cut off from Moon, 4.843
Susa, under Pisces, 4.804
Swan : *see* Cycnus
swimming, styles of, 5.423-430
Sylvester II, Pope : *see* Gerbert
symbols, list of astronomical, xvi

Symplegades, 5.45n
Syracuse, Athenian defeat at, 5.50
SYRIA:
on the map, 4.624
curly-haired, 4.722
under Aries, 4.750
Syrian scents, 5.264
Syrtes, 4.600

Tanais=river Don, divides Europe and Asia, 4.677
Tarquinius Superbus, Lucius, 1.778
Tartarus, 2.46, 794
Tarutius Firmanus, Lucius, 4.773a
TAURUS (1), the Bull:
listing, xxiv; 1.264
stellar magnitudes, ci
location, 1.361
interpolated mention, 1.371
characteristics, 2.150-269*
conjunctions, 2.270-432*
under Venus, 2.439
rules neck, 2.457f.; 4.704
relationships, 2.466-692*
myth, 2.489f.
dodecatemories, 2.693-748*
rising and setting times, 3.203-509*
years given by, 3.568f.
influences on native, 4.140-151
decans, 4.316-319
loves the plough, 4.380
injurious degrees, 4.449-453
special degrees, 4.518-524
lands influenced by, 4.753f.
paranatellonta, 5.140-156
TAURUS (2), the mountain range:
allusion to Aratus, 1.402
on the map, 4.623, 675
teeth, gnashing of, 5.225
Telamon, of line of Aeacus, 1.762a
Tellus, mother of Typhon, 2.875
temple-slaves, 5.345n
temples, the twelve, lviif.; 2.856-967
Tenedos, 4.638
Terra, mother of Giants, 1.421
terrene signs, 2.226-229
Teucer, repels Trojans, 5.298-301
Teutoburgiensis, disaster of Saltus, xii; 1.898-903
textual tradition of *Astronomica*: *see* interpolations, lacunas, manuscripts, transpositions
Thales, on the origin of the universe, xviii; 1.135f.
THEBES:
war against, 3.16n; 5.463f.
famed for its gods, 4.688n
Themistocles, defeats Persians, 1.776n
Theocritus, works of, 2.40n
Theophrastus, on the Milky Way, 1.718-720
Thessaly, 4.614, 690
Thoas, 2.585b
THRACE:
aider of Troy, 1.769
counts Mars a citizen, 4.691
under Gemini, 4.756
Thrascias, compass-point: map
Thunderer: *see* Jupiter (1)
Thyestes, eats sons, 3.18c; 5.462n
Tiberius: *see* Caesar (3)
tides, influenced by Moon and Sun, 2.90-92
tightrope-walking, 5.652-655
Tigris, under Pisces, 4.806
Timaeus of Tauromenium, on Alexander, 3.22e
times of life influenced, 2.841-855*
Timocharis, astronomer, lxxxii
Timosthenes, 12-point compass of, xc
Tiphys, helmsman of the Argo, 5.45
Titanian planet=Sun, *q.v.*, 1.869
Titans, 2.15
Tolumnius, Lars, king of Veii, 1.788k
Tonans: *see* Jupiter (1)
Torquatus, Titus Manlius Imperiosus, 5.107
tragedy, themes of, 5.461-469
TRANSPOSITIONS in the text:
1.30f., 154, 167, 355-611 (*see* cvi and table 7), 641, 743, 766, 805-808, 849;
2.18 with 23, 126, 232, 374, 570-578, 583-585 with 589-591, 622f., 642, 683-686, 764, 935ff.;
3.268-270, 411f., 473f.;
4.38f., 190, 257f., 343f., 388, 725-727, 806;
5.30f. (*see* xcviii), 260, 362f., 444, 454, 480-482, 510, 514, 1f., 728

INDEX

Trasimene, battle of Lake, wrongly dated, 4.39n, 566n
Trebia, battle of, 4.566, 660
Triangle, *Trigonum*: *see* Deltoton
trigonal signs, fig. 3 (xlii) ; 2.273-286,* 673-682
Trinacria = Sicily, *q.v.*, 4.632, 787
Trivia = Diana, *q.v.*, 4.788
tropic degree, lxxxi
tropic of Cancer/Capricorn, celestial : *see* summer/winter tropic
tropic signs, lxxxi ; 2.178f. ; 3.618-682
TROY :
  at war, 1.501, 765 ; 2.1-3 ; 3.7f. ; 4.65 ; 5.47, 302
  reborn as Rome, 1.508-512 ; 4.24
  ruined by Paris, 4.80
  Pella's aid to, 4.689
  as a theme for drama, 5.485
Tullius (1) = Cicero, *q.v.*, 1.794
Tullius (2), Servius, 4.66n
tunny (*thynni*), 5.663
Twins : *see* Gemini
Tydides = Diomedes, 1.763
TYPHON (*Typhoeus*) :
  seat of, Temples 2, 8 : 2.874f.
  wars against heaven, 2.874n
  chases Venus, 4.581, 801
Tyre, under Aquarius, 4.798
Tyrian dyes, 5.258
Tyrian navigators, 1.301
Tyrrhenian sea (*Italiae undae*), 4.781

Ulysses, 1.764n ; 2.4n ; 5.126
UNIVERSE :
  origin of, xviii ; 1.118-254
  Stoic view of, 1.247-254 ; 2.60-83 ; 3.48-55 ; 4.888-890
upright signs, 2.200n
Uranus, planet, c
Urna = Aquarius, *q.v.*, 2.542, 561
Ursa (Major) = Helice, *q.v.*, 1.619
Ursa Minor : *see* Cynosura

Valerius (1) Corvinus, Marcus, 1.782
Valerius (2) Flaccus, echoes Manilius, xiv
Varro (1), Gaius Terentius, 4.38
Varro (2), Marcus Terentius, on *Iugula*, 5.175c
Varus, Publius Quinctilius, 1.899 ; 4.794a
Vela : *see* Argo
VENUS (1), goddess :
  gens Julia descended from, 1.798
  rules Taurus, 2.439
  goddess of love, 2.926 ; 4.258
  sexual love, 3.655 ; 5.143
  gives Romans complexion, 4.719
  flees Typhon, 4.801
  birds of = doves, 5.384
VENUS (2), planet :
  abode in Temple 10 : 2.922
  characteristics, xcviiif.
  both Morning and Evening Star, 1.178n
  listing, 1.808 ; 5.7
  orbit of, 1.872a
Vesta, rules Capricorn, 2.445 ; 4.243
videntia signa, fig. 7 (xlvii) ; 2.466ff.
Virdomarus, 1.788k
VIRGIL :
  Aeneid 6 imitated, xxxv
  interpolation based on, 1.435b
  imitations of, 1.792a ; 4.24a, 64f, 814j
  causes misunderstanding, 1.910d
  Georgics 4 imitated, 299
VIRGO, the Virgin (*Erigone*) :
  listing, xxivf. ; 1.266
  stellar magnitudes, ci
  interpolated mention, 1.565A
  myth, 2.32 ; 4.542
  characteristics, 2.150-269*
  conjunctions, 2.270-432*
  under Ceres, 2.442
  rules belly, 2.461 ; 4.706
  relationships, 2.466-692*
  dodecatemories, 2.693-748*
  rising and setting times, 3.203-509*
  years given by, 3.572f.
  influences on native, 4.189-202
  decans, 4.334-337
  a teacher, 4.382
  injurious degrees, 4.469-472
  special degrees, 4.542-546
  lands influenced by, 4.763-768
  paranatellonta, 5.251-292
viticulture, 4.736f. ; 5.238-242

Volturnus, compass-point, xc
Vulcanus, rules Libra, 2.443

Waterman : *see* Aquarius
water-model of universe, 4.267c
week, origin of, lxxviii
weights and measures, 4.205
Whale : *see* Cetus
winter signs, 2.265-269
winter tropic, celestial, 1.582-588
word of greeting (*ave!*). 5.65n
world, geographical description of: map; lxxxixf.; 4.585-695

Xenophanes, on the origin of the universe, xviii; 1.122-124
Xerxes, 3.19d; 4.65n; 5.49

years of paradise (*aetherii*), 1.761

Zephyrus, west: map; 4.592
ZODIAC (*see also* Signs of):
a celestial circle, xxxiv; 1.666-680
development, lxxxiif.
today's is artificial, lxxxiiif.
rightward, 2.273n
Zoroaster (Zarathustra), xviif.; 1.42a

Printed in Great Britain by R. & R. Clark, Limited, *Edinburgh*

THE SKIES OF MANILIUS

These star-charts are designed to illustrate Manilius's conception of the skies as expressed in his poem, whether or no he derived it from a celestial globe so marked. The frame adopted is that of the star-charts in the Loeb Aratus, centred on the north and south poles respectively, i.e. northern and southern hemispheres are projected on to the plane of the celestial equator as located in the time of Hipparchus, with the equinoxes at 0°♈ and 0°♎ (projection on to the plane of the ecliptic, though convenient for the plotting of Ptolemaic coordinates, carries unfortunately with it the disadvantage of bisecting every sign of the zodiac).

Manilius's **Constellations,** detailed in his poem at 1.256-455 (a list of which is given on pages xxiv-xxxi), are figured as closely to his descriptions as is consistent with astronomical accuracy; all **Stars of the first three magnitudes** visible from Rome (as registered in the catalogue on pages ci-cv) are plotted together with Bayer's identification letters.

The parallel circles are described at 1.539-602. The **Arctic circle** is located 54° north of the celestial equator (see page xxxii). The **Summer tropic** is a parallel circle drawn through the summer solstitial point, located according to the poet's rough reckoning at latitude 24°N. Next comes the **Equator** (represented by the outer circle of

each star-chart), a parallel circle midway between the two poles and at right-angles to the polar axis. The **Winter tropic** and the **Antarctic circle** are symmetrically located at 24° S and 54° S respectively.

The **Equinoctial colure** is the great circle connecting the equinoxes and the poles; the **Solstitial colure,** that connecting the solstices and the poles; on the star-charts these are represented by horizontal and vertical straight lines respectively. The poet describes their paths in detail at 1.603-630.

The **Ecliptic** is the path of the Sun's apparent annual circuit through the skies: at 1.666-680 the zodiac is represented as a zone, and it is possible that Manilius's globe, like the Farnese, was marked with parallel circles at a distance of 6° on either side of the ecliptic.

A three-dimensional representation of all these circles is given in figure 2, page xxxiii.

The poet's mapping of the **Milky Way** at 1.684-698 is clearly a versified form of something much more detailed. Ptolemy goes to great lengths to give a precise account (*Syntaxis* 8.2, volume 2, pages 170-179 Heiberg), and the boundaries of the Milky Way here given are based on his.

Frontispiece and star charts printed by Edward Stanford Ltd., London.

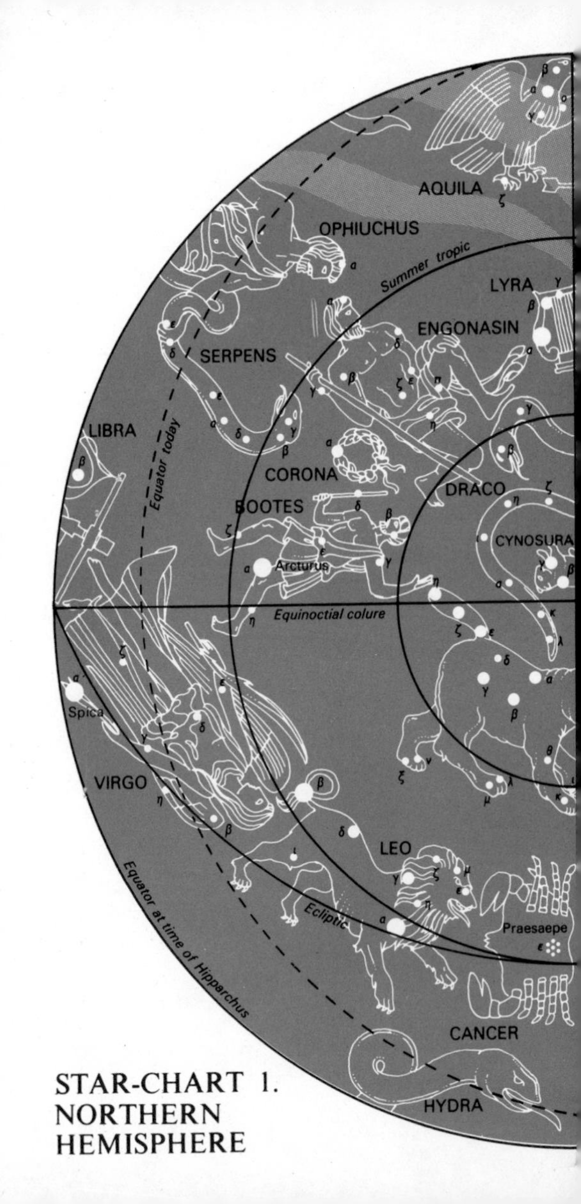

STAR-CHART 1.
NORTHERN
HEMISPHERE

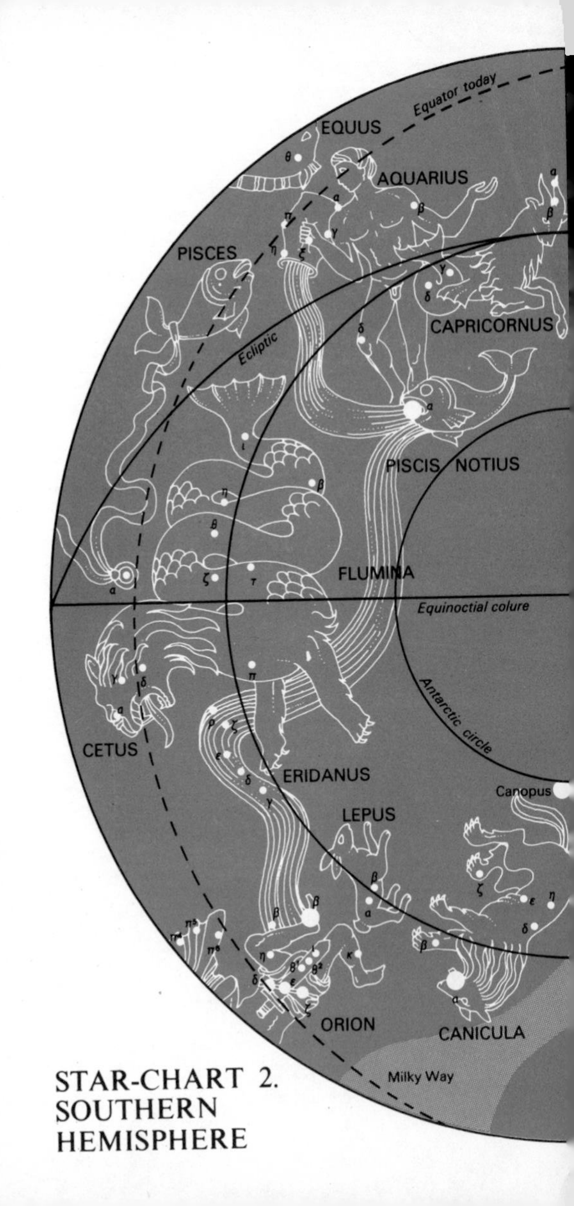

STAR-CHART 2.
SOUTHERN
HEMISPHERE

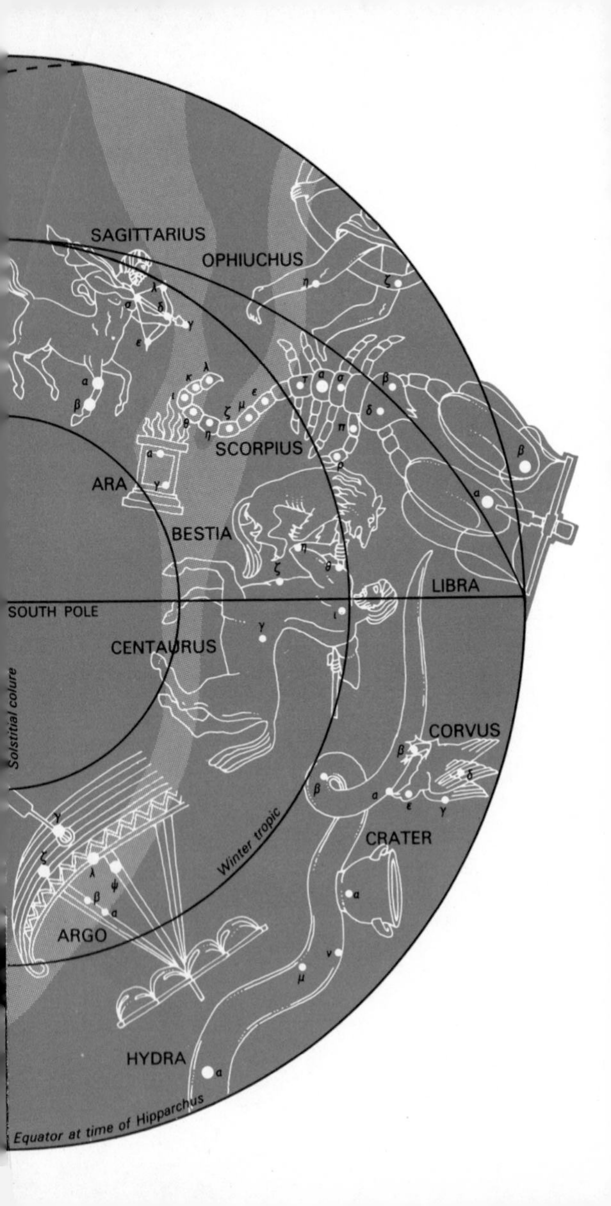
SAGITTARIUS
OPHIUCHUS
SCORPIUS
ARA
BESTIA
LIBRA
SOUTH POLE
CENTAURUS
Solstitial colure
CORVUS
CRATER
Winter tropic
ARGO
HYDRA
Equator at time of Hipparchus

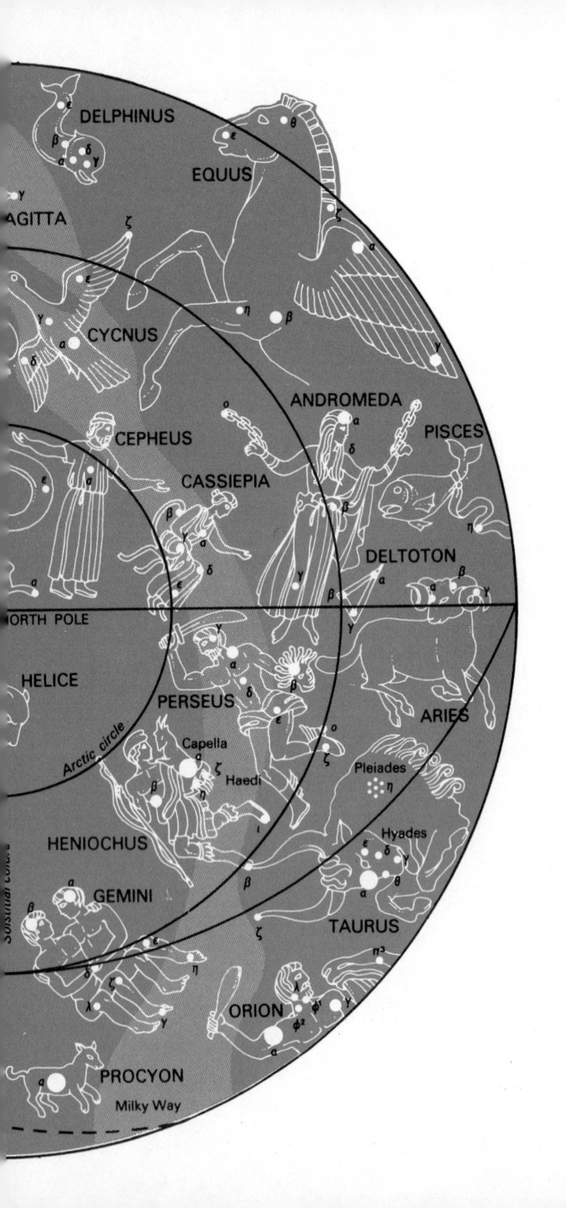

DELPHINUS
EQUUS
AGITTA
CYCNUS
ANDROMEDA
PISCES
CEPHEUS
CASSIEPIA
DELTOTON
ORTH POLE
HELICE
PERSEUS
ARIES
Arctic circle
Capella
Haedi
Pleiades
HENIOCHUS
Hyades
GEMINI
TAURUS
ORION
PROCYON
Milky Way

THE LOEB CLASSICAL LIBRARY

VOLUMES ALREADY PUBLISHED

LATIN AUTHORS

Ammianus Marcellinus. J. C. Rolfe. 3 Vols.
Apuleius: The Golden Ass (Metamorphoses). W. Adlington (1566). Revised by S. Gaselee.
St. Augustine: City of God. 7 Vols. Vol. I. G. E. McCracken. Vol. II. W. M. Green. Vol. III. D. Wiesen. Vol. IV. P. Levine. Vol. V. E. M. Sanford and W. M. Green. Vol. VI. W. C. Greene. Vol. VII. W. M. Green.
St. Augustine, Confessions of. W. Watts (1631). 2 Vols.
St. Augustine: Select Letters. J. H. Baxter.
Ausonius. H. G. Evelyn White. 2 Vols.
Bede. J. E. King. 2 Vols.
Boethius: Tracts and De Consolatione Philosophiae. Rev. H. F. Stewart and E. K. Rand. Revised by S. J. Tester.
Caesar: Alexandrian, African and Spanish Wars. A. G. Way.
Caesar: Civil Wars. A. G. Peskett.
Caesar: Gallic War. H. J. Edwards.
Cato and Varro: De Re Rustica. H. B. Ash and W. D. Hooper.
Catullus. F. W. Cornish; Tibullus. J. B. Postgate; and Pervigilium Veneris. J. W. Mackail.
Celsus: De Medicina. W. G. Spencer. 3 Vols.
Cicero: Brutus and Orator. G. L. Hendrickson and H. M. Hubbell.
Cicero: De Finibus. H. Rackham.
Cicero: De Inventione, etc. H. M. Hubbell.
Cicero: De Natura Deorum and Academica. H. Rackham.
Cicero: De Officiis. Walter Miller.
Cicero: De Oratore, etc. 2 Vols. Vol. I: De Oratore, Books I and II. E. W. Sutton and H. Rackham. Vol. II: De Oratore, Book III; De Fato; Paradoxa Stoicorum; De Partitione Oratoria. H. Rackham.
Cicero: De Republica, De Legibus. Clinton W. Keyes.

Cicero: De Senectute, De Amicitia, De Divinatione. W. A. Falconer.
Cicero: In Catilinam, Pro Murena, Pro Sulla, Pro Flacco. New version by C. Macdonald.
Cicero: Letters to Atticus. E. O. Winstedt. 3 Vols.
Cicero: Letters to his Friends. W. Glynn Williams, M. Cary, M. Henderson. 4 Vols.
Cicero: Philippics. W. C. A. Ker.
Cicero: Pro Archia, Post Reditum, De Domo, De Haruspicum Responsis, Pro Plancio. N. H. Watts.
Cicero: Pro Caecina, Pro Lege Manilia, Pro Cluentio, Pro Rabirio. H. Grose Hodge.
Cicero: Pro Caelio, De Provinciis Consularibus, Pro Balbo. R. Gardner.
Cicero: Pro Milone, In Pisonem, Pro Scauro, Pro Fonteio, Pro Rabirio Postumo, Pro Marcello, Pro Ligario, Pro Rege Deiotaro. N. H. Watts.
Cicero: Pro Quinctio, Pro Roscio Amerino, Pro Roscio Comoedo, Contra Rullum. J. H. Freese.
Cicero: Pro Sestio, In Vatinium. R. Gardner.
[Cicero]: Rhetorica ad Herennium. H. Caplan.
Cicero: Tusculan Disputations. J. E. King.
Cicero: Verrine Orations. L. H. G. Greenwood. 2 Vols.
Claudian. M. Platnauer. 2 Vols.
Columella: De Re Rustica, De Arboribus. H. B. Ash, E. S. Forster, E. Heffner. 3 Vols.
Curtius, Q.: History of Alexander. J. C. Rolfe. 2 Vols.
Florus. E. S. Forster; and Cornelius Nepos. J. C. Rolfe.
Frontinus: Stratagems and Aqueducts. C. E. Bennett and M. B. McElwain.
Fronto: Correspondence. C. R. Haines. 2 Vols.
Gellius. J. C. Rolfe. 3 Vols.
Horace: Odes and Epodes. C. E. Bennett.
Horace: Satires, Epistles, Ars Poetica. H. R. Fairclough.
Jerome: Select Letters. F. A. Wright.
Juvenal and Persius. G. G. Ramsay.
Livy. B. O. Foster, F. G. Moore, Evan T. Sage, A. C. Schlesinger and R. M. Geer (General Index). 14 Vols.
Lucan. J. D. Duff.
Lucretius. W. H. D. Rouse. Revised by M. F. Smith.
Manilius. G. P. Goold.
Martial. W. C. A. Ker. 2 Vols. Revised by E. H. Warmington.
Minor Latin Poets: from Publilius Syrus to Rutilius Namatianus, including Grattius, Calpurnius Siculus,

Nemesianus, Avianus, with "Aetna," "Phoenix" and other poems. J. Wight Duff and Arnold M. Duff.
Ovid: The Art of Love and other Poems. J. H. Mozley.
Ovid: Fasti. Sir James G. Frazer.
Ovid: Heroides and Amores. Grant Showerman. Revised by G. P. Goold.
Ovid: Metamorphoses. F. J. Miller. 2 Vols.
Ovid: Tristia and Ex Ponto. A. L. Wheeler.
Petronius. M. Heseltine; Seneca: Apocolocyntosis. W. H. D. Rouse. Revised by E. H. Warmington.
Phaedrus and Babrius (Greek). B. E. Perry.
Plautus. Paul Nixon. 5 Vols.
Pliny: Letters, Panegyricus. B. Radice. 2 Vols.
Pliny: Natural History. 10 Vols. Vols. I-V. H. Rackham. Vols. VI-VIII. W. H. S. Jones. Vol. IX. H. Rackham. Vol. X. D. E. Eichholz.
Propertius. H. E. Butler.
Prudentius. H. J. Thomson. 2 Vols.
Quintilian. H. E. Butler. 4 Vols.
Remains of Old Latin. E. H. Warmington. 4 Vols. Vol. I (Ennius and Caecilius). Vol. II (Livius, Naevius, Pacuvius, Accius). Vol. III (Lucilius, Laws of the XII Tables). Vol. IV (Archaic Inscriptions).
Sallust. J. C. Rolfe.
Scriptores Historiae Augustae. D. Magie. 3 Vols.
Seneca: Apocolocyntosis. *Cf.* Petronius.
Seneca: Epistulae Morales. R. M. Gummere. 3 Vols.
Seneca: Moral Essays. J. W. Basore. 3 Vols.
Seneca: Naturales Quaestiones. T. H. Corcoran. 2 Vols.
Seneca: Tragedies. F. J. Miller. 2 Vols.
Seneca the Elder. M. Winterbottom. 2 Vols.
Sidonius: Poems and Letters. W. B. Anderson. 2 Vols.
Silius Italicus. J. D. Duff. 2 Vols.
Statius. J. H. Mozley. 2 Vols.
Suetonius. J. C. Rolfe. 2 Vols.
Tacitus: Agricola and Germania. M. Hutton; Dialogus. Sir Wm. Peterson. Revised by R. M. Ogilvie, E. H. Warmington, M. Winterbottom.
Tacitus: Histories and Annals. C. H. Moore and J. Jackson. 4 Vols.
Terence. John Sargeaunt. 2 Vols.
Tertullian: Apologia and De Spectaculis. T. R. Glover; Minucius Felix. G. H. Rendall.
Valerius Flaccus. J. H. Mozley.
Varro: De Lingua Latina. R. G. Kent. 2 Vols.

Velleius Paterculus and Res Gestae Divi Augusti. F. W. Shipley.
Virgil. H. R. Fairclough. 2 Vols.
Vitruvius: De Architectura. F. Granger. 2 Vols.

GREEK AUTHORS

Achilles Tatius. S. Gaselee.
Aelian: On the Nature of Animals. A. F. Scholfield. 3 Vols.
Aeneas Tacticus, Asclepiodotus and Onasander. The Illinois Greek Club.
Aeschines. C. D. Adams.
Aeschylus. H. Weir Smyth. 2 Vols.
Alciphron, Aelian and Philostratus: Letters. A. R. Benner and F. H. Fobes.
Apollodorus. Sir James G. Frazer. 2 Vols.
Apollonius Rhodius. R. C. Seaton.
The Apostolic Fathers. Kirsopp Lake. 2 Vols.
Appian: Roman History. Horace White. 4 Vols.
Aratus. *Cf.* Callimachus: Hymns and Epigrams.
Aristides. C. A. Behr. 4 Vols. Vol. I.
Aristophanes. Benjamin Bickley Rogers. 3 Vols. Verse trans.
Aristotle: Art of Rhetoric. J. H. Freese.
Aristotle: Athenian Constitution, Eudemian Ethics. Virtues and Vices. H. Rackham.
Aristotle: The Categories. On Interpretation. H. P. Cooke; Prior Analytics. H. Tredennick.
Aristotle: Generation of Animals. A. L. Peck.
Aristotle: Historia Animalium. A. L. Peck. 3 Vols. Vols. I and II.
Aristotle: Metaphysics. H. Tredennick. 2 Vols.
Aristotle: Meteorologica. H. D. P. Lee.
Aristotle: Minor Works. W. S. Hett. "On Colours," "On Things Heard," "Physiognomics," "On Plants," "On Marvellous Things Heard," "Mechanical Problems," "On Invisible Lines," "Situations and Names of Winds," "On Melissus, Xenophanes, and Gorgias."
Aristotle: Nicomachean Ethics. H. Rackham.
Aristotle: Oeconomica and Magna Moralia. G. C. Armstrong. (With Metaphysics, Vol. II.)
Aristotle: On the Heavens. W. K. C. Guthrie.

THE LOEB CLASSICAL LIBRARY

Aristotle : On the Soul, Parva Naturalia, On Breath. W. S. Hett.
Aristotle : Parts of Animals. A. L. Peck ; Movement and Progression of Animals. E. S. Forster.
Aristotle : Physics. Rev. P. Wicksteed and F. M. Cornford. 2 Vols.
Aristotle : Poetics ; Longinus on the Sublime. W. Hamilton Fyfe ; Demetrius on Style. W. Rhys Roberts.
Aristotle : Politics. H. Rackham.
Aristotle : Posterior Analytics. H. Tredennick ; Topics. E. S. Forster.
Aristotle : Problems. W. S. Hett. 2 Vols.
Aristotle : Rhetorica ad Alexandrum. H. Rackham. (With Problems, Vol. II.)
Aristotle : Sophistical Refutations. Coming-to-be and Passing-away. E. S. Forster ; On the Cosmos. D. J. Furley.
Arrian : History of Alexander and Indica. 2 Vols. Vol. I. P. Brunt. Vol. II. Rev. E. Iliffe Robson.
Athenaeus : Deipnosophistae. C. B. Gulick. 7 Vols.
Babrius and Phaedrus (Latin). B. E. Perry.
St. Basil : Letters. R. J. Deferrari. 4 Vols.
Callimachus : Fragments. C. A. Trypanis ; Musaeus : Hero and Leander. T. Gelzer and C. Whitman.
Callimachus : Hymns and Epigrams, and Lycophron. A. W. Mair ; Aratus. G. R. Mair.
Clement of Alexandria. Rev. G. W. Butterworth.
Colluthus. *Cf.* Oppian.
Daphnis and Chloe. *Cf.* Longus.
Demosthenes I : Olynthiacs, Philippics and Minor Orations : I-XVII and XX. J. H. Vince.
Demosthenes II : De Corona and De Falsa Legatione. C. A. and J. H. Vince.
Demosthenes III : Meidias, Androtion, Aristocrates, Timocrates, Aristogeiton. J. H. Vince.
Demosthenes IV-VI : Private Orations and In Neaeram. A. T. Murray.
Demosthenes VII : Funeral Speech, Erotic Essay, Exordia and Letters. N. W. and N. J. DeWitt.
Dio Cassius : Roman History. E. Cary. 9 Vols.
Dio Chrysostom. 5 Vols. Vols. I and II. J. W. Cohoon. Vol. III. J. W. Cohoon and H. Lamar Crosby. Vols. IV and V. H. Lamar Crosby.
Diodorus Siculus. 12 Vols. Vols. I-VI. C. H. Oldfather. Vol. VII. C. L. Sherman. Vol. VIII. C. B. Welles. Vols.

IX and X. Russel M. Geer. Vols. XI and XII. F. R. Walton. General Index. Russel M. Geer.
Diogenes Laertius. R. D. Hicks. 2 Vols. New Introduction by H. S. Long.
Dionysius of Halicarnassus: Critical Essays. S. Usher. 2 Vols.
Dionysius of Halicarnassus: Roman Antiquities. Spelman's translation revised by E. Cary. 7 Vols.
Epictetus. W. A. Oldfather. 2 Vols.
Euripides. A. S. Way. 4 Vols. Verse trans.
Eusebius: Ecclesiastical History. Kirsopp Lake and J. E. L. Oulton. 2 Vols.
Galen: On the Natural Faculties. A. J. Brock.
The Greek Anthology. W. R. Paton. 5 Vols.
The Greek Bucolic Poets (Theocritus, Bion, Moschus). J. M. Edmonds.
Greek Elegy and Iambus with the Anacreontea. J. M. Edmonds. 2 Vols.
Greek Mathematical Works. Ivor Thomas. 2 Vols.
Herodes. *Cf.* Theophrastus: Characters.
Herodian. C. R. Whittaker. 2 Vols.
Herodotus. A. D. Godley. 4 Vols.
Hesiod and the Homeric Hymns. H. G. Evelyn White.
Hippocrates and the Fragments of Heracleitus. W. H. S. Jones and E. T. Withington. 4 Vols.
Homer: Iliad. A. T. Murray. 2 Vols.
Homer: Odyssey. A. T. Murray. 2 Vols.
Isaeus. E. S. Forster.
Isocrates. George Norlin and LaRue Van Hook. 3 Vols.
[St. John Damascene]: Barlaam and Ioasaph. Rev. G. R. Woodward, Harold Mattingly and D. M. Lang.
Josephus. 9 Vols. Vols. I-IV. H. St. J. Thackeray. Vol. V. H. St. J. Thackeray and Ralph Marcus. Vols. VI and VII. Ralph Marcus. Vol. VIII. Ralph Marcus and Allen Wikgren. Vol. IX. L. H. Feldman.
Julian. Wilmer Cave Wright. 3 Vols.
Libanius: Selected Works. A. F. Norman. 3 Vols. Vols. I and II.
Longus: Daphnis and Chloe. Thornley's translation revised by J. M. Edmonds; and Parthenius. S. Gaselee.
Lucian. 8 Vols. Vols. I-V. A. M. Harmon. Vol. VI. K. Kilburn. Vols. VII and VIII. M. D. Macleod.
Lycophron. *Cf.* Callimachus: Hymns and Epigrams.
Lyra Graeca. J. M. Edmonds. 3 Vols.
Lysias. W. R. M. Lamb.

Manetho. W. G. Waddell; Ptolemy: Tetrabiblos. F. E. Robbins.
Marcus Aurelius. C. R. Haines.
Menander. F. G. Allinson.
Minor Attic Orators. 2 Vols. K. J. Maidment and J. O. Burtt.
Musaeus : Hero and Leander. *Cf.* Callimachus : Fragments.
Nonnos : Dionysiaca. W. H. D. Rouse. 3 Vols.
Oppian, Colluthus, Tryphiodorus. A. W. Mair.
Papyri. Non-Literary Selections. A. S. Hunt and C. C. Edgar. 2 Vols. Literary Selections (Poetry). D. L. Page.
Parthenius. *Cf.* Longus.
Pausanias : Description of Greece. W. H. S. Jones. 4 Vols. and Companion Vol. arranged by R. E. Wycherley.
Philo. 10 Vols. Vols. I-V. F. H. Colson and Rev. G. H. Whitaker. Vols. VI-X. F. H. Colson. General Index. Rev. J. W. Earp.
Two Supplementary Vols. Translation only from an Armenian Text. Ralph Marcus.
Philostratus : The Life of Apollonius of Tyana. F. C. Conybeare. 2 Vols.
Philostratus : Imagines ; Callistratus : Descriptions. A. Fairbanks.
Philostratus and Eunapius : Lives of the Sophists. Wilmer Cave Wright.
Pindar. Sir J. E. Sandys.
Plato: Charmides, Alcibiades, Hipparchus, The Lovers, Theages, Minos and Epinomis. W. R. M. Lamb.
Plato : Cratylus, Parmenides, Greater Hippias, Lesser Hippias. H. N. Fowler.
Plato : Euthyphro, Apology, Crito, Phaedo, Phaedrus. H. N. Fowler.
Plato : Laches, Protagoras, Meno, Euthydemus. W. R. M. Lamb.
Plato : Laws. Rev. R. G. Bury. 2 Vols.
Plato : Lysis, Symposium, Gorgias. W. R. M. Lamb.
Plato : Republic. Paul Shorey. 2 Vols.
Plato : Statesman, Philebus. H. N. Fowler ; Ion. W. R. M. Lamb.
Plato : Theaetetus and Sophist. H. N. Fowler.
Plato : Timaeus, Critias, Clitopho, Menexenus, Epistulae. Rev. R. G. Bury.
Plotinus. A. H. Armstrong. 6 Vols. Vols. I-III.

PLUTARCH: MORALIA. 17 Vols. Vols. I-V. F. C. Babbitt. Vol. VI. W. C. Helmbold. Vol. VII. P. H. De Lacy and B. Einarson. Vol. VIII. P. A. Clement, H. B. Hoffleit. Vol. IX. E. L. Minar, Jr., F. H. Sandbach, W. C. Helmbold. Vol. X. H. N. Fowler. Vol. XI. L. Pearson, F. H. Sandbach. Vol. XII. H. Cherniss, W. C. Helmbold. Vol. XIII, Parts 1 and 2. H. Cherniss. Vol. XIV. P. H. De Lacy and B. Einarson. Vol. XV. F. H. Sandbach.
PLUTARCH: THE PARALLEL LIVES. B. Perrin. 11 Vols.
POLYBIUS. W. R. Paton. 6 Vols.
PROCOPIUS: HISTORY OF THE WARS. H. B. Dewing. 7 Vols.
PTOLEMY: TETRABIBLOS. *Cf.* MANETHO.
QUINTUS SMYRNAEUS. A. S. Way. Verse trans.
SEXTUS EMPIRICUS. Rev. R. G. Bury. 4 Vols.
SOPHOCLES. F. Storr. 2 Vols. Verse trans.
STRABO: GEOGRAPHY. Horace L. Jones. 8 Vols.
THEOPHRASTUS: CHARACTERS. J. M. Edmonds; HERODES, etc. A. D. Knox.
THEOPHRASTUS: DE CAUSIS PLANTARUM. G. K. K. Link and B. Einarson. 3 Vols. Vol. I.
THEOPHRASTUS: ENQUIRY INTO PLANTS. Sir Arthur Hort. 2 Vols.
THUCYDIDES. C. F. Smith. 4 Vols.
TRYPHIODORUS. *Cf.* OPPIAN.
XENOPHON: ANABASIS. C. L. Brownson.
XENOPHON: CYROPAEDIA. Walter Miller. 2 Vols.
XENOPHON: HELLENICA. C. L. Brownson.
XENOPHON: MEMORABILIA AND OECONOMICUS. E. C. Marchant; SYMPOSIUM AND APOLOGY. O. J. Todd.
XENOPHON: SCRIPTA MINORA. E. C. Marchant and G. W. Bowersock.

CAMBRIDGE, MASS. HARVARD UNIV. PRESS

LONDON WILLIAM HEINEMANN LTD.